D1521051

THE CLASSICAL DYNAMICS OF PARTICLES

THE CLASSICAL DYNAMICS OF PARTICLES

Galilean and Lorentz Relativity

RONALD A. MANN

Department of Physics
University of Dayton
Dayton, Ohio

1974

ACADEMIC PRESS New York San Francisco London
A Subsidiary of Harcourt Brace Jovanovich, Publishers

Copyright © 1974, by Academic Press, Inc.
ALL RIGHTS RESERVED.
NO PART OF THIS PUBLICATION MAY BE REPRODUCED OR TRANSMITTED IN ANY FORM OR BY ANY MEANS, ELECTRONIC OR MECHANICAL, INCLUDING PHOTOCOPY, RECORDING, OR ANY INFORMATION STORAGE AND RETRIEVAL SYSTEM, WITHOUT PERMISSION IN WRITING FROM THE PUBLISHER.

ACADEMIC PRESS, INC.
111 Fifth Avenue, New York, New York 10003

United Kingdom Edition published by
ACADEMIC PRESS, INC. (LONDON) LTD.
24/28 Oval Road, London NW1

Library of Congress Cataloging in Publication Data

Mann, Ronald A
The classical dynamics of particles: Galilean and Lorentz relativity.

Includes bibliographies.
1. Dynamics of a particle. 2. Relativity (Physics) 3. Groups, Theory of. I. Title.
QA852.M36 531′.163 73-18937
ISBN 0-12-469250-8

PRINTED IN THE UNITED STATES OF AMERICA

PHYSICS 1418378
QA
852
.M36

To Mom and Dad

CONTENTS

4 The Canonical Formalism

5 Many-Particle Dynamics

6 Group Theory

PREFACE

This is a text on the classical dynamics of particles intended for advanced undergraduate or graduate students of physics. It is assumed that the reader has had an undergraduate course in mechanics and the usual undergraduate mathematics preparation including differential equations and matrix theory. Some exposure to elementary tensors and group theory would be helpful but is not essential.

The rapid evolution of physics invariably produces major and minor curriculum adjustments. This text has been designed to serve either as an independent graduate course in dynamics or as a segment of a graduate theoretical physics course.

Each of the first four chapters is divided into three sections: a presentation of the basic theory, an extension and application of the basic theory, and at least one solved problem for each major topic in the first two sections. Thus for a rapid survey of the basic dynamical ideas required for electricity and magnetism and quantum mechanics, the reader may restrict his attention to the first sections of the first four chapters.

A major effort has been made to show the vitality of this ancient field of physics by incorporating recent developments, emphasizing the contemporary perspective, and presenting a parallel development of dynamics compatible with Galilean and Einstein relativity.

Chapter 1 provides a general introduction and a rather extensive discussion of the special theory of relativity including a section on tachyons. Chapter 2 presents the variational derivation of Lagrangian dynamical equations of charged particle motion and spin angular momentum. Chapter 3 is devoted to a variational derivation of Noether's theorem. The canonical formalism and Dirac's extension of Hamiltonian dynamics and treatment of constraints is presented in Chapter 4. The "No-Interaction Theorem" of Wigner and Van Dam and various efforts to construct a many-particle dynamics compatible with the special theory of relativity are developed in Chapter 5. Chapter 6 presents two applications of group theory in classical mechanics: the factorization of the dynamical matrix and the construction of a canonical formalism from a symmetry group.

The bulk of this text has been presented to my students at the University of Dayton over the past five years. I am deeply indebted to them for their helpful criticisms.

I would also like to express my sincere gratitude to four wonderful teachers who did so much to open my eyes to the beauty, unity, joy, and challenge of physics: Sister Mary Eleanor Fox, Professor Solomon Schwebel, the late Boris Podolsky, and the late Msgr. John Shuler.

Finally, thanks to Sherry Mann who typed much of the manuscript.

1

CLASSICAL DYNAMICS OF PARTICLES

1.1. PRELIMINARY CONCEPTS

A. Introduction

Among our most primitive concepts of the physical world are those of the location of objects in space and their relative change of position, which we call motion. It is the function of mechanics to determine the crucial features and construct the laws and consequences of motion. Granted the infinite richness of nature, it is not surprising that many approaches to the subject, usually differing in their point of emphasis, have developed. What is surprising is that despite having been extensively beaten generation after generation, the old horse refuses to die and remains today a viable research area.

The science of mechanics seeks to define those quantities that are vital to the description of motion, to discover the laws governing that motion, and to establish a class of observers who agree on the laws governing physical phenomena. We shall call these quests kinematics, dynamics, and relativity, respectively.

B. Kinematics

No attempt will be made to define such primitive concepts as mass, time, and so forth, unless they are being used in a context that differs from that common to undergraduate physics. This text is devoted exclusively to particle mechanics, i.e., we consider the fundamental object of motion to be the point mass. Fields are considered only through the role they play in the motion of mass points; field mechanics deserves its own development. Continuum mechanics, which studies the bulk properties of matter, is completely ignored. Recent developments in many-body particle-field mechanics renew hope that the proper approach to the bulk properties of matter is microscopic physics. That this hope is not universally popular is evident from the humorously acid comments of Truesdell [1]. (This reference may be consulted for a most elegant presentation of continuum mechanics.)

In the particle mechanics we shall develop, space-time serves as the arena in which physical phenomena occur. This arena is endowed a priori with a geometry and a metric, thus leaving the nature of space-time independent of dynamical phenomena. Inclusion of the metrical geometry as a dynamical variable is the point of departure of general relativity and various unified field theories that are beyond the scope of this book. We shall postulate a "flat" space-time and either a Euclidean metric

$$ds^2 = g_{mn}\, dx^m\, dx^n = (dx^1)^2 + (dx^2)^2 + (dx^3)^2 \tag{1.1}$$

or a Lorentz metric

$$ds^2 = g_{\mu\nu}\, dx^\mu\, dx^\nu = (dx^0)^2 - (dx^1)^2 - (dx^2)^2 - (dx^3)^2 \tag{1.2}$$

where

$$g_{mn} = \begin{bmatrix} 1 & 0 & 0 \\ 0 & 1 & 0 \\ 0 & 0 & 1 \end{bmatrix}, \qquad g_{\mu\nu} = \begin{bmatrix} 1 & 0 & 0 & 0 \\ 0 & -1 & 0 & 0 \\ 0 & 0 & -1 & 0 \\ 0 & 0 & 0 & -1 \end{bmatrix}$$

and $(ds)^2$ is the square of the interval (distance) associated with the points x^k and $x^k + dx^k$, $x^0 = ct$. Note that the Euclidean metric is positive definite with signature $+3$, whereas the Lorentz metric is indefinite with signature -2. The crucial distinction is not the difference in dimensionality of the spaces but the difference in signature; i.e., time could enter into (1.1) as $+(dx^0)^2$, yielding a Euclidean four-space whose properties would differ considerably from those of Lorentz space.

Unless otherwise stated, we assume the summation convention; i.e., a repeated index in a product implies a sum over the allowed values of the index. For example,

$$x_m y^m = x_1 y^1 + x_2 y^2 + x_3 y^3 \tag{1.3}$$

Throughout the book, it will be assumed that the Latin indices i, j, k,

l, m, n may take on the values 1, 2, 3, whereas the Greek indices λ, μ, ν, ϱ, σ, τ range over 0, 1, 2, 3. For the Euclidean metric, there is no essential difference between a quantity with a superscript (contravariant index) and that same quantity with a subscript (covariant index); for the Lorentz metric, the covariant and contravariant indices are related by

$$x^0 = x_0, \qquad x^k = -x_k$$

For more details concerning the metric and some related tensor properties, consult Appendix A or the references at the end of the chapter.

The use of a flat space-time greatly simplifies the formalism and the structure of mechanics; e.g., we can carry over intact the definitions common to introductory treatments such as vectors, parallelism, differentiation, and so forth.

Furthermore, it will be assumed that there is no intrinsic limit to the accuracy with which we may simultaneously measure the kinematical and dynamical quantities introduced throughout the text. Thus, we restrict ourselves to classical rather than quantum phenomena; and thus the observer has no dynamical influence on the system being observed.

A most important concept is that of the free particle. Historically, Aristotle postulated that a free particle was a particle that was at rest. He was thus led to introduce a force, i.e., a "cause," for all motion. This very simple definition of a free particle requires an exceedingly complex concept of a force; and we have seen in our previous mechanics courses that a much simpler dynamics can be constructed if we use the Newtonian concept of free particle, a statement of which is contained in his first Law.

We shall, throughout the book, define a free particle as a particle whose trajectory is that of a geodesic of the space-time in which it exists.

The *geodesic* is a generalization of a property of the straight line; viz., it is the extremum distance (maximum or minimum) between two points. It can be shown [2] that the equation of the geodesic is

$$\frac{d^2x^\varrho}{d\lambda^2} + \Gamma^\varrho_{\mu\nu}\frac{dx^\mu}{d\lambda}\frac{dx^\nu}{d\lambda} = 0 \tag{1.4}$$

where the $x^\varrho(\lambda)$ are the coordinates of the points lying on a curve joining two points in space-time and $\Gamma^\varrho_{\mu\nu}$ is the *Christoffel symbol of the second kind* defined in terms of the metric by

$$\Gamma^\varrho_{\mu\nu} = \tfrac{1}{2}\, g^{\varrho\sigma}(g_{\mu\sigma,\nu} + g_{\sigma\nu,\mu} - g_{\mu\nu,\sigma}) \tag{1.5}$$

Throughout the text a comma appearing in a subscript or superscript of tensor indices signifies the partial derivative with respect to the subsequent indices, e.g.,

$$g_{\mu\sigma,\nu} \equiv \partial g_{\mu\sigma}/\partial x^\nu, \qquad T^{\mu\nu\lambda,\varrho\lambda}{}_{\nu} \equiv \partial T^{\mu\nu\lambda}/\partial x_\varrho \, \partial x_\lambda \, \partial x^\nu \tag{1.6}$$

It has already been mentioned that we will consider only two types of metrics. For the Euclidean and Lorentz metrics, it can be shown by substituting (1.1) and (1.2) into (1.5) that

$$\Gamma^{\varrho}_{\mu\nu} = 0 \tag{1.7}$$

and, therefore, (1.4) is simply the equation of a straight line. Hence, for the Euclidean and Lorentz metrics, a free particle has the trajectory of a straight line. If a particle does not follow the trajectory of a geodesic, we conclude that there must be some force which causes it to deviate from the geodesic trajectory.

C. Relativity

The fundamental postulate of relativity is that there exists a class of observers, called *inertial* observers, for whom the laws of physics are the same. Suppose, now, that two observers express the dynamical laws relative to their own coordinate reference frame. To verify that these laws are in agreement it is necessary to know the rule for transforming the coordinates of one frame into the coordinates of the other. Since they are expressed in terms of the coordinates, the laws may be compared by performing the coordinate transformation.

At this point, one may simply postulate the functional form of the coordinate transformations or, alternatively, the allowed transformations may be derived by requiring a particular dynamical law to be valid in two given coordinate systems. We now illustrate the latter procedure.

Suppose we have an observer in each of the two inertial frames S and S′. We now apply the principle of relativity to Newton's second law. The S observer would write

$$F_i = ma_i = d^2x_i/dt^2 \tag{1.8}$$

Similarly, for the S′ observer,

$$F_i' = m'a_i' = d^2x_i'/dt'^2 \tag{1.9}$$

Our object is to determine the transformation equations relating the primed and unprimed coordinates such that (1.8) and (1.9) satisfy the principle of relativity. Since S and S′ are Newtonian inertial observers, the first law must hold. If $\mathbf{F} = 0$ in S, $\mathbf{F}'$ must likewise be zero in S′ so that both observers will report uniform motion (i.e., geodesic motion). It follows immediately that the most general *linear* relationship satisfying the zero force requirement is

$$F_i' = \omega_{ij} F_j \tag{1.10}$$

where ω_{ij} is the transformation matrix which maps a vector in the S frame into a vector in the S′ frame. To determine the nine elements of this matrix, additional assumptions are necessary.

Let us first assume that S and S′ would measure the same length for a given vector. Specifically, measurement of the force would yield the same number whether performed by S or S′. Therefore,

$$F'^2 = F^2 \tag{1.11}$$

or

$$F_i' F_i' = F_j F_j \tag{1.12}$$

Substituting (1.10) in (1.12)

$$\omega_{ij} \omega_{ik} F_j F_k = F_j F_j \tag{1.13}$$

Therefore, it is necessary that

$$\omega_{ij} \omega_{ik} = \delta_{jk} \tag{1.14}$$

where δ_{jk} is the Kronecker delta

$$\delta_{jk} = \begin{cases} 0 & \text{for } j \neq k \\ 1 & \text{for } j = k \end{cases} \tag{1.15}$$

Equation (1.14) is the definition of an *orthogonal transformation.*

Furthermore, we require that both observers measure the same value for the mass of the particle.

$$m = m' \tag{1.16}$$

Finally, we assume that the time duration of a given process as measured by S is identical to that measured by S′, i.e.,

$$\Delta t' = \Delta t \tag{1.17}$$

where $\Delta t = t_1' - t_2'$. Hence, we may write

$$t' = t + \tau \tag{1.18}$$

where τ is a constant.

Substituting (1.8) and (1.9) into (1.10) and using (1.16) and (1.17), we obtain

$$d^2x_i'/dt^2 = \omega_{ij}\, d^2x_j/dt^2 \tag{1.19}$$

The solution of (1.19) is

$$x_i' = \omega_{ij}x_j + \alpha_i t + \beta_i \tag{1.20}$$

and

$$\dot{x}_i' = \omega_{ij}\dot{x}_j + \alpha_i \tag{1.21}$$

where α_i and β_i are integration constants whose physical significance is yet to be determined and $\dot{x}_i \equiv dx_i/dt$.

A particle attached to the origin of the S frame would be recorded by S as being located at x_i : (0, 0, 0), whereas at $t = 0$ the S′ observer would locate the particle at x_i' : (x_1', x_2', x_3'). But the primed and unprimed coordinates are connected by (1.20). Hence

$$x_i' = \beta_i \tag{1.22}$$

i.e., the β_i are the coordinates of the origin of the S frame as recorded by the S′ frame at $t = 0$.

Now the S observer would record the velocity of the particle fixed at the origin as zero ($\dot{x}_j = 0$); however, the S′ observer would record the velocity as $\dot{x}_i'$: $(\dot{x}_1', \dot{x}_2', \dot{x}_3')$ and from (1.21) we find

$$\dot{x}_i' = \alpha_i \tag{1.23}$$

i.e., the α_i are the components of the velocity of the origin of the S frame as measured by the S′ observer.

We may rewrite (1.20) as

$$x_i' = \omega_{ij}x_j + v_i t + x_{0_i}, \qquad t' = t + \tau \tag{1.24}$$

It remains to identify the nine elements of the transformation matrix.

The orthogonality condition (1.14) provides six independent equations relating the nine matrix elements, thus we need but three more conditions to uniquely specify the transformation matrix. These three additional conditions are simply the equations describing the relative orientation of the S and S′ frames. In fact, taking the determinant of both sides of (1.14) yields

$$\det \omega_{ij} = \pm 1 \tag{1.25}$$

Thus, when the determinant is $+1(-1)$, ω_{ij} is the rotation (inversion) matrix which at $t = 0$ maps the unprimed frame into an orientation parallel to the primed frame. The three independent parameters describing the relative orientation of S and S′ may be chosen to be the Euler angles, the direction cosines of the primed axes relative to the unprimed axes, and so forth.

Thus, the transformation equation (1.24) is specified uniquely by x_{0_i} (the location of the S origin relative to the S′ frame), by v_j (the velocity of the S frame relative to the S′ frame), and by ω_{ij} (the rotation matrix describing the orientation of the S frame relative to the S′ frame). The transformation (1.24) is called the *Galilean transformation* and it represents the relativity transformation that is compatible with Newton's laws of motion and the principle of relativity. Its derivation has been presented so as to make clear the assumptions involved in its use. The set of transformations (1.24) form a group called the Galilean group.

Now Newton's laws are not the only laws of physics. Another important class of laws are those describing electromagnetic phenomena, viz., Maxwell's equations. That these equations are incompatible with Galilean relativity is demonstrated in Section 1.2. For a more detailed account, consult any of the texts on relativity listed at the end of the chapter.

We now derive a set of coordinate transformations which permit Maxwell's equations to satisfy the principle of relativity. The derivation is simplified considerably if, instead of Maxwell's equations, we use as our invariant law the law describing the propagation of an electromagnetic wave emitted from a point source. From Maxwell's equations it follows immediately [2] that such a wave will propagate as a spherical wave with the source as the center of the sphere. The S observer would write for the equation of the spherical wave front

$$c^2(t - t_0)^2 - (x_i - x_{0_i})^2 = 0 \qquad \text{or} \qquad (x^\mu - x_0{}^\mu) g_{\mu\nu} (x^\nu - x_0{}^\nu) = 0 \tag{1.26}$$

where c is the velocity of light, x_{0_i} the location of the source, and t_0 the time of emission all of which are measured relative to S. The x_i are the locus of points comprising the spherical wave front at a time t. $g_{\mu\nu}$ is the metric tensor. S′ would write

$$c'^2(t' - t_0')^2 - (x_i' - x_{0_i}')^2 = 0 \qquad \text{or} \qquad (x'^\mu - x_0'^\mu) g_{\mu\nu} (x^{\nu\prime} - x_0^{\nu\prime}) = 0 \tag{1.27}$$

Proceeding now as we did for Newton's laws, we ask whether there exists a set of coordinate transformations such that the laws described by (1.27) and (1.26) are the same (i.e., transform into one another). It should be noted that the starting point in the Newtonian case was a differential equation, whereas here we start with an algebraic equation obtained from the differential equation describing wave propagation.

It is clear that we do not wish to obtain the Galilean transformations, so we must eliminate or revise one or more of the assumptions contained in the previous derivation. Mass does not enter into (1.26) and (1.27); however, time does. Let us, therefore, lift the assumption expressed in (1.17) leaving the relation between time in the two frames to be determined in the same manner as the coordinate relations. Furthermore, we assume as before that the transformation equations are linear. (It should, perhaps, be remarked that the introduction of nonlinearity into physical theories leads, in general, to such nightmarish complexities that it is done only when all reasonable linear possibilities have been exhausted. That correct dynamical laws must be nonlinear is argued quite succinctly by Infeld [3].) Thus we write

$$x^{\mu\prime} = L^{\mu}{}_{\nu}\, x^{\nu} + a^{\mu} \tag{1.28}$$

Substituting (1.28) into (1.27)

$$L^{\mu}{}_{\nu}\,(x^{\nu} - x_0^{\nu}) g_{\mu\lambda} L^{\lambda}{}_{\varrho}\,(x^{\varrho} - x_0^{\varrho}) = 0 \tag{1.29}$$

According to the principle of relativity (1.29) must be identical with (1.26) thus we conclude that

$$L^{\mu}{}_{\nu}\, g_{\mu\lambda} L^{\lambda}{}_{\varrho} = g_{\nu\varrho} \tag{1.30}$$

Equation (1.30) is the *orthogonality condition*; it provides ten independent equations reducing the number of parameters comprising $L^{\mu}{}_{\nu}$ from sixteen to six. Taking the determinant of both sides of (1.30) yields

$$\det(L^{\mu}{}_{\nu}) = \pm 1 \tag{1.31}$$

Thus for the $+1(-1)$, $L^{\mu}{}_{\nu}$ is a rotation (inversion) matrix in the four-dimensional linear space of coordinates and time. From (1.28) it is clear that a^{μ} is the location of the origin of the S frame as measured by the S′ frame. The remaining six parameters of $L^{\mu}{}_{\nu}$ are determined by the relative orientation of the S and S′ frames. An explicit evaluation of all the parameters is presented in Section 1.2. The set of transformations of (1.28), called the *inhomogeneous Lorentz transformations*, form a group called the Poincaré group.

The principal differences between the Galilean and Lorentz transformations are: (a) The G.T. are defined on a four-dimensional Euclidean space, whereas the L.T. are defined on a four-dimensional space having the Lorentz metric

$$g_{\mu\nu} = \begin{bmatrix} 1 & & & 0 \\ & -1 & & \\ & & -1 & \\ 0 & & & -1 \end{bmatrix}$$

(b) The principle of relativity and Newton's mechanics are compatible under the Galilean group but not under the Poincaré group; vice versa for Maxwell's electromagnetic theory.

The question of the compatibility between the dynamical formalism, the principle of relativity, and the relativity group is most important and will be discussed more fully in a subsequent chapter. Here the relativity group has been derived, but if, as often happens, it is postulated at the outset for a dynamical theory, the question of compatibility must be investigated.

Whether the dynamics constructed on the kinematics and relativity presented above is rich enough to describe the phenomena of nature remains to be seen. It is sufficient at this point to note that we have already made a number of crucial assumptions that will greatly restrict the nature and content of the dynamics we are about to construct.

1.2. SALIENT FEATURES OF THE SPECIAL THEORY OF RELATIVITY

A. Lorentz Space

A Lorentz observer at the origin of the reference system shown in Fig. 1.1 may divide Lorentz space as follows. Let S_{21} be the interval between two events (a "point" in Lorentz space is usually called an *event*) in Lorentz space, i.e.,

$$S_{21}^2 \equiv ({}_2x^\mu - {}_1x^\mu)({}_2x_\mu - {}_1x_\mu)$$

To simplify the discussion let point 1 lie at the origin. If $S_{21}^2 > 0$, point 2 lies within the cone and the interval is said to be *timelike*; if $({}_2x^0 - {}_1x^0) > 0$, then point 2 is said to be in the *absolute future* of point 1; if $({}_2x^0 - {}_1x^0) < 0$,

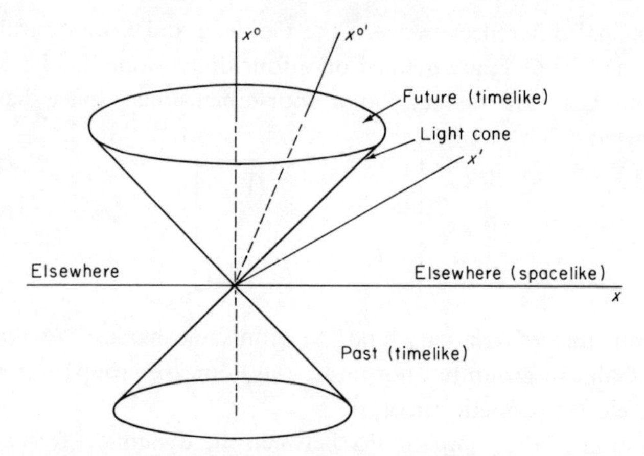

Fig. 1.1. Lorentz diagram illustrating the characteristic regions of Lorentz space. Point 1 is located at the origin.

then point 2 lies in the *absolute past* of point 1. The region within the cone is called the timelike region. If $S_{21}^2 < 0$, point 2 lies outside the cone in the *elsewhere* or *spacelike* region and the interval is said to be *spacelike*. If $S_{21}^2 = 0$, point 2 lies on the *light cone* of point 1 and the interval is said to be *lightlike* or *null*. The light cones provide the boundary separating the elsewhere from the absolute past and future.

A vector A^μ is *timelike* if $A^2 \equiv A^\mu A_\mu > 0$. If $A^2 < 0$, the vector is spacelike. If $A^2 = 0$, A^μ is a *null* vector.

Recall that a tensor is defined by its transformation properties and that a vector is a tensor of rank 1. The space-time arena employed in the description of physical phenomena is chosen to be a "representation" space of the relativity group. Thus the vectors (or in general the tensors) transform according to the relativity group. In Euclidean space the tensors transform according to (1.24) and in Lorentz space according to (1.28).

Consider now two events recorded by an inertial observer P as having an interval

$$S_{12}^2 = (x_{12}^0)^2 - l_{12}^2 \tag{1.32}$$

where $x_{12}^0 = {}_1x^0 - {}_2x^0$ and $l_{12}^2 = ({}_1x^k - {}_2x^k)^2$, the spatial separation of the events. In order that a second inertial observer P' record the events as occurring simultaneously, i.e., $x_{12}^0 = 0$, it is necessary that

$$S_{12}^{\prime 2} = -l_{12}^{\prime 2} < 0 \tag{1.33}$$

Since both observers are inertial there exists a Lorentz transformation connecting them, and since S^0_{12} is a scalar we conclude

$$S'^2_{12} = S^2_{12} = (x^0_{12})^2 - l^2_{12} < 0 \tag{1.34}$$

i.e., if the interval is *spacelike* the inertial observers will not agree on the time ordering of events which of course is crucial to the establishment of cause-effect relationships. A similar analysis shows that *timelike* intervals do preserve the time ordering of events under a Lorentz transformation, and this is why the region within the light cone is designated as absolute past or future.

It is not difficult to show that a Lorentz observer P′ moving with a three-velocity v relative to P may be represented on the Lorentz diagram in Fig. 1.1 by the dotted axes labeled x_0' and x'. The angle which x' makes with x is just the $\tan^{-1}(v/c)$. x_0' makes the same angle with x_0.

B. Geodesics

For Lorentz space the geodesic equation (1.4) becomes

$$d^2x^\varrho/d\lambda^2 = 0 \tag{1.35}$$

and the solutions are

$$dx^\varrho/d\lambda = k^{\varrho\prime} \tag{1.36a}$$

$$x^\varrho - \underset{0}{x^\varrho} = k^{\varrho\prime}\lambda \tag{1.36b}$$

where $\underset{0}{x^\varrho}$ and $k^{\varrho\prime}$ are integration constants. It is particularly advantageous to use the interval, i.e., the arc length along the curve, as the parameter appearing in (1.35) and (1.36). Then (1.36) becomes

$$dx^\varrho/ds = k^\varrho \tag{1.37a}$$

$$x^\varrho - \underset{0}{x^\varrho} = k^\varrho s \tag{1.37b}$$

But from (1.2) it follows that

$$\frac{dx^\mu}{ds}\frac{dx_\mu}{ds} = 1 = k^\varrho k_\varrho \tag{1.38}$$

Thus the tangent to the curve, $dx^\varrho/ds = k^\varrho$, is a unit vector and (1.38) represents a supplementary condition which restricts the solutions of the geodesic equations.

There exists a very important class of geodesics which are not subject to the supplementary condition (1.38). We have already seen that in Lorentz space there are four-vectors that are null vectors. Now if the interval is null, i.e., if

$$(ds)^2 = dx^\mu \, dx_\mu = 0 \tag{1.39}$$

the interval s can no longer be used to characterize the events x^μ because the arc length between distinct events having $x^\mu x_\mu = 0$ is zero. Thus in this case we must return to (1.36) as our geodesic equations but with the restriction that $\lambda \neq s$. From (1.36a) and (1.39) it follows that

$$\frac{dx^\varrho}{d\lambda} \frac{dx_\varrho}{d\lambda} = 0 = k^\varrho k_\varrho \tag{1.40}$$

Geodesics that satisfy (1.40) are called *null geodesics.* We conclude that the geodesics of Lorentz space may be separated into two classes, viz., null geodesics satisfying (1.40) and regular geodesics satisfying (1.38). The null geodesics do not appear in Euclidean space because the norm of a Euclidean vector is always greater than zero if anyone of the components is nonzero.

From (1.40) it follows that

$$\left(\frac{dx^0}{d\lambda}\right)^2 \left[1 + \frac{dx^j}{dx^0} \frac{dx_j}{dx^0}\right] = 0 \tag{1.41}$$

and therefore

$$(dx^j/dx^0)^2 = (v^j/c)^2 = 1 \tag{1.42}$$

where v^j is the usual three velocity dx^j/dt. From (1.42) we see that along null geodesics the three-velocity is a constant. Moreover, since (1.40) is Lorentz invariant (i.e., null geodesics transform into null geodesics under a Lorentz transformation), (1.42) or the constancy of the three-velocity is Lorentz invariant and has the value of c for all Lorentz observers. It seems quite natural, then, to identify the null geodesics as the trajectories of free particles moving with the speed of light (e.g., photons or neutrinos).

A similar consideration of the geodesics represented by (1.37) and (1.38) leads to

$$(dx^0/ds)^2[1 - (\beta^j)^2] = 1 \tag{1.43}$$

where

$$\beta^j = dx^j/dx^0 = v^j/c \tag{1.44}$$

Proceeding as we did for null geodesics we identify the geodesics represented by (1.37) and (1.38) as the trajectories followed by free particles having a three-velocity v^j.

We define the four-velocity of a particle as

$$u^\mu = dx^\mu/ds \tag{1.45}$$

whose components are, according to (1.43) and (1.44),

$$u^0 = 1/[1 - \beta^2]^{1/2} \equiv \gamma \tag{1.46a}$$

$$u^j = (dx^0/ds)(dx^j/dx^0) = \gamma v^j/c = \gamma\beta^j \tag{1.46b}$$

Comparing (1.37) and (1.45) we see that

$$k^\mu = u^\mu \tag{1.47}$$

and from (1.38), (1.46), and (1.47)

$$k^2 = (k^j)^2 = (k^0)^2 - 1 = \beta^2/(1 - \beta^2) \tag{1.48a}$$

$$\beta^2 = k^2/(1 + k^2) \tag{1.48b}$$

From (1.48) it follows that

$$\text{if } 0 \leq \beta^2 < 1, \quad \text{then } 0 \leq k^2 < \infty \text{ and } 1 \leq (k^0)^2 < \infty \tag{1.49a}$$

$$\text{if } 1 < \beta^2 \leq \infty, \quad \text{then } -\infty < k^2 \leq -1 \text{ and } -\infty < (k^0)^2 \leq 0 \tag{1.49b}$$

$$\text{if } -\infty \leq \beta^2 < 0, \quad \text{then } -1 \leq k^2 < 0 \text{ and } 0 < (k^0)^2 < 1 \tag{1.49c}$$

Although mathematically the geodesics represented by (1.49b) and (1.49c) may exist, it does not seem that there are any particles with the properties necessary to follow these trajectories. The trajectories of (1.49b) correspond to the "tachyon" motion which will be discussed later. At the present time (1.49c) must be rejected as possible free particle trajectories because it requires "imaginary" velocities ($\beta^2 < 0$). At least it can safely be said that only (1.49a) which yields free particle trajectories having speeds less than that of light and the null geodesics which correspond to particles moving with the speed of light are compatible with observed particle motion.

C. Energy Momentum

It is common to define the *four-momentum* p^μ of a particle of mass m, called the rest mass or invariant mass, as

$$p^\mu = mcu^\mu \tag{1.50a}$$

where the velocity of light c has been introduced to provide the four-momentum with the usual dimensions of mass X velocity (u^μ is dimensionless). Some problems associated with this definition will be discussed in subsequent chapters. From (1.38), (1.45), and (1.50) it follows that

$$p^\mu p_\mu = m^2c^2 \tag{1.51a}$$

It is common practice to associate p^0 with the total energy of the system (this will be discussed more fully in Chapter 3) so that (1.51) may be written

$$E^2 - (p^k)^2c^2 = m^2c^4 \tag{1.52}$$

And the energy of a particle at rest is

$$E = mc^2 \tag{1.53}$$

For a system of free particles the total momentum of the system is simply the sum of the momenta of the individual particles, i.e.,

$$P^\mu = {}_1p^\mu + {}_2p^\mu + \cdots \tag{1.50b}$$

The *invariant mass* of the system is defined by the equation

$$P^\mu P_\mu = M^2C^2 \tag{1.51b}$$

D. Evaluation of the Lorentz Parameters

The homogeneous Lorentz transformation is

$$x^{\mu\prime} = L^\mu{}_\nu \, x^\nu \tag{1.54}$$

The inverse follows immediately by multiplying both sides of (1.54) by $L_\lambda{}^\varrho g_{\varrho\mu}$ and using (1.30) to obtain

$$g_{\lambda\nu}x^\nu = x_\lambda = L^\varrho{}_\lambda \, x_\varrho{}' \tag{1.55}$$

Consider two inertial observers S and S′ whose coordinate axes coincide at $x^0 = 0$. S observes S′ to be moving with a speed v^i (see Fig. 1.2). Thus at $x^0 = 0$ a physical point P attached to the origin of S′ would have the coordinates (0, 0, 0) in both S′ and S. At a later time the point P would still have the coordinates (0, 0, 0) in S′ but in S the coordinates are $x^i = \beta^i x^0$,

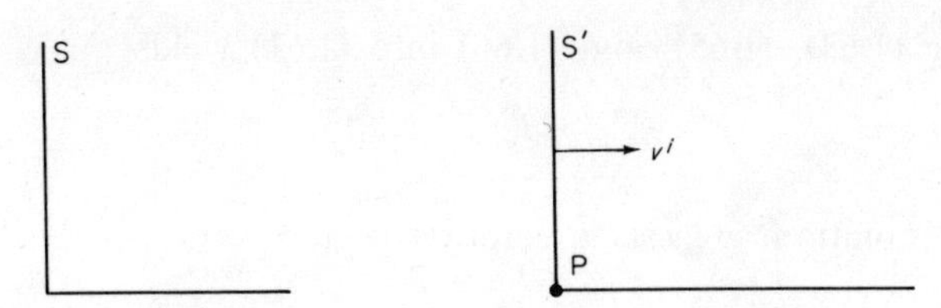

Fig. 1.2. A point P is attached to the origin of the S′ system which moves with a speed v^i relative to S.

where $\beta^i = v^i/c$. Plugging these values into (1.55) we find

$$x_0 = L^0{}_0\, x_0{}' \tag{1.56a}$$

$$x_j = L^0{}_j\, x_0{}' \tag{1.56b}$$

Eliminating $x_0{}'$ from these equations and using $x_j = \beta_j x^0$, we find

$$L^0{}_j = L^0{}_0\, \beta_j \tag{1.57}$$

Similarly the point Q attached to the origin of S would have the spatial coordinates (0, 0, 0) in S and $x^{j\prime} = -\beta^j x^{0\prime}$ in S′. Proceeding as before we find from (1.54) that

$$L^j{}_0 = -L^0{}_0\, \beta^j \tag{1.58}$$

The remaining coefficients may be obtained from the orthogonality conditions which may be written as

$$(L^0{}_0)^2 - (L^j{}_0)^2 = 1 \tag{1.59}$$

$$(L^0{}_k)^2 - (L^j{}_k)^2 = -1 \qquad \text{(no sum on } k) \tag{1.60}$$

$$L^0{}_0\, L^0{}_k - L^j{}_0\, L^j{}_k = 0 \tag{1.61}$$

$$L^0{}_k\, L^0{}_m - L^j{}_k\, L^j{}_m = 0 \qquad (\text{for } k \neq m) \tag{1.62}$$

Using (1.58) in (1.59) we obtain

$$L^0{}_0 = \pm \frac{1}{(1 - \beta^2)^{1/2}} \equiv \pm \gamma \tag{1.63}$$

where

$$\beta^2 = (\beta^1)^2 + (\beta^2)^2 + (\beta^3)^2 \tag{1.64}$$

and therefore

$$L^0{}_j = \gamma \beta_j \tag{1.65}$$

$$L^j{}_0 = -\gamma \beta^j = -u^j \tag{1.66}$$

Substituting (1.63), (1.65), and (1.66) into (1.61) yields

$$\gamma\beta^j L^j{}_k = -\gamma^2\beta_k \tag{1.67}$$

The tensor notation suggests a solution of the form

$$L^j{}_k = g^j{}_k + b\gamma\beta^j\beta_k/\beta^2 \tag{1.68}$$

where b is a constant to be determined. Actually (1.67) suggests the second term on the right side of (1.68), whereas the $g^j{}_k$ is inserted to provide a transformation connection between $x^{j\prime}$ and x^j even when $\beta^j = 0$. Inserting (1.68) into (1.67) we find

$$b = -(\gamma - 1)/\gamma \tag{1.69}$$

Thus

$$L^j{}_k = g^j{}_k - (\gamma - 1)\beta^j\beta_k/\beta^2 = g^j{}_k - u^j u_k/(1 + \gamma) \tag{1.70}$$

Inserting (1.70), (1.66), (1.65), and (1.63) into (1.60) and (1.62) we obtain an identity thereby confirming the validity of our expressions for the Lorentz parameters. When S′ is moving with constant velocity along the x' axis the Lorentz transformations are quite simple and may be displayed conveniently in matrix form

$$L^\mu{}_\nu = \begin{bmatrix} \gamma & -\beta\gamma & 0 & 0 \\ -\beta\gamma & \gamma & 0 & 0 \\ 0 & 0 & 1 & 0 \\ 0 & 0 & 0 & 1 \end{bmatrix} \tag{1.71}$$

where $\beta = \beta^1$.

E. Relativistic Addition of Velocities

The velocity of an object as measured by different Lorentz observers could be obtained by differentiating the Lorentz transformations, a process which is complicated somewhat by the need to differentiate the primed coordinates by t' and the unprimed coordinates by t. An alternate procedure follows.

Consider three Lorentz frames S, S′, and S″ orientated and moving in such a way that the coordinate transformations take the form of (1.71) (i.e., as in Fig. 1.3). We write in matrix notation

$$x'' = Lx, \qquad x' = L_1 x, \qquad x'' = L_2 x' = L_2 L_1 x = Lx \tag{1.72}$$

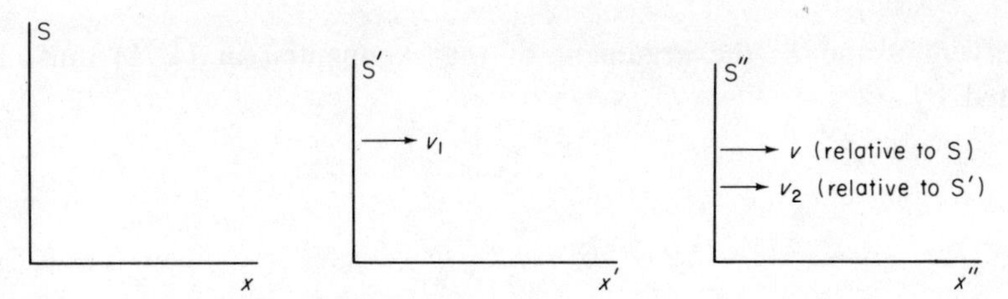

Fig. 1.3. Three Lorentz frames moving along the x axis. S′ has a speed v_1 relative to S′. S″ has a speed v relative to S and v_2 relative to S′.

Therefore

$$L_2 L_1 = L \tag{1.73}$$

Using (1.71) in (1.73)

$$\begin{bmatrix} \gamma_2 & -\beta_2\gamma_2 & 0 & 0 \\ -\beta_2\gamma_2 & \gamma_2 & 0 & 0 \\ 0 & 0 & 1 & 0 \\ 0 & 0 & 0 & 1 \end{bmatrix} \begin{bmatrix} \gamma_1 & -\beta_1\gamma_1 & 0 & 0 \\ -\beta_1\gamma_1 & \gamma_1 & 0 & 0 \\ 0 & 0 & 1 & 0 \\ 0 & 0 & 0 & 1 \end{bmatrix} = \begin{bmatrix} \gamma & -\beta\gamma & 0 & 0 \\ -\beta\gamma & \gamma & 0 & 0 \\ 0 & 0 & 1 & 0 \\ 0 & 0 & 0 & 1 \end{bmatrix} \tag{1.74}$$

Equation (1.74) yields after some algebra

$$\beta = (\beta_1 + \beta_2)/(1 + \beta_1\beta_2) \tag{1.75}$$

It follows immediately from (1.75) that if S′ observes S″ to be moving with the velocity of light, S will also observe S″ to move with the velocity of light.

F. Doppler Effect and Aberration

The Lorentz transformation was developed from a consideration of the laws of light propagation. It can be shown that the electromagnetic wave equation admits a plane wave solution

$$\phi = A \exp(ik^\mu x_\mu) \tag{1.76}$$

where the four-vector k^μ is a null vector

$$k^\mu : (\omega/c, \mathbf{k}) \tag{1.77}$$

ω is the angular frequency and $\mathbf{k}$ the wave vector of the wave. For two observers S and S′ the argument of the exponential in (1.76) must be related by

$$k^{\mu\prime} x_{\mu}{}' = k^{\mu} x_{\mu} \tag{1.78}$$

But

$$k^{\mu\prime} = L^{\mu}{}_{\nu}\, k^{\nu} \tag{1.79}$$

It follows from (1.77) that

$$k^{\mu} k_{\mu} = 0 \tag{1.80}$$

and therefore

$$k^{0} = |\,\mathbf{k}\,| = 2\pi/\lambda \tag{1.81}$$

If we assume that S and S′ are related by (1.78) and that (1.71) is valid, (1.79) becomes

$$k^{0\prime} = \gamma k^{0} - \beta\gamma k^{2} \tag{1.82}$$

$$k^{1\prime} = -\beta\gamma k^{0} + \gamma k^{1} \tag{1.83}$$

We define α as the angle that the propagation direction of the wave makes with the x^1 axis (i.e., the angle between $\mathbf{k}$ and x^1). Similarly α' is the angle between $\mathbf{k}'$ and $x^{1\prime}$. Thus

$$k^{1} = |\,\mathbf{k}\,| \cos\alpha = k^{0} \cos\alpha \tag{1.84}$$

$$k^{1\prime} = |\,\mathbf{k}'\,| \cos\alpha' = k^{0\prime} \cos\alpha' \tag{1.85}$$

Using (1.84) in (1.82), we obtain the Doppler frequency shift

$$\omega' = \omega\gamma(1 - \beta\cos\alpha) \tag{1.86}$$

Using (1.84) and (1.85) in (1.83) we obtain the aberration formula

$$\cos\alpha' = (\cos\alpha - \beta)/(1 - \beta\cos\alpha) \tag{1.87}$$

G. Covariance and Invariance

The principle of relativity requires that the laws of physics be invariant under transformations comprising the relativity group. When the laws are constructed from tensors defined on the relativity group, the transformation properties are immediately evident and equations formulated in this manner are called *covariant*.

It is important to note that equations may be relativistically invariant but not covariant. For example, Maxwell's equations may be written in either form.

Relativistically invariant

$$\begin{aligned} \nabla \cdot \mathbf{B} &= 0, \qquad \nabla \cdot \mathbf{E} = \varrho \\ \partial_t \mathbf{B} + \nabla \times \mathbf{E} &= 0, \qquad \nabla \times \mathbf{B} = \varrho \mathbf{v} + \partial_t \mathbf{E} \end{aligned} \tag{1.88}$$

Covariant

$$\begin{aligned} F^{\mu\nu}{}_{,\nu} &= -j^\mu/c \\ F_{\nu\varrho,\sigma} + F_{\sigma\nu,\varrho} + F_{\varrho\sigma,\nu} &= 0 \end{aligned} \tag{1.89}$$

where

$$F^{\mu\nu} = \begin{bmatrix} 0 & -E_1 & -E_2 & -E_3 \\ E_1 & 0 & -B_3 & B_2 \\ E_2 & B_3 & 0 & -B_1 \\ E_3 & -B_2 & B_1 & 0 \end{bmatrix}, \qquad F^{jk} = \varepsilon^{jkl} B_l \tag{1.90}$$

$$j^\mu = \varrho\, dx^\mu/ds \tag{1.91}$$

H. Transformation from the Rest (Proper) Frame to the Lab Frame

Transformations from rest frames to laboratory frames are essential to discussions of the relativistic motions of particles and scattering theory. We shall consider transformations from the rest frame of the particle to the lab frame and transformations from the rest frame of the system to the lab frame.

I. Rest Frame of the Particle

These transformations are important because the properties of the particle, e.g., mass, spin, charge, and so forth, are defined in the rest frame. Consider a particle following a world line which is described in the laboratory by (Fig. 1.4). Its rest frame is defined as the inertial frame in which the particle is at rest, i.e., the frame which is moving with the same velocity as that of the particle. If the particle is accelerating the velocity of the particle is changing; nevertheless, at each instant there is an inertial rest

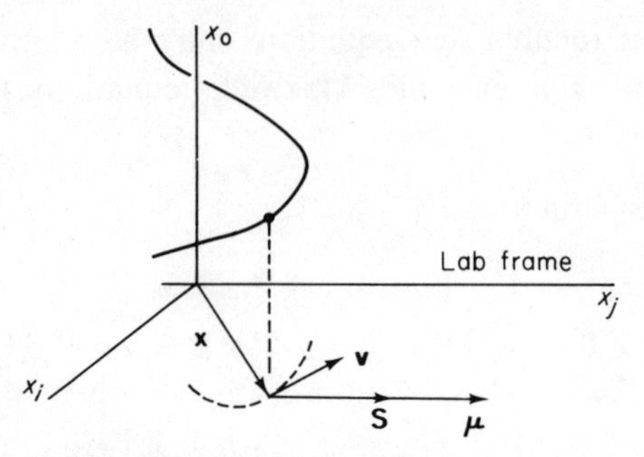

Fig. 1.4. World line of relativistic particle with spin **S** and magnetic moment μ. The rest frame of the particle has instantaneous velocity **v** relative to laboratory frame.

frame which in the next instant is rejected in favor of a new one having the new particle velocity. Thus for accelerating particles there is always an instantaneous inertial (Lorentz) rest frame and the transformation from the rest frame to the lab frame is just the Lorentz transformation.

We may associate a spatial orientation with particles that have intrinsic angular momentum, i.e., spin, $\mathbf{S}'$. The three-vector $\mathbf{S}'$ is defined within the rest system of the particle. To determine its orientation in the lab frame we construct a four-vector S^μ whose components in the rest frame of the particle are $(0, \mathbf{S}')$. It follows immediately that S^μ is a spacelike vector and

$$S^\mu S_\mu = -(S^{k\prime})^2 \tag{1.92}$$

$$u^\mu S_\mu = 0 \tag{1.93}$$

where u^μ is the four-velocity of the particle. The orientation of the particle can be separated from the magnitude of the spin by introducing the polarization four-vector $\mathcal{P}^\mu$ whose components in the rest frame are $(0, \hat{S}')$ where $\hat{S}'$ is the unit three-vector $\mathbf{S}'/|\mathbf{S}'|$. Then

$$\mathcal{P}^\mu \mathcal{P}_\mu = -1 \tag{1.94}$$

$$u^\mu \mathcal{P}_\mu = 0 \tag{1.95}$$

We now develop a covariant equation of motion for the spin compatible with (1.92) and (1.93) and subject to the restriction that within the rest frame of the particle there is no torque acting on the particle causing its spin to change, i.e.,

$$d\mathbf{S}'/dx^{0\prime} = 0 \tag{1.96}$$

We speculate that if the nonrest frame observers detect a change in the spin, that change may depend on the spin itself and on the parameters characteristic of the motion of the particle, viz., position, velocity, and so forth. Therefore, we write quite generally

$$dS^\mu/ds = AS^\mu + Bx^\mu + Du^\mu + E\,du^\mu/ds \tag{1.97}$$

where A, B, D, and E are Lorentz scalars that remain to be determined. Higher derivatives of u^μ could be included but it can be shown that their coefficients must be zero. Consider now the requirement that in the rest frame (1.96) must be valid. Thus

$$\frac{dx^{0'}}{ds}\frac{dS^{j'}}{dx^{0'}} = AS^{j'} + Bx^{j'} + Du^{j'} + E\frac{du^{j'}}{ds} = 0 \tag{1.98}$$

Since $S^{j'}$, $x^{j'}$, $u^{j'}$, and $du^{j'}/ds$ are independent, (1.98) must be satisfied only if each term is zero. Thus we conclude that

$$A = B = E = 0 \tag{1.99}$$

The value of D is not determined because $u^{j'} = 0$ (the velocity is zero in the rest frame). Thus (1.97) reduces to

$$dS^\mu/ds = Du^\mu \tag{1.100}$$

To determine D we differentiate (1.93) with respect to the proper time and obtain

$$(du^\mu/ds)S_\mu + u^\mu\, dS_\mu/ds = 0 \tag{1.101}$$

But taking the scalar product of (1.100) with u_μ and using the fact that $u^\mu u_\mu = 1$ we find that

$$D = u_\mu\, dS^\mu/ds \tag{1.102}$$

Substituting (1.102) into (1.101) we find

$$D = -S_\mu\, du^\mu/ds \tag{1.103}$$

and the covariant equation of motion for the spin is

$$dS^\mu/ds = -S_\nu(du^\nu/ds)u^\mu \tag{1.104}$$

We now determine the rate of change of $\mathbf{S}'$ as observed by the lab (unprimed) frame, i.e., $d\mathbf{S}'/dx^0$. The lab frame spin is related to the rest frame spin by the inverse Lorentz transformation

$$S_\mu = L^\lambda{}_\mu\, S_\lambda' \tag{1.105}$$

Differentiating both sides of this equation with respect to laboratory time we find

$$\frac{dS_\mu}{dx^0} = \left(\frac{d}{dx^0} L^\lambda{}_\mu\right) S_\lambda' + L^\lambda{}_\mu \frac{dS_\lambda'}{dx^0} \tag{1.106}$$

Substituting (1.106) into (1.104) we obtain

$$L_\lambda{}^\mu \frac{dS^{\lambda\prime}}{dx^0} = -L^\lambda{}_\nu S_\lambda' \frac{du^\nu}{dx^0} u^\mu - \left(\frac{d}{dx^0} L^{\lambda\mu}\right) S_\lambda' \tag{1.107}$$

where x^0 and u^μ are the laboratory time and velocity of the particle, S_λ' is the rest frame spin, and $L^\mu{}_\lambda$ is the Lorentz transformation connecting the rest frame and the lab frame. To isolate $dS^{\lambda\prime}/dx^0$ we operate on both sides of (1.107) with the inverse of $L^\mu{}_\lambda$, viz., $L^\varrho{}_\mu$, and obtain

$$\frac{dS^{\varrho\prime}}{dx^0} = -L^\lambda{}_\nu S_\lambda' \frac{du^\nu}{dx^0} u^{\varrho\prime} - \left(\frac{d}{dx^0} L^\lambda{}_\mu\right) S_\lambda' L^{\varrho\mu} \tag{1.108}$$

where $u^{\varrho\prime} = L^\mu{}_\varrho u^\mu$ is the rest frame velocity. Since $u^{j\prime} = 0$ and $S^{0\prime} = 0$, (1.108) becomes

$$\frac{dS^{j\prime}}{dx^0} = -\left(\frac{d}{dx^0} L^k{}_\mu\right) S_k' L^{j\mu} \tag{1.109}$$

Inserting (1.66) and (1.70) into (1.109) we find

$$\frac{d\mathbf{S}'}{dt} = \frac{\gamma^2}{(\gamma+1)c^2} [(\mathbf{S}' \cdot \mathbf{a})\mathbf{v} - (\mathbf{S} \cdot \mathbf{v})\mathbf{a}] = \frac{\gamma^2}{(\gamma+1)c^2} \mathbf{S}' \times (\mathbf{v} \times \mathbf{a}) \tag{1.110}$$

where $\mathbf{a}$ is the acceleration of the particle as measured in the lab frame, i.e., $d\mathbf{v}/dt$. Equation (1.110) is known as the *Thomas Precession Equation* since it predicts that in the laboratory frame $\mathbf{S}'$ will precess. To see this recall [4] that the change of a radius vector is given by

$$d\mathbf{r} = \mathbf{r} \times d\mathbf{\Omega} \tag{1.111}$$

where $d\Omega$ is the element of solid angle swept out by $d\mathbf{r}$ in the time dt. By comparing the form of (1.111) with (1.110) we see that the differential element of precession angle for $\mathbf{S}'$ is

$$d\mathbf{\Omega}_{s'} = \frac{\gamma^2}{(\gamma+1)c^2} (\mathbf{v} \times \mathbf{a})\, dt = (\gamma - 1) \frac{(\mathbf{v} \times \mathbf{a})\, dt}{v^2} \tag{1.112}$$

or

$$\boldsymbol{\omega}_{s'} = d\mathbf{\Omega}_{s'}/dt = (\gamma - 1)(\mathbf{v} \times \mathbf{a})/v^2 \tag{1.113}$$

where $\boldsymbol{\omega}_{s'}$ is the rate of precession of the spin $\mathbf{S}'$.

From (1.111) it follows that

$$\mathbf{v} = \mathbf{r} \times d\mathbf{\Omega}/dt = \mathbf{r} \times \boldsymbol{\omega} \tag{1.114}$$

where $\mathbf{r}$ is a spatial radius vector locating the particle in the lab frame, $\boldsymbol{\omega}$ is the angular velocity, and $\mathbf{v}$ is the linear velocity of the particle in the lab frame. Now, if the particle is executing uniform circular motion in the lab frame, $d\boldsymbol{\omega}$ is zero and

$$d\mathbf{v} = d\mathbf{r} \times \boldsymbol{\omega} \tag{1.115}$$

Taking the vector product of (1.115) with $\mathbf{v}$, we obtain

$$\mathbf{v} \times d\mathbf{v} = \mathbf{v} \times (d\mathbf{r} \times \boldsymbol{\omega}) = (\mathbf{v} \cdot \boldsymbol{\omega})\, d\mathbf{r} - (\mathbf{v} \cdot d\mathbf{r})\boldsymbol{\omega} \tag{1.116}$$

But it follows from (1.114) that $\mathbf{v} \cdot \boldsymbol{\omega} = 0$. Thus dividing (1.116) by dt we find

$$\boldsymbol{\omega} = \mathbf{a} \times \mathbf{v}/v^2 \tag{1.117}$$

Thus for uniform circular motion we may write (1.113) as

$$\boldsymbol{\omega}_{s'} = -(\gamma - 1)\boldsymbol{\omega} \tag{1.118}$$

The minus sign indicates that the spin precesses in a direction opposite to that of the particle; i.e., if the orbit of the particle is counterclockwise the precession of the spin vector is clockwise, and vice versa. The most famous illustration of this relativistic precession of spin is the precession of the spin of the electron while in orbit about the nucleus of an atom.

J. Rest Frame of the System (Center-of-Momentum Frame)

In Newtonian physics it is often advantageous to employ a center-of-mass reference frame when investigating the motion of a system of particles. In Einsteinian dynamics the center-of-mass system is not at all advantageous, but an analogous system, the center of momentum, is. The *center-of-momentum* frame or equivalently the *system rest frame* is defined as the reference system for which the total three-momentum of the particles comprising the system is zero, i.e.,

$$\hat{P}^j = \sum_\alpha {}_\alpha p^j = 0 \tag{1.119}$$

Since a discussion of dynamics is reserved for Chapter 2, we consider here only *collision* or *scattering* problems in which there is initially a system of noninteracting, i.e., free, particles which eventually interact (collide) with one another over a finite period of time and then once again become free (noninteracting). It is shown in Chapter 3 that the four-momentum P^μ of the system should be conserved. This represents a conservation of both energy and three-momentum. The essential features of this type of problem are illustrated by the two-particle system represented in Fig. 1.5.

Fig. 1.5. (a) Collision of particles 1 and 2 as observed in laboratory frame. Particles 3 and 4 emerge from the collision. (b) Collision of particles 1 and 2 as observed in the center-of-momentum frame.

Figure 1.5a represents the collision of particles 1 and 2 as observed by an observer in the "laboratory frame"; after collision he observes the free particles 3 and 4. The total momentum before and after collision is

$$_BP^\mu = {}_1p^\mu + {}_2p^\mu \tag{1.120a}$$

$$_AP^\mu = {}_3p^\mu + {}_4p^\mu \tag{1.120b}$$

Let us now introduce the rest frame of the system where the total momentum is

$$_BP^{\mu'} = {}_1p^{\mu'} + {}_2p^{\mu'} \tag{1.121a}$$

$$_Bp^{j'} = {}_1p^{j'} + {}_2p^{j'} = 0 \tag{1.121b}$$

and from the conservation of momentum it follows that

$$_AP^{j'} = {}_3p^{j'} + {}_4p^{j'} = 0 \tag{1.122}$$

Thus in the rest system

$$_3p^{j'} = -{}_4p^{j'} \tag{1.123a}$$

$$_2p^{j'} = -{}_1p^{j'} \tag{1.123b}$$

as shown in Fig. 1.5b.

If the rest frame is an inertial frame it is related to the laboratory frame by a Lorentz transformation, viz.,

$$_BP^{\mu'} = L^{\mu}{}_{\nu}\, {}_BP^{\mu} \tag{1.124}$$

That there is such a frame follows from the fact that $_BP^{\mu}$ is a timelike vector, since one can always find an inertial frame for which a timelike vector has all of its spatial components equal to zero. (See Problem 1.5.) From (1.124) and (1.121a) one can determine the velocity of the rest frame relative to the lab frame. The process is facilitated by assuming that the lab axes are parallel to the rest system axes and that the x^1 axis has the same direction as $_BP^j$. It is left to the reader to verify that this is possible. (See Problem 1.19.) We find that

$$\beta^1 = {}_BP^1/{}_BP^0, \qquad \beta^2 = \beta^3 = 0 \tag{1.125}$$

i.e., the rest frame is moving in the same direction as the total momentum.

Quantities as measured in the lab frame are related to quantities as measured in the center-of-momentum frame and vice versa by the Lorentz transformation connecting the two frames. It is often simpler, however, to make use of the invariance of the scalar products of four-vectors to determine the relation between quantities. Thus to determine the total energy of a system as measured by the center-of-momentum frame we write

$$(P^{\mu'})^2 = (P^{0'})^2 - (\mathbf{P}')^2 = (P^{\mu})^2 = (P^0)^2 - (\mathbf{P})^2 \tag{1.126}$$

where $P^{\mu'}$ is the four-momentum as measured by the center-of-momentum frame and P^{μ} is the four-momentum as measured by the lab frame. But in the center-of-momentum frame, $\mathbf{P}' = 0$ and (1.126) becomes

$$P^{0'} = [(P^0)^2 - (\mathbf{P})^2]^{1/2} \tag{1.127}$$

K. Distribution Functions and Scattering Cross Sections

Many experiments are designed to determine the distribution functions of physical variables. Thus in a scattering experiment in which a beam of protons is incident on a target one is usually interested in the energy-momentum distribution of the emerging particles, i.e., in how many particles were scattering through an angle θ_1, with an energy E_1, how many were scattered through an angle θ_2 with energy E_2, and so forth. If the measured variable is x, then the *distribution function* $f(x)$ is defined as that function which when integrated from x_1 to x_2 yields the number of events for which the variable has a measured value in the range x_1 to x_2. In other words $f(x)\,dx$ is the number of events in which the variable has a value between x and $x + dx$. If we normalize to a single event $f(x)$ is the probability density of x. Our concern is with how a distribution function transforms from one Lorentz frame to another.

If we integrate the distribution function over all possible values of the variable, we obtain the total number of events (or one if we are dealing with the probability density) which should be the same for all Lorentz observers. Thus we can require

$$f'(x')\,dx' = f(x)\,dx \tag{1.128}$$

where the prime designates variables and distributions as measured by the primed observer. The variables of interest are often the components of four-momentum for which (1.128) becomes

$$N(p^\mu)\,d^4p = N'(p^{\mu\prime})\,d^4p' \tag{1.129}$$

where $d^4p = dp^0\,dp^1\,dp^2\,dp^3$ is the four-dimensional volume element and $N(p^\mu)\,d^4p$. It can be shown [5] that a coordinate transformation produces a change in the volume element such that the original and new volume elements are related by

$$d^4p' = J\,d^4p \tag{1.130}$$

where J is the Jacobian of the transformation from p^μ to $p^{\mu\prime}$, i.e.,

$$J \equiv \frac{\partial(p^{0\prime}p^{1\prime}p^{2\prime}p^{3\prime})}{\partial(p^0p^1p^2p^3)} = \begin{vmatrix} \frac{\partial p^{0\prime}}{\partial p^0} & \frac{\partial p^{1\prime}}{\partial p^0} & \frac{\partial p^{2\prime}}{\partial p^0} & \frac{\partial p^{3\prime}}{\partial p^0} \\ \frac{\partial p^{0\prime}}{\partial p^1} & \frac{\partial p^{1\prime}}{\partial p^1} & \cdot & \vdots \\ \frac{\partial p^{0\prime}}{\partial p^2} & \cdot & & \\ \frac{\partial p^{0\prime}}{\partial p^3} & & \cdots & \frac{\partial p^{3\prime}}{\partial p^3} \end{vmatrix} \tag{1.131}$$

It is not difficult to show that the Jacobian of an orthogonal transformation is unity. Thus for the Lorentz transformation (1.130) becomes

$$d^4p' = d^4p \tag{1.132}$$

and from (1.12a) it follows that $N(p^\mu)$ transforms as a scalar function. The restriction on the possible values the momentum may have is expressed by (1.51) and may be explicitly inserted into our discussion of the momentum distribution by means of the Dirac delta function (the properties of the Dirac delta function are summarized by Butkov [6]). Thus we may write

$$N(p^\mu)\,\delta(p^\mu p_\mu - m^2c^2)\,d^4p = N'(p')\,\delta(p^{\mu\prime} p_\mu{}' - m^2c^2)\,d^4p' \qquad (1.133)$$

Integrating both sides of (1.133) over $dp_0(0 \leq p_0 \leq \infty$ since p^μ is a timelike vector), we obtain

$$N(E/c, \mathbf{p})\,d\mathbf{p}/E = N'(E'/c, \mathbf{p})\,d\mathbf{p}'/E' \qquad (1.134)$$

where $d\mathbf{p} = dp^1\,dp^2\,dp^3$ is the three-dimensional volume element and E is the total energy, i.e.,

$$E = (\mathbf{p}^2 + m^2c^2)^{1/2}$$

From our previous results we know that

$$N = N' \qquad (1.135)$$

and therefore

$$d\mathbf{p}/E = d\mathbf{p}'/E' \qquad (1.136)$$

or the Jacobian of the transformation is just

$$J = \partial(p^1p^2p^3)/\partial(p^{1\prime}p^{2\prime}p^{3\prime}) = E/E' \qquad (1.137)$$

In many experiments it is advantageous to express distribution functions in terms of spherical coordinates rather than cartesian coordinates. As in (1.128) we may write

$$g(y)\,dy = f(x)\,dx \qquad (1.138)$$

where $g(y)$ is the distribution function expressed in terms of the new coordinates, y, which are related to the original coordinates by

$$y = y(x) \qquad (1.139)$$

Thus for a momentum distribution which is to be expressed in terms of spherical coordinates we have

$$f(\mathbf{p})\,d\mathbf{p} = f(\mathbf{p}(p, \Omega))p^2\,dp\,d\Omega \equiv g(p, \Omega)\,dp\,d\Omega \qquad (1.140)$$

where

$$p^2 = (p^1)^2 + (p^2)^2 + (p^3)^2 \tag{1.141a}$$

$$d\Omega = \text{element of solid angle} = \sin^2\theta \, d\theta \, d\phi \tag{1.141b}$$

The $f(\mathbf{p})$ that appears in (1.140) corresponds to the N/E that appears in (1.134). Thus in terms of f, (1.135) may be written

$$f(\mathbf{p})E = f'(\mathbf{p}')E' \tag{1.142}$$

or using (1.140)

$$g(p, \Omega)\, E/p^2 = g'(p', \Omega')\, E'/p'^2 \tag{1.143}$$

where (1.143) relates the spherical distribution functions as measured by the primed and unprimed Lorentz observers. A similar analysis in which the volume element in (1.133) is expressed as $dp^0 \, dp \, d\Omega$ and the integration is performed over p, the magnitude of the three-momentum, yields an energy distribution function for which

$$h(E, \Omega)/p = h'(E', \Omega')/p' \tag{1.144}$$

The *cross section* σ provides a measure of the probability of a given event and in scattering problems it is usually defined as the number of a certain kind of process divided by the number of particles incident on the target. Thus in an experiment in which high-energy protons strike a hydrogen rich target, one may observe, among other things, the production of antiprotons. The number of antiprotons produced in a given time divided by the number of incident protons in that same time is the cross section for antiproton production. The *differential cross section* for the scattering of particles having momentum between p and $p + dp$ into a solid angle between Ω and $\Omega + d\Omega$ is defined in terms of the total cross section for scattering σ by the equation

$$\sigma = \iint dp \, d\Omega \left(\frac{d^2\sigma}{dp \, d\Omega} \right) \tag{1.145}$$

where the integrand is the differential scattering cross section. It is clear from its definition that the differential cross section is a distribution function and therefore it satisfies (1.143). Thus

$$\frac{d^2\sigma}{dp \, d\Omega} \frac{E}{p^2} = \frac{d^2\sigma'}{dp' \, d\Omega'} \frac{E'}{p'^2} \tag{1.146}$$

and

$$\frac{d^2\sigma}{dE \, d\Omega} \frac{1}{p} = \frac{d^2\sigma'}{dE' \, d\Omega'} \frac{1}{p'} \tag{1.147}$$

L. Tachyons

A particle moving with a speed greater than the speed of light is called a *tachyon*. It has been asserted for years [4] that the existence of tachyons is not compatible with Einstein's relativity. Recently, however, arguments were advanced claiming that tachyons may indeed exist [7, 8] in harmony with relativity. This has stimulated a lively debate and the physics literature has been inundated with papers for and against the existence of tachyons [9–12]. As this is written a definitive answer to the question is not available, but it would appear that opponents to their existence have the upper hand. In any event, the very disagreement concerning this old fundamental question testifies to the enormous vitality of every branch of physics.

Tachyon advocates agree with the conventional argument that it is impossible to *accelerate* a particle with a speed less than that of light (subluminal) to a speed greater than that of light (superluminal) because the mass approaches infinity as $v \to c$. They argue that just as photons and neutrinos travel with the speed of light without being accelerated to c, is it not possible that there exist particles whose speed is *always* greater than c?

The concept of free tachyons is compatible with our assertion that free particle trajectories are geodesics; they correspond to the class of geodesics designated by (1.49b).

Some of the implications of a tachyon theory are illustrated in Fig. 1.6a, which presents a graphical representation of (1.52). It is not difficult to show that the slope of the hyperbola at a given point yields the velocity of the particle, i.e.,

$$v_x = dE/dp_x \tag{1.148}$$

The two intersecting lines are the degenerate hyperbolas corresponding to particles of zero rest mass moving with the velocity of light (they are called *luxons*). The hyperbolas opening to the top and to the bottom have their slopes everywhere smaller than the slope of the luxons and therefore these hyperbolas represent particles moving with speeds less than the speed of light, called *tardyons*. The upper curve corresponds to positive energies and the lower one corresponds to negative energies which are unobservable. Each point on the hyperbola gives the energy and momentum of the particle in a given reference frame. Thus Lorentz transformations connect the various points comprising the hyperbola but no Lorentz transformation connects the branch of positive energies with the branch of negative energies.

The hyperbolas opening to the right and left correspond to the tachyons since the slope is everywhere greater than the slope of the luxons. Now it is

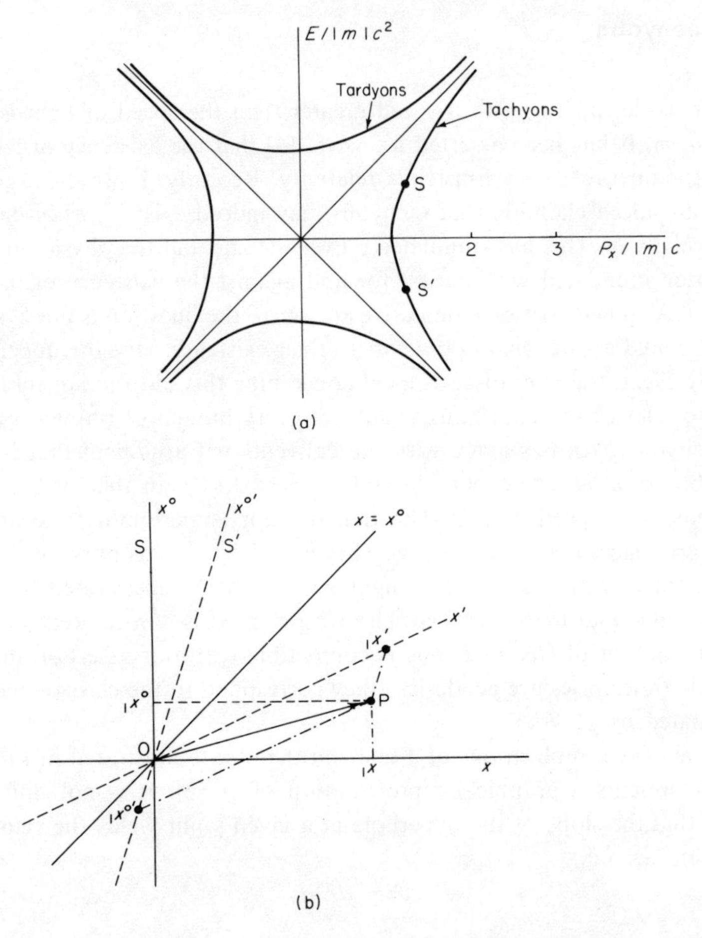

Fig. 1.6. (a) The relativistic energy versus momentum curves. The slope of the curves is the velocity of the particles. (b) The Lorentz diagram shows the reference system for S and S′ of Fig. 1.3a and forms the basis of the reinterpretation principle.

immediately obvious that for some Lorentz observers the tachyons would have positive energy and be observable; whereas, for other Lorentz observers they would presumably have negative energies and be unobservable, a most unpleasant situation. To avoid this conclusion the resourceful tachyon advocates have developed the *reinterpretation principle*, the basis of which is Fig. 1.6b.

The observers S and S' appearing in Fig. 1.6a are represented on the Lorentz diagram of Fig. 1.3b by the unprimed and primed reference axes as shown. To the S observer, a tachyon event at the origin, e.g., the emission of a tachyon at $x = 0$, $t = 0$, would at some subsequent time be observed at the event P which may for the sake of discussion be the absorption of the tachyon at $x = x_1$, $t = t_1$. Now, however, to the S' observer the event P occurs *before* the event 0 and this forms the basis of the reinterpretation principle. According to this principle S' would observe the tachyon to be emitted at P and absorbed at 0. Thus he would disagree with the S observer on what was the emitter and what was the observer and also on the direction of propagation of the tachyon, i.e., S says the tachyon is moving away from the origin and S' says it is moving toward the origin. These differences of appearance are no more unpalatable than "length contraction" or "time dilation" and thus it appears that the tachyon hypothesis is compatible with relativity.

The tachyon hypothesis has several interesting implications. From (1.46) and (1.50) it is clear that for the momentum p_k to be real (measurable) it is necessary that the mass m must be imaginary to counteract the imaginary γ, i.e.,

$$m^2 < 0 \tag{1.149}$$

That this is acceptable follows from the fact that the mass of the tachyon m is not directly measurable because there is no "proper" or rest inertial frame for a tachyon which follows from the fact that the Lorentz transformation from a Lorentz observer to an observer moving with a speed greater than that of light would yield imaginary coordinates (because of the γ appearing in the transformations).

Most of the efforts to destroy the tachyon hypothesis have centered around attempts to prove that the hypothesis leads to violations of causality. To represent this type of attack, we present a situation developed by Pirani [13].

The experimental arrangement is represented in Fig. 1.7. The four observers A, B, C, and D move in the xy plane. A emits tachyon 1. B receives tachyon 1 and at once emits tachyon 2. C receives tachyon 2 and at once emits tachyon 3. D receives tachyon 3 and at once emits tachyon 4. A receives tachyon 4 before his emission of tachyon 1. If we arrange that reception of a tachyon by A triggers the destruction of his emitter we have an untenable result and Pirani concludes that the tachyon hypothesis cannot be sustained.

Although arguments such as the one just presented are impressive, they will probably not convince a tachyon enthusiast who will question the concept of causality embodied in the argument or the reception–emission mechanisms required in the experiment. The debate will undoubtably

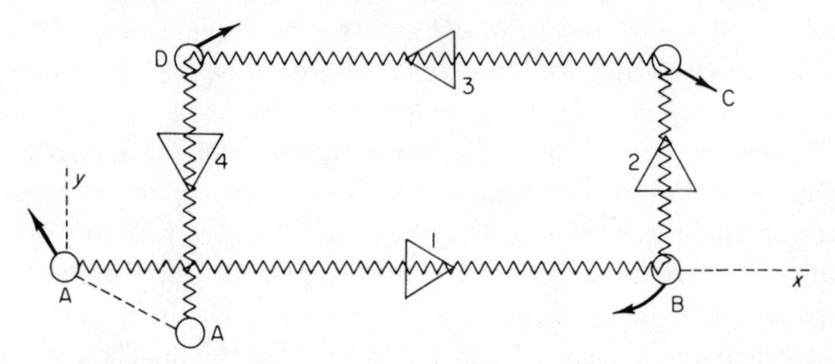

Fig. 1.7. Experimental arrangement for causality violations with tachyons. Wiggly lines are tachyon paths and solid arrows denote observer three-velocities.

continue, but it would seem that the existence of tachyons is tenable only if we are willing to relinquish cherished concepts about causality and/or emission–absorption processes.

1.3. SOLVED EXAMPLES

Example 1.1. An astronaut on an inertial space platform observes a space ship moving at a speed of $0.6c$. In checking his space ship identification manual, he finds a ship, the "Galaxy", that resembles the approaching ship but the manual lists the length of the ship as 120 m whereas he measures it to be 96 m long. The manual says that the "Galaxy" emits light pulses at 80-sec intervals but he measures the interval between pulses to be 100 sec. Could the approaching ship be the "Galaxy"?

Solution To answer the question posed by this problem, one must determine the relationship between space and time intervals as measured by the two inertial systems. An inertial system in which an object is at rest is called the *proper* reference system and times and lengths measured in that system are called *proper times* and *proper lengths.* In this problem the proper system is a system attached to the space ship and we shall designate this system as the S′ system. The identification manual lists proper times and lengths. We shall designate the astronaut's system as the S system and assume an orientation of S and S′ such that (1.71) is valid. Now

$$X = L^{-1}X' \tag{1.150}$$

where it follows from (1.71) that

$$L^{-1} = \begin{bmatrix} \gamma & \beta\gamma & 0 & 0 \\ \beta\gamma & \gamma & 0 & 0 \\ 0 & 0 & 1 & 0 \\ 0 & 0 & 0 & 1 \end{bmatrix} \tag{1.151}$$

Therefore,

$$x^0 = \gamma x^{0'} + \beta\gamma x^{1'}, \qquad x^1 = \beta\gamma x^{0'} + \gamma x^{1'} \tag{1.152}$$

Now the length of the space ship in the S frame is

$$l = {}_2x^1 - {}_1x^1 \tag{1.153}$$

where ${}_2x^1$ and ${}_1x^1$ are the positions of the front and rear of the spaceship at the instant x^0. Thus

$$l = \beta\gamma({}_2x^{0'} - {}_1x^{0'}) + \gamma({}_2x^{1'} - {}_1x^{1'}) = \beta\gamma\, \Delta x^{0'} + \gamma l'$$

But from the x^0 equation and the fact that $\Delta x^0 = 0$ it follows that

$$\beta\gamma\, \Delta x^{0'} = -\beta^2\gamma l'$$

Therefore

$$l = -\beta^2\gamma l' + \gamma l' = (1 - \beta^2)^{1/2} l' \tag{1.154}$$

This is the famous "length contraction" formula and for $\beta = 0.6$ and $l' = 120$ m we find

$$l = (0.8)(120) = 96 \quad \text{m}$$

To investigate the time interval between light pulses we simply use the x^0 equation and obtain

$$\Delta x^0 = \gamma\, \Delta x^{0'} \tag{1.155}$$

since $\Delta x^{1'}$ (the change in the position of the pulsing light) is zero in the S′ frame. This is the famous *time dilation* formula and in our case

$$\Delta x^0 = 100 \quad \text{sec}$$

We conclude that the ship could be the "Galaxy."

Example 1.2. Observer A finds that D is moving away from him in the horizontal direction with a speed of $0.6c$. A also notes that B is moving toward him with a speed of $0.8c$. A, B, and D are arranged as shown in Fig. 1.8. D decides to emit a continuous beam of green light ($\lambda = 0.54\ \text{m}\mu$) at an angle of 170° to his direction of motion. The beam reflects off B to A.

(a) With what speed does B observe D to be moving?
(b) What is the wavelength of the light seen by A?
(c) At what angle can A expect to observe the beam?

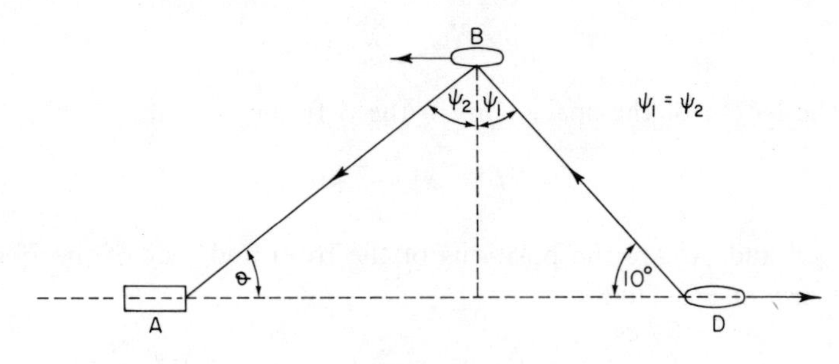

Fig. 1.8

Solution. In this problem there is aberration and frequency shift as the light goes from D to B and also as it goes from B to A. Let us first calculate the speed of D relative to B. We can use (1.75) if we identify D with S″, B with S′, and A with S. Then β is the velocity of D relative to A, β_1 is the velocity of B relative to A, and β_2 is the velocity of D relative to B. Therefore, we wish to solve (1.75) for β_2.

$$\beta_2 = (\beta - \beta_1)/(1 - \beta_1\beta) = 1.40/1.48 = 0.946$$

Now we determine ψ_1 and λ_B the angle and wavelength of light incident on B. The primed quantities in (1.86) and (1.87) correspond to the frequency and angle as measured by D whereas the unprimed quantities are those measured by B. Therefore, we solve (1.86) and (1.87) for the unprimed quantities and obtain

$$\omega = \gamma\omega'(1 + \beta \cos \alpha') \tag{1.156}$$

$$\cos \alpha = (\cos \alpha' + \beta)/(1 + \beta \cos \alpha') \tag{1.157}$$

Therefore,

$$2\pi c/\lambda_B = (2\pi c/\lambda_D)\gamma_2(1 + \beta_2 \cos \alpha')$$

$$\lambda_D = (1 - \beta_2{}^2)^{1/2}\lambda_D/(1 + \beta_2 \cos \alpha')$$

$$\lambda_B = (1 - (0.946)^2]^{1/2}(0.54\ \text{m}\mu)/(1 + 0.946 \cos 170°) = 2.56\ \text{m}\mu$$

The incident angle ψ_1 is given by

$$\cos(90° + \psi_1) = -\sin\psi_1 = \frac{\cos(170°) + (0.946)}{1 + (0.946 \cos 170°)} = -0.5674$$

Therefore,

$$\psi_1 = 34°34'$$

By the laws of reflection $\psi_2 = \psi_1$. Thus the beam makes an angle of 235°26′ with the horizontal as it leaves B (i.e., as measured by B). A receives the light at the angle θ which is obtained from

$$\cos\alpha = \frac{\cos 235°26' - 0.8}{1 + (-0.8)\cos 235°26'} = -0.9418$$

Therefore,

$$\alpha = 199°39' \qquad \text{or} \quad \theta = 19°39'$$

The wavelength observed by A is

$$\lambda_A = (1 - \beta_1{}^2)^{1/2}\lambda_B/(1 + \beta_1 \cos\alpha')$$
$$\lambda_A = (1 - 0.64)^{1/2}(2.56\ \text{m}\mu)/[1 + (0.8)(0.5674)] = 1.056\ \text{m}\mu\ \text{(infrared)}$$

Example 1.3. A negative beam of pions while passing through a target of liquid hydrogen interacts with a proton to produce, among other things, a neutral kaon K°. The kaon moves downstream with constant velocity. While passing through a spark chamber it decays into a pair of pions. What is the lifetime of the kaon if event #1 (kaon production) occurred at $x = y = z = t = 0$ and event #2 (kaon decay) occurred at $x = 1.4$ m, $y = 0$, $z = 0$, and $t_2 = 5.3 \times 10^{-9}$ sec.

Solution. This problem deals with the relativity of time in different reference frames. In the laboratory frame the lifetime of the kaon is measured to be 5.3×10^{-9} sec $(t_2 - t_1)$. However, in the rest frame of the kaon, the proper frame, the lifetime will be different and it is this proper lifetime that we wish to determine, i.e., we want to know the lifetime of the kaon according to its clock. To find this we simply use the invariance of the interval. In the lab frame

$$-s^2 = (\Delta x)^2 + (\Delta y)^2 + (\Delta z)^2 - (\Delta ct)^2 = 1.96 + 0 + 0 - 2.53 = -.57\ \text{m}^2$$

In the proper frame

$$-(s')^2 = (\Delta x')^2 + (\Delta y')^2 + (\Delta z')^2 - c^2(\Delta t')^2 = 0 + 0 + 0 - c^2(\Delta t')^2$$

But $(s')^2 = s^2$. Therefore,

$$\Delta t' = 2.5 \times 10^{-9} \text{ sec.}$$

Example 1.4. (a) The Weston accelerator is designed to produce 200 GeV protons. Determine the energy of the proton–proton system in the center-of-momentum frame if the proton beam is incident on a proton target that is at rest in the lab frame. (b) Compare the results in (a) with the center-of-momentum energy for the proton–proton colliding beam system at CERN. Each proton beam is accelerated to 28 GeV in the lab frame prior to their head-on collision.

Solution. (a) The center-of-momentum energy is expressed in terms of lab quantities in (1.127). Thus

$$cP^{0\prime} = E' = c[(P^0)^2 - (\mathbf{P})^2]^{1/2}$$

where

$$P^0 = m_1 c\gamma_1 + m_2 c\gamma_2, \qquad \mathbf{P} = m_1 c\gamma_1\, {}_1\boldsymbol{\beta} + m_2 c\gamma_2\, {}_2\boldsymbol{\beta}$$

But the target mass m_2 is equal to the beam particle mass m_1. Also the target particle is at rest. Therefore,

$$P^0 = mc(\gamma + 1), \qquad \mathbf{P} = mc\gamma\boldsymbol{\beta}, \qquad \text{or} \qquad \gamma = (P^0 - mc)/mc$$

Inserting these relations into our energy expression we obtain

$$E' = (2mc^2E)^{1/2} = [2(0.938)(200)]^{1/2} \text{ GeV} = 19.4 \text{ GeV}$$

(b) In the colliding beam experiment the lab frame is also the center-of-momentum frame. Thus

$$E' = E = E_1 + E_2 = 28 \text{ GeV} + 28 \text{ GeV} = 56 \text{ GeV}$$

The important point is that the maximum energy available for the production of a new particle is just the center-of-momentum energy of the proton–proton system (i.e., the new particle can have a rest mass energy equal to E').

(Why?) Thus the CERN accelerator can create more massive particles than the Weston accelerator. Can you suggest an advantage of the Weston-type accelerator?

Example 1.5. In the reaction

$$\pi^- + p^+ \rightarrow \mathrm{K}^0 + \Lambda^0$$

the target proton is at rest in the lab frame. (a) What is the energy of the pion at threshold? (b) In an experiment in which the pions have a momentum 2.50 GeV/c the Λ^0 are observed to have a monentum 0.60 GeV/c at an angle of 45° with respect to the incident pion. What is the velocity of the center-of-momentum frame? What is the momentum of the K^0 in the lab frame and the center-of-momentum frame?

Solution. (a) Threshold is defined as the energy to make the particles K^0 and Λ° at rest in the center-of-momentum frame. Because of the conservation of four-momentum, we have in the center-of-momentum frame

$$P_{\mathrm{B}}^{0\prime} = P_{\mathrm{A}}^{0\prime} \qquad \text{and} \qquad \mathbf{P}_{\mathrm{B}}{}' = \mathbf{P}_{\mathrm{A}}{}' = 0$$

But

$$P_{\mathrm{B}}^{\mu\prime} P'_{\mu\mathrm{B}} = (P_{\mathrm{B}}^{0\prime})^2 = P_{\mathrm{B}}{}^{\mu} P_{\mu\mathrm{B}} = (P_{\mathrm{B}}{}^0)^2 - (\mathbf{P}_{\mathrm{B}})^2 = (P_{\mathrm{A}}^{0\prime})^2$$

where $P_B{}^\mu$ is the total four-momentum in the lab frame before collision. Now

$$\mathbf{P}_{\mathrm{B}} = \mathbf{p}_\pi + \mathbf{p}_p = \mathbf{p}_\pi, \qquad P_B{}^0 = p_\pi{}^0 + p_p{}^0, \qquad p_p{}^0 = m_p c$$

But

$$p_\pi{}^\mu p_{\mu\pi} = (p_\pi{}^0)^2 - (\mathbf{p}_\pi)^2 = m_\pi{}^2 c^2$$

or

$$(\mathbf{p}_\pi)^2 = (p_\pi{}^0)^2 - m_\pi{}^2 c^2$$

Thus

$$(P_{\mathrm{A}}^{0\prime})^2 = (p_\pi{}^0 + p_p{}^0)^2 - (p_\pi{}^0)^2 + m_\pi{}^2 c^2$$

$$E_\pi = \frac{(E_{\mathrm{A}}{}')^2 - (m_\pi c^2)^2 - (m_p c^2)^2}{2 m_p c^2}$$

$$= \frac{(1.614)^2 - (0.140)^2 - (0.938)^2}{2(0.938)} \text{ GeV} = 0.909 \text{ GeV}$$

(b) The speed of the center-of-momentum frame is given by (1.125). Thus

$$\mathbf{B} = \mathbf{P}_B/P_B{}^0 = c\mathbf{P}_B/E_B$$

But

$$P_B{}^0 = p_\pi{}^0 + p_p{}^0 = [m_\pi{}^2c^2 + (\mathbf{p}_\pi)^2]^{1/2} + m_p{}^2c^2$$

Therefore,

$$\beta = \frac{c \mid \mathbf{p}_\pi \mid}{[(m_\pi c^2)^2 + (c \mid \mathbf{p}_\pi \mid)^2]^{1/2} + m_p c^2}$$

$$= \frac{2.50}{[(0.140)^2 + (2.50)^2]^{1/2} + (0.938)^2} = 0.739$$

$$\gamma = 1/(1 - \beta^2)^{1/2} = 1.484$$

The momentum of the K^0 in the lab frame follows directly from the conservation of momentum. Thus

$$\mathbf{p}_p + \mathbf{p}_\pi = \mathbf{p}_{K^0} + \mathbf{p}_\Lambda \Rightarrow \mathbf{p}_{K^0} = \mathbf{p}_\pi - \mathbf{p}_\Lambda$$

But

$$\mathbf{p}_\pi = 2.50 \ (\mathrm{GeV}/c)\, \hat{\imath}$$

and

$$\mathbf{p}_\Lambda = p_\Lambda \cos 45^\circ\, \hat{\imath} + p_\Lambda \sin 45^\circ\, \hat{\jmath} = (0.60 \ \mathrm{GeV/c})(\cos 45^\circ\, \hat{\imath} + \sin 45^\circ\, \hat{\jmath})$$

Therefore,

$$\mathbf{p}_K = 2.08 \ (\mathrm{GeV}/c)\, \hat{\imath} - 0.42 \ (\mathrm{GeV}/c)\, \hat{\jmath}$$

or

$$\mid \mathbf{p}_K \mid = 2.12 \ \mathrm{GeV}/c$$

and

$$\tan\theta = -0.42/2.08 = 0.2019 \quad \Rightarrow \quad \theta = 11^\circ 25'$$

To determine the momentum of K^0 in the center-of-momentum frame we may transform the lab frame momentum to the center-of-momentum frame by using the Lorentz transformation to a frame moving with a speed of $\beta = 0.739$. Thus

$$p^{\mu'}_K = L^\mu{}_\nu\, p_K{}^\nu$$

and

$$\mathbf{p}_K{}' = (-\beta\gamma p_K{}^0 + \gamma p_K{}^1)\,\hat{\mathbf{i}} + p_K{}^2\,\hat{\mathbf{j}}$$

Now

$$p_K{}^0 = \frac{1}{c}\,[(m_K c^2)^2 + (c\mathbf{p}_K)^2]^{1/2} = [(0.498)^2 + (2.12)^2]^{1/2} = 2.18 \quad \text{GeV}/c$$

Therefore

$$\begin{aligned}\mathbf{p}_K{}' &= [(-0.739)(1.484)(2.18) + (1.484)(2.08)]\,\hat{\imath} - 0.42\,\hat{\jmath} \\ &= (0.696\,\hat{\imath} - 0.42\,\hat{\jmath}) \quad \text{GeV}/c\end{aligned}$$

and

$$|\,\mathbf{p}_K{}'\,| = 0.813 \quad \text{GeV}/c, \quad \theta' = \tan^{-1}\frac{-0.42}{0.696} = \tan^{-1}(-0.6034) = -31°6'$$

Example 1.6. A beam of pions having a momentum of 10.0 GeV/c serves as a source of neutrinos by decaying into a vacuum. In the rest frame of the pions, the neutrino momentum distribution is given by $g'(p', \Omega') = (1/4\pi)\,\delta(p' - a)$, where $\delta(p' - a)$ is the Dirac delta function and $a = 29.8 \times 10^{-3}$ GeV/c. Determine the momentum distribution of the neutrinos in the lab frame.

Solution. A delta function distribution implies that the distribution is isotropic in the center-of-momentum frame. From (1.143) we find that

$$g(p, \Omega) = (p/p')^2(E'/E)g'(p', \Omega')$$

Since the neutrino is massless, its four-momentum is a null vector and therefore

$$E = cp, \qquad E' = cp'$$

Thus

$$g(p, \Omega) = (p/p')(1/4\pi)\,\delta(p' - a)$$

To express $g(p, \Omega)$ properly we must express p' in terms of p and Ω. This is accomplished by noting that

$$p^{0\prime} = L^0{}_\nu\, p^\nu = L^0{}_0\, p^0 + L^0{}_3\, p^3$$

We have chosen our axes such that the momentum is along the polar axis, i.e., the x^3 axis. Thus $L^0{}_1 = L^0{}_2 = 0$ and $p^3 = p\cos\theta$. Since the neutrinos are massless

$$p^{0\prime} = p' = \gamma p^0 + (-\beta\gamma)p\cos\theta = \gamma p(1 - \beta\cos\theta)$$

To determine β and γ we use (1.125) (the rest frame of the pion is the center-of-momentum frame) and the fact that $p_\pi^0 = [(m_\pi c^2)^2 + (c\mathbf{p}_\pi)^2]^{1/2}$ to obtain

$$\beta = \frac{10}{[(0.140)^2 + (10)^2]^{1/2}} = 0.9999, \qquad \gamma = 71.42$$

The distribution in the lab frame is

$$g(p, \Omega) = \frac{1}{4\pi\gamma(1 - \beta\cos\theta)}\,\delta([1 - \beta\cos\theta]\gamma p - a)$$

i.e., the distribution is not isotropic in the lab frame. The delta function tells us that there is a contribution to the momentum distribution only when the argument of the delta function is zero, i.e.,

$$p = a/\gamma(1 - \beta\cos\theta)$$

Thus the maximum and minimum momenta in the distribution occur at $\theta = 0^0$ and $\theta = 180°$, respectively, and have the values

$$p_{\max} = \frac{a}{\gamma(1-\beta)} = \frac{29.8\times 10^{-3}}{71.42(10^{-4})} = 4.17 \quad \mathrm{GeV}/c$$

$$p_{\min} = \frac{a}{\gamma(1+\beta)} = \frac{29.8\times 10^{-3}}{(71.42)(2)} = 0.209 \quad \mathrm{MeV}/c$$

Now

$$g(\Omega) = \int_0^\infty dp\, p^2\, g(p, \Omega) = \frac{1}{4\pi}\,\frac{a^2}{\gamma^2(1 - \beta\cos\theta)^2}$$

gives the neutrino angular distribution and we see that it is rather sharply peaked in the forward direction. On the other hand,

$$g(p) \equiv \iint d\varphi\, d\theta\, \sin\theta\, g(p, \Omega) = \int_0^{2\pi} d\varphi \int d(\cos\theta)\,\frac{\delta[(1 - \beta\cos\theta)\gamma p - a]}{4\pi\gamma(1 - \beta\cos\theta)}$$

Letting $w = p\gamma(1 - \beta\cos\theta)$ we find

$$g(p) = \int_0^{2\pi} d\varphi \int \frac{dw}{-p\gamma\beta}\,\frac{1}{4\pi}\,\frac{p}{v}\,\delta(w - a) = \frac{1}{2\gamma\beta a}$$

i.e., in the lab frame the momentum distribution is flat between $p_{\min}$ and $p_{\max}$ (Fig. 1.9).

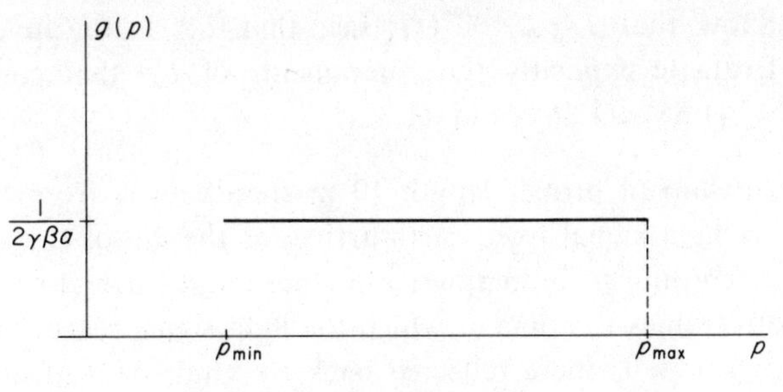

Fig. 1.9

PROBLEMS

1.1 Show that $\Gamma^{\varrho}_{\mu\nu} = 0$ for a Euclidean and Lorentz metric.

1.2 Construct a matrix which rotates the coordinate axes about the z axis through an angle ϕ and show that the determinant is $+1$.

1.3 Construct the matrix which inverts the x axis and show that the determinant is -1.

1.4 Determine the metric tensor for (a) a three-dimensional Euclidean space when cylindrical coordinates are used, and for (b) the two-dimensional space on the surface of a sphere.

1.5 Carry out the analysis necessary to show that timelike intervals permit cause–effect relationships to be established.

1.6 Show that if A^{μ} is timelike, then any other vector orthogonal to it is spacelike.

1.7 An event A occurs at (0, 0, 0, 0). Another event B occurs at (3, 3, 4, 0) cm. Find a frame S′ in which B precedes A in time by 1.0×10^{-10} sec. Is there a frame in which B and A occur at the same place?

1.8 Show that the electromagnetic wave equations are not invariant under a Galilean transformation.

1.9 Show that an observer P′ moving with a three-velocity v relative to an observer P may be represented on the Lorentz diagram as indicated in Fig. 1.1, where $\alpha = \tan^{-1}(v/c)$.

1.10 Derive (1.75) by differentiating the transformation.

1.11 (a) Show that $L^{\varrho}{}_{\mu} L_{\lambda}{}^{\mu} = g^{\varrho}{}_{\lambda}$, i.e., that $L_{\lambda}{}^{\mu}$ is the inverse of $L^{\varrho}{}_{\mu}$. (b) Evaluate explicitly the components of $L_{\lambda}{}^{\mu}$ that correspond to (1.63), (1.65), (1.66), and (1.70).

1.12 A spaceship of proper length 10 m speeds away from the earth at $0.8c$. A light signal from earth arrives at the tail of the ship at time zero according to both spaceship clocks and earth clocks. Calculate in both frames the time at which the light signal reaches the head of the rocket. It is there reflected back by a mirror. Calculate in both frames the time at which the light signal again reaches the tail. Compare the elapsed time in the two frames for the round trip tail-head-tail.

1.13 A particle at the origin executes simple harmonic motion along the y direction, with a 60 Hz frequency and 1.0 cm amplitude. Determine the position and velocity as functions of time in a reference system S′ moving with a velocity $0.8c$ in the x direction. Calculate the time and distance between successive maxima of y'. Sketch the motion as observed by S′.

1.14 A man on an inertial space platform sees two spaceships approaching each other at a rate of $\frac{7}{5}c$. A man in one of the spaceships sees the other spaceship approaching him with a velocity $\frac{35}{37}c$. What are the velocities of the spaceships with respect to the station?

1.15 A light ray makes an angle of 60° with the x^1 axis in the S frame. What angle will the ray make with the $x^{1\prime}$ axis of the S′ frame which is moving with a speed of $\frac{1}{2}c$ relative to S in such a way that (1.71) is valid?

1.16 Relative to the earth, the stars A, B, and D are moving in the x direction with a common velocity $0.6c$. At the same time in the earth frame, they are at the vertices of an equilateral triangle of side $L = 100$ light years. Determine the sides and orientation of the triangle ABC in the rest frame of the three stars. The earth is in the direction $\hat{\jmath}$ relative to the three stars at a distance $\gg 100$ light years. In what direction in the rest frame of the stars is the light emitted that travels along $\hat{\jmath}$, in the earth frame? What frequency does the H_β line ($\nu = 0.616 \times 10^{14}$ Hz) from a star have in the earth frame?

1.17 How fast must you travel to see a red light (6300 Å) as green (5400 Å)?

1.18 Show that (1.89) is equivalent to (1.88).

1.19 Show that in going from a lab system to a rest system via Eq. (1.124) one may use (1.71) rather than (1.70) and (1.66) for $L^{\mu}{}_{\nu}$.

1.20 Determine the rate of precession of the spin of a particle whose motion relative to the lab system is governed by the equations

$$d^2z/dt^2 = \ddot{z} = k_1 \qquad \varphi = -k_2 z, \qquad \varrho = k_3$$

where $\{z, \varphi, \varrho\}$ are the cylindrical coordinates of the particle in the lab frame and k_1, k_2, and k_3 are constants.

1.21 Derive (1.144) from (1.133) by the method suggested in the text immediately preceeding (1.144).

1.22 A K^0 meson decays into two pions

$$K^0 \rightarrow \pi^+ + \pi^-$$

(a) Determine the energy and momentum of the pions in the rest frame of K^0 and in the rest frame of π^+.
(b) The angular distribution in the K^0 rest frame is isotropic. What is the energy distribution of π^+ in the lab frame?
(c) If K^0 has a momentum 3.0 GeV/c in the lab frame, what are the maximum and minimum values of the energy of the positive pion in the lab frame?

1.23 A proton having a lab momentum of 70 GeV/c collides with a beam of "target" protons which are moving toward the incident beam with a lab momentum of 0.20 GeV/c. (a) Determine the velocity of the center-of-momentum frame. (b) Determine the energy and momentum of each proton in the center-of-momentum frame. (c) Determine the total invariant mass of the proton–proton systems and compare this result with the case when the target proton is at rest. (d) In the proton–proton collision a hyperon is created and emitted at right angles to the direction of the protons with a momentum in the center-of-momentum frame of 2.0 GeV/c. It lives for 2.0×10^{-10} sec in its rest frame. How long does it live in the center-of-momentum frame? What is its momentum and direction in the lab frame? How far does it travel in the lab frame before it disintegrates? Rest energy of proton = 0.938 GeV. Rest energy of Λ hyperon = 1.116 GeV.

1.24 A 20 GeV photon strikes a proton producing a pair of particles W and $\overline{\mathrm{W}}$ of equal invariant mass in the reaction

$$\gamma + p \rightarrow p + \mathrm{W} + \overline{\mathrm{W}}$$

What is the maximum invariant mass the W particle can have and what are the laboratory momenta of the three particles if the reaction just barely occurs (i.e., 20 GeV is the threshold for W production)?

1.25 A free neutron is produced in the Crab Nebula 5000 light years from earth. If the half-life of a free neutron at rest is 13 min and its rest energy is 940 MeV, determine the energy the neutron must have to have a fifty–fifty chance to reach the earth before it decays.

REFERENCES

1. C. Truesdell and W. Noll, Non-linear field theories of mechanics. *In* "Handbuch der Physik" (S. Flügge, ed.), Vol. 3, Springer-Verlag, Berlin and New York, 1965.
2. J. Anderson, "Principles of Relativity Physics." Academic Press, New York, 1967.
3. L. Infeld and J. Plebanski, "Motion and Relativity." Pergamon, Oxford, 1960.
4. Y. P. Terletskii, "Paradoxes in the Theory of Relativity." Plenum, New York, 1968.
5. R. D. Sard, "Relativistic Mechanics." Benjamin, New York, 1970.
6. E. Butkov, "Mathematical Physics." Addison-Wesley, Reading, Massachusetts, 1968.
7. O. M. P. Bilaniuk, V. K. Deshpande, and E. C. Sudarshan, *Amer. J. Phys.* **30**, 718 (1962).
8. Physics Today **22**, 43 (1969).
9. G. Feinberg, *Phys. Rev.* **159**, 1089 (1967).
10. T. Alvager and M. N. Kreisler, *Phys. Rev.* **171**, 1357 (1968).
11. G. A. Benford, D. L. Book, and W. A. Newcomb, *Phys. Rev.* **2**, 263 (1970).
12. R. G. Cawley, *Phys. Rev. D* **2**, 276 (1970).
13. F. A. E. Pirani, *Phys. Rev. D* **1**, 3224 (1970).

BIBLIOGRAPHY

H. Arzelies, "Relativistic Point Dynamics." Pergamon, Oxford, 1972.
A. O. Barut, "Electrodynamics and Classical Theory of Fields and Particles." Macmillan, New York, 1964.
P. G. Bergmann, The Special Theory of Relativity. "Handbuch der Physik" (S. Flügge, ed.), Vol. 4. Springer-Verlag, Berlin and New York, 1962.

D. Bohm, "The Special Theory of Relativity." Benjamin, New York, 1965.

O. Costa de Beauregard, "Precis of Special Relativity." Academic Press, New York, 1966.

W. R. Davis, "Classical Fields, Particles, and the Theory of Relativity." Plenum, New York, 1968.

H. Goldstein, "Classical Mechanics." Addison-Wesley, Reading, Massachusetts, 1950.

R. H. Good and T. J. Nelson, "Classical Theory of Electric and Magnetic Fields." Academic Press, New York, 1971.

R. Hagedorn, "Relativistic Kinematics." Benjamin, New York, 1964.

H. M. Schwartz, "Introduction to Special Relativity." McGraw-Hill, New York, 1968.

J. L. Synge, "Relativity: The Special Theory." Wiley, New York, 1956.

2

LAGRANGIAN DYNAMICS

2.1. BASIC THEORY

In Chapter 1 we presented the kinematical and relativistic framework within which our dynamical theory will be constructed. The actual form that our theory takes depends on our choice of the fundamental dynamical principle. One could follow Newton and take his second law as the starting point. We choose, however, to present our dynamical principle as an "extremum principle" because (a) almost all physical theories may be presented in this form and thus extremal principles serve as a natural vehicle to unify the diverse areas of physics; (b) the resulting dynamics has a most elegant mathematical structure; (c) they are very useful as the starting point for various approximations necessary to obtain solutions to actual physical systems. For a dissenting opinion see Kilmister [1].

A. Hamilton's Principle

Let the configuration of a physical system be specified by the set of coordinates $\{q_a\}$, where $a = 1, 2, \ldots, N$ and N is the number of degrees of freedom of the system. We assume that there exists a state function, the Lagrangian,

$$L = L(q_a, \dot{q}_a, \tau) \tag{2.1}$$

where $\dot{q}_a = dq_a/d\tau$ and τ is the independent variable. Then Hamilton's principle states that the initial configuration $\{q_a(\tau_1)\}$ evolves into the configuration $\{q_a(\tau_2)\}$ in such a way that the integral

$$I = \int_{\tau_1}^{\tau_2} L(q_a, \dot{q}_a, \tau)\, d\tau \tag{2.2}$$

is an extremum. I is called the *action*.

The physical evolution of the system is described by the function $q_a = q_a(\tau)$. The action I is a *functional*. A functional, defined on the space of functions $q_a(\tau)$, assigns to each *function* $q_a(\tau)$ a real or complex number I; whereas a function assigns to each number τ a real or complex number.

The problem posed by Hamilton's principle is to determine from all possible functions connecting τ_1 and τ_2 (e.g., see Fig. 2.1a where each curve corresponds to a different function) that function for which (2.2) is an extremum. For functions one has the analogous problem of determining those values of the independent variables for which the function is an extremum (i.e., a maximum or minimum). This problem is disposed of in ordinary calculus as follows. Consider a function

$$f = f(x, y, z, \ldots) \tag{2.3}$$

We select an arbitrary point, defined by the set of independent variables $\{x_0, y_0, z_0, \ldots\}$. Now we introduce arbitrary infinitesimal changes $\{dx, dy, dz, \ldots\}$ and form the total differential of $f(x, y, z, \ldots)$,

$$df = \frac{\partial f}{\partial x} dx + \frac{\partial f}{\partial y} dy + \frac{\partial f}{\partial z} dz + \cdots \tag{2.4}$$

where all partial derivatives are to be evaluated at the point $\{x_0, y_0, z_0, \ldots\}$. Since $df = 0$ at an extremum[2], it is necessary that at the extremum point

$$\left.\frac{\partial f}{\partial x}\right|_{\{x_0, \ldots\}} = \left.\frac{\partial f}{\partial y}\right|_{\{\,\}} = \cdots = 0 \tag{2.5}$$

Solving (2.5), we obtain the point (or points) $\{x_0, y_0, \ldots\}$, where $f(x, y, \ldots)$ is an extremum.

The situation for functionals, while employing the same idea, is more complicated. Variational Calculus is that branch of mathematics which considers the problem of extremizing functionals. We begin by choosing an arbitrary point, not in the space of real or complex numbers, but rather

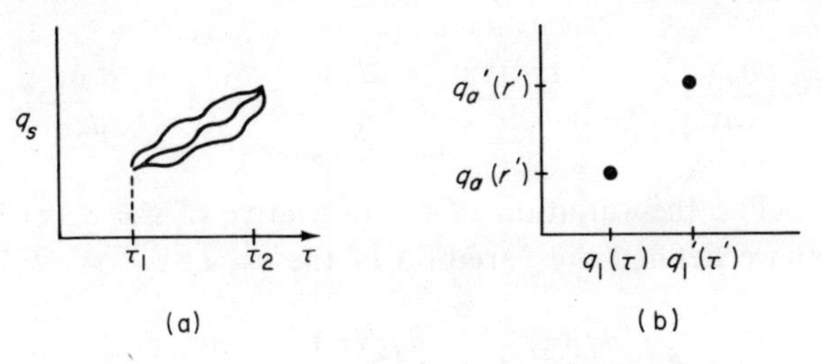

Fig. 2.1

in the space of functions. The coordinates of this point in function space will be designated by the set of functions $\{q_a(\tau)\}$ (see Fig. 2.1b). Consider now the new set of functions $\{q_a'(\tau)\}$ related to the original set by

$$q_a'(\tau') = q_a(\tau) + \varepsilon u_a(\tau) \equiv q_a(\tau) + \delta q_a(\tau) \tag{2.6}$$

$$\tau' = \tau + \varepsilon'\eta(\tau) = \tau + \delta\tau \tag{2.7}$$

$$\dot{q}_a'(\tau') \equiv dq_a'(\tau')/d\tau' \tag{2.8}$$

where ε and ε' are infinitesmal numbers and $\eta(\tau)$ and $u_a(\tau)$ are arbitrary functions of τ. $\delta q_a(\tau)$, defined by (2.6), is called the *variation* of $q_a(\tau)$.

The variation of the q_a and τ induces a variation of the functional, i.e., the action, defined by

$$\delta I \equiv \int_{\tau_1'}^{\tau_2'} L'(q_a'(\tau'), \dot{q}_a'(\tau'), \tau')\, d\tau' - \int_{\tau_1}^{\tau_2} L(q_a(\tau), \dot{q}_a(\tau), \tau)\, d\tau \tag{2.9}$$

In other words, just as a change of location in the space of real numbers (i.e., a change in the numerical value of the argument of the function) produces a change in the value of a function defined on that space, so also does a change of location in the space of functions produce a change in the value of a functional defined on that space.

Before simplifying (2.9), we develop a more useful form of (2.8):

$$\frac{dq_a'(\tau')}{d\tau'} = \dot{q}_a'(\tau') = \frac{dq_a(\tau)}{d\tau'} + \frac{d}{d\tau'}\,\delta q_a(\tau) = \frac{d\tau}{d\tau'}\left(\frac{dq_a}{d\tau} + \frac{d}{d\tau}\,\delta q_a\right) \tag{2.10}$$

But from (2.7)

$$\frac{d\tau}{d\tau'} = 1 - \frac{d\,\delta\tau}{d\tau'} \tag{2.11}$$

Thus (2.10) becomes

$$\frac{dq_a'(\tau')}{d\tau'} = \left(\frac{dq_a(\tau)}{d\tau} + \frac{d}{d\tau}\,\delta q_a(\tau)\right)\left(1 - \frac{d\,\delta\tau}{d\tau'}\right) \tag{2.12}$$

We now define the variation of the derivative of the $q_a(\tau)$ in the same manner that we defined the variation of the q_a, i.e.,

$$\delta\left(\frac{dq_a(\tau)}{d\tau}\right) \equiv \frac{dq_a'(\tau')}{d\tau'} - \frac{dq_a(\tau)}{d\tau} \tag{2.13}$$

Comparing (2.12) and (2.13) we conclude that to first order in the infinitesmals

$$\frac{d}{d\tau}\,\delta q_a(\tau) = \delta\,\frac{dq_a}{d\tau} + \frac{dq_a}{d\tau}\,\frac{d(\delta\tau)}{d\tau'} \tag{2.14}$$

i.e., the operators of differentiation and variation do not in general commute.

We now impose the additional restriction of *form invariance* on our Lagrangian

$$L'(q_a, \dot{q}_a, \tau) = L(q_a, \dot{q}_a, \tau) \tag{2.15}$$

Using (2.11), (2.13), and (2.15), (2.9) becomes

$$\delta I = \int_{\tau_1}^{\tau_2} L(q_a(\tau) + \delta q_a(\tau), \dot{q}_a + \delta\dot{q}_a, \tau + \delta\tau)(1 + d\,\delta\tau/d\tau)\,d\tau$$
$$- \int_{\tau_1}^{\tau_2} L(q_a(\tau), \dot{q}_a(\tau)\,\tau)\,d\tau \tag{2.16}$$

Performing a Taylor expansion of the integrand of the first integral of (2.16) about the point $\{q_a(\tau)\}$ yields

$$\delta I = \int_{\tau_1}^{\tau_2} \Big\{ L(q_a, \dot{q}_\alpha, \tau) + \frac{\partial L}{\partial q_a}\Big|_{\{q_a,\, \dot{q}_a,\, \tau\}} \delta q_a + \frac{\partial L}{\partial \dot{q}_a}\Big|_{\{\,\}} \delta\dot{q}_a$$
$$+ \frac{\partial L}{\partial T}\Big|_{\{\,\}} \delta\tau + \text{H.O.}\Big\}\left(1 + \frac{d\,\delta\tau}{d\tau}\right) d\tau - \int_{\tau_1}^{\tau_2} L(q_a, \dot{q}_a, \tau)\,d\tau \tag{2.17}$$

where H.O. represents the infinitesimal terms of "higher order." Neglecting the H.O., (2.17) becomes

$$\delta I = \int_{\tau_1}^{\tau_2} \left[\frac{\partial L}{\partial q_a}\,\delta q_a + \frac{\partial L}{\partial \dot{q}_a}\,\delta\dot{q}_a + \frac{\partial L}{\partial T}\,\delta\tau + L\,\frac{d\,\delta\tau}{d\tau} \right] d\tau \tag{2.18}$$

Using (2.14) we obtain

$$\delta I = \int_{\tau_1}^{\tau_2} \left[\left(\frac{\partial L}{\partial q_a} + \frac{\partial L}{\partial \dot{q}_a}\,\frac{d}{d\tau} \right) \delta q_a \right.$$
$$\left. + \frac{\partial L}{\partial \tau}\,\delta\tau - \dot{q}_a\,\frac{\partial L}{\partial \dot{q}_a}\,\frac{d\,\delta\tau}{d\tau} + L\,\frac{d\,\delta\tau}{d\tau} \right] d\tau \tag{2.19}$$

Equation (2.19) may be written as

$$\delta I = \int_{\tau_1}^{\tau_2} \frac{d}{d\tau}\left(\frac{\partial L}{\partial \dot{q}_a}\,\delta q_a - \dot{q}_a \frac{\partial L}{\partial \dot{q}_a}\,\delta\tau + L\,\delta\tau\right) d\tau$$
$$+ \int_{\tau_1}^{\tau_2} \left(\frac{\partial L}{\partial q_a} - \frac{d}{d\tau}\frac{\partial L}{\partial \dot{q}_a}\right) \delta q_a\, d\tau$$
$$+ \int_{\tau_1}^{\tau_2} \left[\frac{\partial L}{\partial \tau} + \frac{d}{d\tau}\left(\dot{q}_a \frac{\partial L}{\partial \dot{q}_a}\right) - \frac{dL}{d\tau}\right] \delta\tau\, d\tau \tag{2.20}$$

However,

$$\frac{\partial L}{\partial \tau} = \frac{dL}{d\tau} - \frac{dL}{\partial q_a}\,\dot{q}_a - \frac{\partial L}{\partial \dot{q}_a}\,\ddot{q}_a \tag{2.21}$$

Substituting (2.21) into (2.20)

$$\delta I = \int_{\tau_1}^{\tau_2} \frac{d}{d\tau}\left[\delta q_a \frac{\partial L}{\partial \dot{q}_a} + \left(L - \dot{q}_a \frac{\partial L}{\partial \dot{q}_a}\right)\delta\tau\right] d\tau$$
$$+ \int_{\tau_1}^{\tau_2} \left(\frac{\partial L}{\partial q_a} - \frac{d}{d\tau}\frac{\partial L}{\partial \dot{q}_a}\right)(\delta q_a - \dot{q}_a\,\delta\tau)\, d\tau \tag{2.22}$$

A necessary condition for an extremum is that

$$\delta I = 0 \tag{2.23}$$

For a detailed discussion of this condition see the references at the end of the chapter. We present here a simple argument for the necessity of (2.23). Suppose that I is a minimum. Displacement of the configuration of the system from the minimum is produced by the variation δq_a. If the resulting $\delta I > 0$ (this is compatible with the minimum property), then $\delta I < 0$ when the sign of δq_a is reversed. But $\delta I < 0$ is not compatible with the minimum property. Thus we conclude that $\delta I = 0$ so that it does not change sign upon reversing the sign of the variation.

Now Hamilton's principle specifies a particular initial and final configuration, viz., $q_a(\tau_1)$ and $q_a(\tau_2)$; hence at the endpoints

$$\delta q_a(\tau_1) = \delta q_a(\tau_2) = 0 = \delta\tau \tag{2.24}$$

As a result the first integration on the right side of (2.22) is zero, and with (2.23), (2.22) becomes

$$\int_{\tau_1}^{\tau_2} \left(\frac{\partial L}{\partial q_a} - \frac{d}{d\tau}\frac{\partial L}{\partial \dot{q}_a}\right)(\delta q_a - \dot{q}_a\,\delta\tau)\, d\tau = 0 \tag{2.25}$$

Thus for arbitrary and independent variations it is necessary that

$$\frac{d}{d\tau}\frac{\partial L}{\partial \dot{q}_a} - \frac{\partial L}{\partial q_a} = 0 \tag{2.26a}$$

where it is understood that the q_a are the extremal paths.

The set of equations (2.26a) are the *Euler–Lagrange* equations of motion describing the evolution of the system from $q_a(\tau_1)$ to $q_a(\tau_2)$. In Galilean physics τ is identified as the time and the Lagrangian is prescribed as

$$L = T - V \tag{2.27}$$

where T is the kinetic energy and V the potential energy of the system.

Defining the *momentum* p_a of the system as

$$p_a = \partial L/\partial \dot{q}_a \tag{2.28}$$

and the *force* Q_a on the ath particle as

$$Q_a = \partial L/\partial q_a \tag{2.29}$$

the Euler–Lagrange equations, (2.26), have the Newtonian form

$$\frac{d}{d\tau} p_a = Q_a \tag{2.30}$$

If $L \neq L(q_a)$, then the Lagrangian is said to be *cyclic* in q_a and it follows that

$$\frac{d}{d\tau} p_a = 0 \quad \Rightarrow \quad p_a = \text{const} \tag{2.31}$$

Thus the momentum corresponding to a cyclic coordinate is conserved. The Euler–Lagrange equations are readily generalized to include nonconservative forces (e.g., friction) by adding the generalized forces associated with the nonconservative forces to the right side of (2.6a); thus

$$\frac{d}{d\tau}\frac{\partial L}{\partial \dot{q}_a} - \frac{\partial L}{\partial q_a} = Q_a \tag{2.26b}$$

where the work done by the nonconservative forces is

$$\delta W = Q_a\, \delta q_a = \sum_{\alpha=1}^{N} \mathbf{F}_\alpha \cdot d\mathbf{r}_\alpha \tag{2.26c}$$

and $\mathbf{F}_\alpha$ is the nonconservative force acting on the αth particle. Equation

(2.26c) may be utilized to obtain the Q_a of (2.26b) when the nonconservative forces $\mathbf{F}_\alpha$ and the transformation equations connecting the inertial cartesian coordinates $\mathbf{r}_\alpha$ and the generalized coordinates q_a [see (2.59)] are known.

The Lagrangian yielding a given set of equations of motion is not unique. It is easy to show that a Lagrangian L' yields the same equations of motion as the Lagrangian L if L' is related to L according to

$$L' = bL + \frac{d}{d\tau} G(q_a, \tau) \tag{2.32}$$

where b is an arbitrary constant and G is an arbitrary function of the coordinates and the independent variable, but may not be a function of the velocities. Thus

$$\frac{d}{d\tau}\frac{\partial L'}{\partial \dot{q}_a} - \frac{\partial L'}{\partial q_a} = b\left(\frac{d}{d\tau}\frac{\partial L}{\partial \dot{q}_a} - \frac{\partial L}{\partial q_a}\right) + \frac{d}{d\tau}\left[\frac{\partial}{\partial \dot{q}_a}\left(\frac{\partial G}{\partial q_b}\dot{q}_b + \frac{\partial G}{\partial \tau}\right)\right]$$
$$- \frac{\partial}{\partial q_a}\left(\frac{\partial G}{\partial q_b}\dot{q}_b + \frac{\partial G}{\partial \tau}\right) = 0 \tag{2.33}$$

The second and third terms cancel leaving

$$\frac{d}{d\tau}\frac{\partial L'}{\partial \dot{q}_a} - \frac{\partial L'}{\partial q_a} = b\left(\frac{d}{d\tau}\frac{\partial L}{\partial q_a} - \frac{\partial L}{\partial q_a}\right) = 0 \tag{2.34}$$

B. Constraints

In the derivation of (2.26) it was assumed that the q_a were all independent. This fact was used explicitly in the going from (2.25) to (2.26a). If, however, there are constraints on the motion of the system, the q_a are no longer all independent and we may not conclude (2.26a) from (2.25). If the constraints are holonomic, i.e., if the constraint conditions can be expressed as equations connecting the coordinates and the time, the equations of constraint may be used to eliminate the dependent variables and the variational procedure proceeds as before with the new set of independent functions.

It often happens that the elimination procedure is either difficult to perform because of the intractability of the constraint equations or undesirable because it removes certain functions that one may for one reason or another prefer to keep. Fortunately, there is another method for handling constraints, the Lagrange multiplier technique, which allows one to follow the previously presented variational program by modifying the Lagrangian.

Consider a system for which the r constraint conditions may be expressed as

$$f_b(q_a, \dot{q}_a, \tau) = 0, \qquad b = 1, 2, \ldots, p \tag{2.35}$$

We now multiply each of the constraints (2.35) by a "Lagrange Multiplier" $\lambda_a(\tau)$, add the equations together and replace the original Lagrangian L by a modified Lagrangian $\bar{L}$:

$$\bar{L} = L + \lambda_b(\tau) f_b(q_a, \dot{q}_a, \tau), \qquad a = 1, 2, \ldots, p \tag{2.36}$$

$\bar{L}$ has the same *magnitude* as L.

The modified Lagrangian is then substituted into the action and the variational procedure is performed as before except that the varied functions are now the $q_a(\tau)$ and the $\lambda_b(\tau)$, all of which are considered independent. Equation (2.25) becomes

$$\int_{\tau_1}^{\tau_2} \left(\frac{\partial \bar{L}}{\partial q_a} - \frac{d}{d\tau} \frac{\partial \bar{L}}{\partial \dot{q}_a} \right) (\delta q_a - \dot{q}_a \, \delta\tau) \, d\tau$$
$$+ \int_{\tau_1}^{\tau_2} \left(\frac{\partial \bar{L}}{\partial \lambda_b} - \frac{d}{d\tau} \frac{\partial \bar{L}}{\partial \dot{\lambda}_b} \right) (\delta \lambda_b - \dot{\lambda}_b \, \delta\tau) \, d\tau = 0 \tag{2.37}$$

Thus the equations of motion become

$$\frac{d}{d\tau} \frac{\partial \bar{L}}{\partial \dot{q}_a} - \frac{\partial \bar{L}}{\partial q_a} = 0 \tag{2.38}$$

and

$$f_b = 0 \tag{2.39}$$

Equations (2.38) and (2.39) determine the motion of the system. What the multiplier technique actually does is to construct a Lagrangian that *derives* the constraint equations and thereby guarantees a motion in conformity with the constraints.

If the constraint is independent of the velocities, (2.38) becomes

$$\frac{d}{d\tau} \frac{\partial L}{\partial \dot{q}_a} = \frac{\partial L}{\partial q_a} + \lambda_b \frac{\partial f_b}{\partial q_a} \tag{2.40}$$

Comparing (2.40) with (2.30) it is clear that $\lambda_b \, \partial f_b / \partial q_a$ may be interpreted as the forces due to the constraints; thus the multiplier technique permits the evaluation of such constraint forces as tensions in ropes, the reactive forces of wires, and so forth.

C. Lagrangian Dynamics in Lorentz Space

A number of serious problems arise when one specifies the Poincaré group as the relativity group. We defer a discussion of the most serious of these problems to Chapter 5; however, the motion of a single particle is not beset by serious difficulties and we discuss it here.

In Minkowski space the q_a correspond to the x^μ. The parameter τ may be taken to be the proper time s. The velocities $\dot{q}_a$ are then

$$\dot{q}_a \sim \dot{x}^\mu = dx^\mu/ds$$

Recall that

$$(ds)^2 = dx^\mu \, dx_\mu$$

Hence

$$\dot{x}^\mu \dot{x}_\mu = \frac{dx^\mu}{ds}\frac{dx_\mu}{ds} = 1 \tag{2.41}$$

Equation (2.41) establishes a constraint on the motion which may be handled by the Lagrange multiplier technique. Thus in this case our modified Lagrangian is

$$\bar{L} = L + \tfrac{1}{2}\lambda(s)(\dot{x}^\mu \dot{x}_\mu - 1) \tag{2.42}$$

The equations of motion corresponding to (2.38) and (2.39) are

$$\frac{d}{ds}\frac{\partial L}{\partial \dot{x}^\mu} - \frac{\partial L}{\partial x^\mu} + \frac{d}{ds}(\lambda \dot{x}_\mu) = 0 \tag{2.43}$$

$$\dot{x}^\mu \dot{x}_\mu = 1 \tag{2.44}$$

Equations (2.43) and (2.44) represent five equations in the five unknowns x^μ and λ. It follows directly from (2.44) that

$$\dot{x}^\mu \ddot{x}_\mu = 0 \tag{2.45}$$

Multiplying (2.43) by $\dot{x}^\mu$ and using (2.44) and (2.45), we have

$$\frac{d\lambda}{ds} = \dot{x}^\mu \frac{\partial L}{\partial x^\mu} - \dot{x}^\mu \frac{d}{ds}\frac{\partial L}{\partial \dot{x}^\mu} \tag{2.46}$$

Thus

$$\lambda = \int \frac{\partial L}{\partial x^\mu}\frac{dx^\mu}{ds}\, ds - \int \dot{x}^\mu \frac{d}{ds}\frac{\partial L}{\partial \dot{x}^\mu}\, ds \tag{2.47}$$

But

$$\frac{dL}{ds} = \frac{\partial L}{\partial x^\mu}\frac{dx^\mu}{ds} + \frac{\partial L}{\partial \dot{x}^\mu}\frac{d\dot{x}^\mu}{ds} + \frac{\partial L}{ds} \tag{2.48}$$

Therefore,

$$\begin{aligned}\lambda &= \int dL - \int \frac{\partial L}{\partial \dot{x}^\mu}\frac{d\dot{x}^\mu}{ds}\,ds - \int \dot{x}^\mu \frac{d}{ds}\frac{\partial L}{\partial \dot{x}^\mu}ds - \int \frac{\partial L}{ds}\,ds \\ &= L - \dot{x}^\mu \frac{\partial L}{\partial \dot{x}^\mu} - \int \frac{\partial L}{ds}\,ds \end{aligned} \tag{2.49}$$

If we assume $L \neq L(s)$, (2.43) becomes

$$\frac{d}{ds}\frac{\partial L}{\partial \dot{x}^\mu} - \frac{\partial L}{\partial x^\mu} + \frac{d}{ds}\left[\left(L - \dot{x}^\nu \frac{\partial L}{\partial \dot{x}^\nu}\right)\dot{x}_\mu\right] = 0 \tag{2.50}$$

Equations (2.50) are the Euler–Lagrange equations of motion for a Lorentz invariant system of one particle. The prescription embodied in (2.27) is not applicable to Lorentz invariant systems. In fact, we are without any clear prescription for determining L except that we know it must be a scalar and have the functional dependence indicated in (2.1). Since the action I has the dimensions of "energy × time," and since the independent variable s has the dimension of length, it follows that L must have the dimensions of "mass × velocity." In practice one must construct (postulate) a Lagrangian for the system by making an educated guess. If the motion predicted by (2.50) agrees with experiment, we consider it a valid model of the system; if it does not we try again.

The limitations imposed on the above development are indeed severe. We might well ask how it is possible to have a nontrivial dynamics with just a single particle. If the system has only one particle, where or how does the force on that particle originate? It is clear that within the realm of pure *particle* mechanics a single particle system must be a free particle mechanics. Recent attempts to construct a legitimate multiparticle dynamics in Lorentz space are discussed in Chapter 5. Nevertheless, it is surprising that almost fifty years expired after the advent of Einstein's special theory of relativity before any significant progress was made in the development of a pure multiparticle relativistic mechanics.

In the interim the problem has been circumvented by combining the particle description with the field description of nature in such a way that the forces acting on the particle are the result of the interaction of the particle with the field; e.g., the electromagnetic or gravitational field. The field is assumed to be a known function of the coordinates; the sources of the field and the interaction of the particle back on these sources are ignored. With these additions (2.50) may be used to describe the motion in Lorentz space of a particle under the influence of a field.

To illustrate the preceding discussion consider the motion of a charged particle in an electromagnetic field. In the limit of zero electromagnetic field the motion and therefore the Lagrangian should be that of a free particle. It follows from (2.50) that the Lagrangian

$$L = mc \tag{2.51}$$

where m is the mass of the particle and c is the velocity of light, yields a geodesic trajectory of Lorentz space; i.e.,

$$d^2x^\mu/ds^2 = 0 \tag{2.52}$$

Therefore, we may consider (2.51) as the free particle Lagrangian. To represent the charged particle moving under the influence of an electromagnetic field we must add to (2.51) a scalar that is constructed from functions characteristic of the particle and the field. The field may be represented by the electromagnetic potentials $A^\mu(x)$, which are defined by

$$F^{\mu\nu} = A^{\nu,\mu} - A^{\mu,\nu} \tag{2.53}$$

where $F^{\mu\nu}$ is the electromagnetic field tensor defined in (1.76). We choose as our Lagrangian

$$L = mc + (e/c)A^\mu\dot{x}_\mu \tag{2.54}$$

where e is the charge of the particle.

Substituting (2.54) in (2.50), we obtain

$$\frac{e}{c}\frac{d}{ds}A^\mu - \frac{e}{c}\dot{x}_\nu A^{\nu,\mu} + \frac{d}{ds}\left[\left(mc + \frac{e}{c}A^\nu\dot{x}_\nu - \frac{e}{c}A^\nu\dot{x}_\nu\right)\dot{x}^\mu\right] = 0$$

$$\frac{e}{c}A^{\mu,\nu}\dot{x}_\nu - \frac{e}{c}A^{\nu,\mu}\dot{x}_\nu + mc\ddot{x}^\mu = 0$$

$$mc\ddot{x}^\mu = \frac{e}{c}\dot{x}_\nu F^{\mu\nu} \tag{2.55}$$

It is left to the reader to verify that (2.55) is indeed the correct equation, i.e., that the right side of (2.55) corresponds to the Lorentz force.

2.2. DEVELOPMENT AND EXTENSION OF THE BASIC THEORY

A. Kinetic Energy (Galilean Physics)

In Galilean physics the Lagrangian is given by (2.27), and therefore one must be able to express the kinetic energy and potential energy in terms of the coordinates being used to describe the motion of the system. It is always a simple, albeit sometimes tedious, process to express the kinetic energy.

For a system containing A particles, the kinetic energy may *always* be written

$$T = \tfrac{1}{2} \sum_{\alpha=1}^{A} m_\alpha ({}_\alpha\dot{x}^j)^2 \tag{2.56}$$

provided that the Cartesian coordinates ${}_\alpha x^j$ are referred to an *inertial* coordinate system. It is often desirable to express the kinetic energy in terms of other coordinates; the recipe for doing this is simply to determine the transformations connecting the desired coordinates with the inertial Cartesian coordinates and substitute them into (2.56). To illustrate we express the kinetic energy in terms of spherical polar coordinates by substituting

$$\begin{aligned} &{}_\alpha x^3 = r_\alpha \cos\theta_\alpha, \qquad {}_\alpha x^2 = r_\alpha \sin\theta_\alpha \sin\varphi_\alpha, \qquad {}_\alpha x^2 = r_\alpha \sin\theta_\alpha \cos\varphi_\alpha \\ &{}_\alpha\dot{x}^3 = \dot{r}_\alpha \cos\theta_\alpha - r_\alpha \dot{\theta}_\alpha \sin\theta_\alpha, \qquad \text{etc.} \end{aligned} \tag{2.57}$$

into (2.56). We obtain after some algebra

$$T = \tfrac{1}{2} \sum_{\alpha=1}^{A} m_\alpha (\dot{r}_\alpha{}^2 + r_\alpha{}^2 \dot{\theta}_\alpha{}^2 + r_\alpha{}^2 \dot{\varphi}_\alpha{}^2 \sin^2\theta_\alpha) \tag{2.58}$$

It is instructive to obtain a more general expression for the kinetic energy. Let ${}_\alpha x^i$ be the *inertial* Cartesian coordinates locating the position of the αth particle of a system having N degrees of freedom. Suppose that it is desired to define the configuration of the system in terms of the coordinates $\{q_a\}$ which are related to the inertial coordinates by

$${}_\alpha x^i = {}_\alpha x^i(q_a, t) \tag{2.59}$$

And

$${}_\alpha\dot{x}^i = \frac{\partial\, {}_\alpha x^i}{\partial q_a} \dot{q}_a + \frac{\partial\, {}_\alpha x^i}{\partial t}, \qquad a = 1, \ldots, N \tag{2.60}$$

Substituting (2.60) into (2.56), we obtain

$$T = \tfrac{1}{2} \sum_{\alpha=1}^{A} m_\alpha \left(\frac{\partial\, {}_\alpha x^i}{\partial q_a} \frac{\partial\, {}_\alpha x^i}{\partial q_b} \dot{q}_a \dot{q}_b + 2 \frac{\partial\, {}_\alpha x^i}{\partial q_a} \frac{\partial\, {}_\alpha x^i}{\partial t} \dot{q}_a + \frac{\partial\, {}_\alpha x^i}{\partial t} \frac{\partial\, {}_\alpha x^i}{\partial t} \right) \tag{2.61}$$

or more compactly

$$T = d_{ab} \dot{q}_a \dot{q}_b + d_a \dot{q}_a + d \tag{2.62}$$

where

$$d_{ab} = \frac{1}{2} \sum_{\alpha=1}^{A} m_\alpha \frac{\partial\, {}_\alpha x^i}{\partial q_a} \frac{\partial\, {}_\alpha x^i}{\partial q_b} \tag{2.63a}$$

$$d_a = \sum_{\alpha=1}^{A} m_\alpha \frac{\partial\, {}_\alpha x^i}{\partial q_a} \frac{\partial\, {}_\alpha x^i}{\partial t} \tag{2.63b}$$

$$d = \sum_{\alpha=1}^{A} m_\alpha \frac{\partial\, {}_\alpha x^i}{\partial t} \frac{\partial\, {}_\alpha x^i}{\partial t} \tag{2.63c}$$

If the transformation equations (2.59) do not depend explicitly on the time, d_a and d are zero and the kinetic energy is quadratic in the velocities.

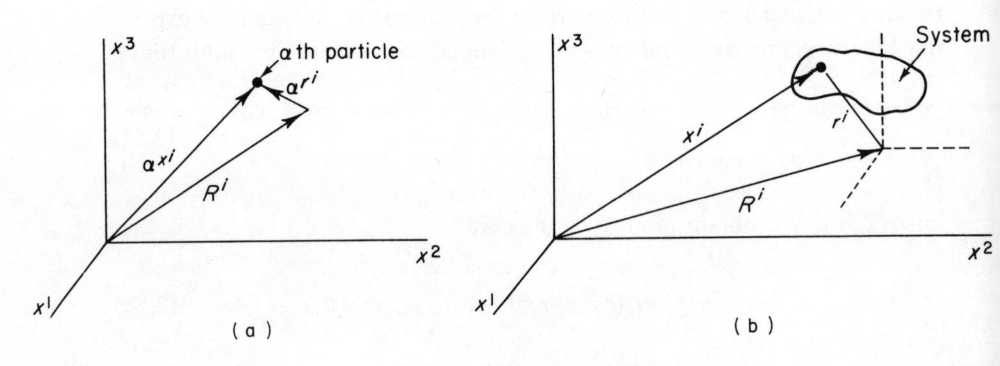

Fig. 2.2

A wide variety of problems utilize the transformation of coordinates presented in Fig. 2.2a, viz.,

$${}_\alpha x^i = R^i + {}_\alpha r^i \tag{2.64}$$

where R^i represents a translation of the origin of the coordinate system and ${}_\alpha r^i$ is the vector defining the location of the αth particle relative to the translated system. A useful generalization of (2.64) is to allow the system to be described by a set of coordinates ${}_\alpha r^{i'}$ which are related to the ${}_\alpha r^i$ by the transformation equations

$$_\alpha r^i = F^{ij}\, _\alpha r^{j'} \tag{2.65}$$

where the F^{ij} may in general be functions of time. Thus (2.64) becomes

$$_\alpha x^i = R^i + F^{ij}\, _\alpha r^{j'} \tag{2.66}$$

To express the kinetic energy in terms of these coordinates we require the time derivative of (2.66).

$$_\alpha \dot{x}^i = \dot{R}^i + \dot{F}^{ij}\, _\alpha r^{j'} + F^{ij}\, _\alpha \dot{r}^{j'} \tag{2.67}$$

Substituting (2.67) into (2.56) we obtain

$$T = \tfrac{1}{2}\{M(\dot{R}^i)^2 + F^{ij}F^{ik}\sum_\alpha m_\alpha\, _\alpha\dot{r}^{j'}\, _\alpha\dot{r}^{k'} + \dot{F}^{ij}\dot{F}^{ik}\sum_\alpha\, _\alpha r^{j'}\, _\alpha r^{k'}$$
$$+2\dot{R}^i F^{ij}\sum_\alpha m_\alpha\, _\alpha\dot{r}^{j'} + 2\dot{R}^i\dot{F}^{ij}\sum_\alpha m_\alpha\, _\alpha r^{j'} + 2F^{ij}\dot{F}^{ik}\sum_\alpha m_\alpha\, _\alpha\dot{r}^{j'}\, _\alpha r^{k'}\} \tag{2.68}$$

where M is the total mass of the system, i.e., $M = \sum_\alpha m_\alpha$.

The *moment of inertia* I^{jk} of a system of masses is defined as

$$I^{jk} = \sum_\alpha m_\alpha\, _\alpha r^j\, _\alpha r^k \tag{2.69a}$$

or for a continuous distribution of mass

$$I^{jk} = \int dm\, r^j r^k \tag{2.69b}$$

With (2.69), (2.68) may be written

$$T = \tfrac{1}{2}\{M(\dot{R}^i)^2 + \dot{F}^{ij}\dot{F}^{ik}I^{jk'} + 2\dot{R}^i\dot{F}^{ij}\sum_\alpha m_\alpha\, _\alpha r^{j'} + F^{ij}F^{ik}\sum_\alpha m_\alpha\, _\alpha\dot{r}^{j'}\, _\alpha\dot{r}^{k'}$$
$$+2\dot{R}^i F^{ij}\sum_\alpha m_\alpha\, _\alpha\dot{r}^{j'} + 2F^{ij}\dot{F}^{ik}\sum_\alpha m_\alpha\, _\alpha\dot{r}^{j'}\, _\alpha r^{k'}\} \tag{2.70}$$

The prime on the moment of inertia indicates that it is the moment of inertia relative to the primed system.

A simple but important example of (2.66) is the transformation to center-of-mass coordinates illustrated in Fig. 2.2b, where R^i now locates the center-of-mass of the system and is defined by

$$R^i = \sum_\alpha m_\alpha\, _\alpha x^i \Big/ \sum_\alpha m_\alpha \tag{2.71}$$

or for continuous distributions of mass

$$R^i = \left[\int dm\, x^i\right] \Big/ \int dm \tag{2.72}$$

The ${}_\alpha r^i$ locate the position of the αth particle relative to the center-of-mass and are therefore called the *relative coordinates*.

To specialize (2.66) to a transformation to center-of-mass coordinates we need only require that $F^{ij} = \delta^{ij}$ and therefore its derivatives are zero. Thus (2.70) becomes

$$T = \tfrac{1}{2}[M(\dot{R}^i)^2 + \sum_\alpha m_\alpha({}_\alpha\dot{r}^i)^2 + 2\dot{R}^i \sum_\alpha m_\alpha\, {}_\alpha\dot{r}^i] \tag{2.73}$$

This can be further simplified by noting that the third term on the right may be written

$$2\dot{R}^i \frac{d}{dt} \sum_\alpha m_\alpha\, {}_\alpha r^i$$

This term is zero since $\sum_\alpha m_\alpha\, {}_\alpha r^i$ is the definition of a radius vector locating the center of mass, but since the ${}_\alpha r^i$ are drawn from the center-of-mass that radius vector is zero. Thus (2.73) is simply

$$T = \tfrac{1}{2}M(\dot{R}^i)^2 + \tfrac{1}{2}\sum_\alpha m_\alpha({}_\alpha\dot{r}^i)^2 \tag{2.74}$$

Therefore the kinetic energy of a system of particles may be expressed as the kinetic energy of a particle of mass M moving with the velocity of the center-of-mass plus the kinetic energy of the motion of the particles relative to the center-of-mass.

Systems containing a large number of particles, i.e., many degrees of freedom, are generally discussed within the framework of statistical or continuum mechanics. A rigid body, however, is a system that is characterized by the fact that the distance between the particles comprising the system remains constant. As a result a rigid body has only six degrees of freedom, viz., three translational and three rotational degrees of freedom. To verify this consider three noncollinear particles. If they were free these particles would have a total of nine degrees of freedom, i.e., the position of each particle is determined by three coordinates. The fact that the distance between the particles may be represented by the three constraint equations

$$d_{12} = \text{const}, \qquad d_{13} = \text{const}, \qquad d_{23} = \text{const} \tag{2.75}$$

where $d_{\alpha\beta}$ is the distance between the αth and βth particles, reduces the number of degrees of freedom from nine to six. For each particle added to our system we would require three additional coordinates. But if the position of this new particle is fixed relative to the first three particles we also have three additional equations of constraint and therefore there is no net increase in the number of degrees of freedom of our system.

Although the choice of the six coordinates to describe rigid body motion is a matter of individual preference, one normally chooses the three coordinates locating the center-of-mass of the system and either the three Euler angles, defined in Appendix D, or any other set of three coordinates defining the orientation of the rigid body relative to the center-of-mass.

The transformation of (2.66) may be specialized to rigid body motion by identifying R^i as a vector locating some point within or on the rigid body. The ${}_\alpha r^i$ are the radius vectors locating the particles of the rigid body relative to an axis system whose origin is at the terminus of R^i and whose axes are parallel to those of the inertial system. The ${}_\alpha r^{i'}$ are the radius vectors locating the particles of the rigid body relative to an axis system whose origin is at the terminus of R^i and whose axes are rigidly attached to the rigid body. The primed axes are called the *body axes* and F^{ij} is the operator which transforms from the primed system to the unprimed system. Now since the body axes are attached to the rigid body, they follow the motion of the body or more specifically the primed coordinates are not time dependent. Therefore,

$$ {}_\alpha\dot{r}^{i'} = 0 \tag{2.76}$$

Since the body is rigid only three coordinates are required to orientate the body (and therefore locate all the particles) relative to the origin fixed by R^i. The location of the body axes at a given instant gives the orientation of the body and we can conveniently choose the Euler angles to specify the orientation of the body system at a given instant. Thus F^{ij} is just the operator which connects the Euler angles with the unprimed system that is parallel to the inertial system.

From (2.76) it follows that (2.70) simplifies to

$$T = \tfrac{1}{2}[M(\dot{R}^i)^2 + \dot{F}^{ij}\dot{F}^{ik}I^{jk'} + 2\dot{R}^i\dot{F}^{ij}\sum_\alpha m_\alpha\, {}_\alpha r^{j'}] \tag{2.77}$$

or using (2.71)

$$T = \tfrac{1}{2}[M(\dot{R}^i)^2 + \dot{F}^{ij}\dot{F}^{ik}I^{jk'} + 2\dot{R}^i\dot{F}^{ij}MR^{j'}] \tag{2.78}$$

where $R^{j'}$ is the location of the center-of-mass relative to the primed system. If R^i locates the center-of-mass of the rigid body then $R^{j'}$ must be the zero vector because $R^{j'}$ is the vector drawn from the terminus of R^i to the center-of-mass and the terminus is the center-of-mass. Equation (2.78) becomes

$$T = \tfrac{1}{2}[M(\dot{R}^i)^2 + \dot{F}^{ij}\dot{F}^{ik}I^{jk'}] \tag{2.79}$$

The *principal axes* of a system are those axes for which the moment of inertia tensor I^{jk} is diagonal. The diagonal elements I^{11}, I^{22}, and I^{33} are called the *principal moments of inertia* and are just the eigenvalues of the characteristic equation

$$I\Phi = \lambda\Phi \tag{2.80}$$

where I is the moment of inertia tensor, λ the eigenvalues, and the eigenvectors Φ are the *principal axes.* A more through discussion of the diagonalization procedure is presented in the next section on small oscillation techniques. Thus if the body system is the principal axis system, (2.79) becomes

$$T = \tfrac{1}{2}\sum_i [M(\dot{R}^i)^2 + \sum_j \dot{F}^{ij}\dot{F}^{ij}I^{j'}] \tag{2.81}$$

where the $I^{j'}$ are the principal moments of inertia; i.e., the diagonal elements of the moment of inertia tensor referred to the principal axes.

B. Small Oscillations

It often occurs that the Euler–Lagrange equations of motion are too difficult to solve without having recourse to some approximation or numerical scheme. Many techniques have been developed to handle a wide variety of problems. We present here a technique that is widely used when the system is performing an oscillatory motion about some equilibrium configuration.

The system is in equilibrium if it is in a configuration for which

$$\partial V/\partial q_a = 0 \qquad \text{for all} \quad a = 1, 2, \ldots, N \tag{2.82}$$

since in this case the net force $Q_a = \partial L/\partial q_a = 0$. But (2.82) is just the condition for the potential energy function to be an extremum; and, therefore, we may state that the equilibrium configurations of a system are those configurations for which the potential energy is an extremum. The equilibrium is *stable* if the potential energy is a minimum, since in this case a small displacement produces a force ($Q_a = -\partial V/\partial q_a$) whose direction is opposite to that of the displacement and, therefore, tends to restore the system to its equilibrium configuration. If the potential energy is a maximum, the equilibrium is *unstable*.

It is our object to determine the motion of a physical system in that portion of configuration space which is in the neighborhood of a point of stable equilibrium. Let the point of stable equilibrium be designated by the set of coordinates $\{q_a'\}$ and let a small displacement from equilibrium be represented by

$$q_a = q_a' + \eta_a, \qquad a = 1, 2, \ldots, N \tag{2.83}$$

where η_a is the small displacement. A Taylor expansion of the potential energy about the equilibrium configuration yields

$$\begin{aligned} V(q_a) &= V(q_a' + \eta_a) \\ &= V(q_a') + [\partial V/\partial q_a]_{q_a'}\eta_a + \tfrac{1}{2}[\partial^2 V/\partial q_a\, \partial q_b]_{q_a' q_b'}\eta_a\eta_b + \cdots \end{aligned} \tag{2.84}$$

where the derivatives in the brackets are to be evaluated at the equilibrium position. $V(q_a')$ is a constant which we may set equal to zero since it will not contribute to the equations of motion. Since $\{q_a'\}$ is the equilibrium configuration, it follows from (2.82) that

$$[\partial V/\partial q_a]_{q_a'} = 0$$

Thus the first nonzero term in the expansion is the second derivative term. Neglecting higher-order terms, we may write the potential energy as

$$V(q_a) = \tfrac{1}{2}[\partial^2 V/\partial q_a\, \partial q_b]_{q_a' q_b'}\eta_a\eta_b \equiv \tfrac{1}{2} V_{ab}\eta_a\eta_b \tag{2.85}$$

where the η_a may now be considered as the coordinates describing the motion of the system, and the V_{ab} are simply constant coefficients. It follows from the properties of the partial derivative that $V_{ab} = V_{ba}$. If the potential energy does not depend explicitly on the time, the equilibrium coordinates are independent of the time, i.e.,

$$\dot{q}_a' = 0 \tag{2.86}$$

and

$$\dot{q}_a = \dot{\eta}_a \tag{2.87}$$

We now assume that the kinetic energy is quadratic in the velocities. From (2.62) and (2.63) it follows that this corresponds to a situation in which the transformation equations connecting the inertial cartesian coordinates and the generalized coordinates do not depend explicitly on the time. This restriction and (2.86) allow us to write for the kinetic energy

$$T = d_{ab}\dot{\eta}_a\dot{\eta}_b \tag{2.88}$$

where d_{ab} is a function of the coordinates only and is given by (2.63a). We now Taylor expand d_{ab} about the equilibrium configuration keeping only the first term in the expansion, since the second term is small relative to the first term. Thus we have

$$d_{ab} = \frac{1}{2}\sum_{\alpha} m_\alpha \left[\frac{\partial\, {}_\alpha x^i}{\partial q_a}\frac{\partial\, {}_\alpha x^i}{\partial q_b}\right]_{q_a', q_b'} \tag{2.89}$$

and the kinetic energy becomes

$$T = \frac{1}{2}\sum_{\alpha} m_\alpha \left[\frac{\partial\, {}_\alpha x^i}{\partial q_a}\frac{\partial\, {}_\alpha x^i}{\partial q_b}\right]_{q_a', q_b'} \dot{\eta}_a\dot{\eta}_b \equiv \frac{1}{2} T_{ab}\dot{\eta}_a\dot{\eta}_b \tag{2.90}$$

The T_{ab} are just constant coefficients, and it follows from the properties of the partial derivatives that

$$T_{ab} = T_{ba} \tag{2.91}$$

From (2.85) and (2.90) the Lagrangian may be written

$$L = T - V = \tfrac{1}{2}[T_{ab}\dot{\eta}_a\dot{\eta}_b - V_{ab}\eta_a\eta_b] \tag{2.92}$$

and the Euler–Lagrange equations of motion are

$$\frac{d}{dt}\frac{\partial L}{\partial \dot{\eta}_c} - \frac{\partial L}{\partial \eta_c} = 0$$

or

$$T_{cb}\ddot{\eta}_b + V_{cb}\eta_b = 0, \qquad c, b = 1, 2, \ldots, N \tag{2.93}$$

In matrix form

$$T\ddot{\eta} + V\eta = 0 \tag{2.94}$$

This in turn may be written as

$$\ddot{\eta} = -A\eta \tag{2.95}$$

where A is an N by N constant matrix called the *dynamical matrix* and is defined by

$$A = T^{-1}V \tag{2.96}$$

Equation (2.95) represents a system of N coupled ordinary differential equations of the second order. The solution of such a system is immensely simplified if the equations can be uncoupled, i.e., if only one dependent variable appears in each equation. It is clear that this will be achieved if A is or can be made a diagonal matrix.

We postulate the existence of a matrix S for which

$$S^{-1}AS = D \tag{2.97}$$

where D is a diagonal matrix. Multiplying (2.95) on the left by S^{-1} we obtain

$$S^{-1}\ddot{\eta} = -S^{-1}ASS^{-1}\eta \tag{2.98}$$

$$\ddot{\psi} = -D\psi \tag{2.99}$$

where $\psi = S^{-1}\eta$. Since D is diagonal, (2.99) represents a system of uncoupled differential equations each of which has the form of the simple harmonic oscillator equation.

Before discussing the solutions of (2.99) let us determine how to construct the matrix S which diagonalizes A. Equation (2.97) may be written

$$AS = SD \tag{2.100}$$

or

$$A_{ab}S_{bc} = S_{ad}\,D_{dc} \tag{2.101}$$

In the sum over d on the right side of (2.101) only the term for which $d = c$ is nonzero since D is diagonal. Thus we may write

$$A_{ab}S_{bc} = D_{cc}S_{ac} \tag{2.102}$$

This equation may be written as

$$A\Phi_c = \lambda_c\Phi_c \qquad \text{(no sum on } c\text{)} \tag{2.103}$$

where Φ_c is the cth column of the S matrix and λ_c is the D_{cc} element of the D matrix. But (2.103) is just the eigenvalue equation with the λ_c being the eigenvalues and the Φ_c the eigenvectors of the dynamical matrix A. Therefore we conclude that the diagonal matrix D is just the matrix having the eigenvalues of A along the main diagonal. The S matrix is constructed by placing the eigenvector Φ_c corresponding to the cth eigenvector in the cth column.

Returning now to (2.99) we see that these equations are of the form

$$\ddot{\psi}_c = -\lambda_c\psi_c \qquad \text{(no sum on } c\text{)} \tag{2.104}$$

The solutions are

$$\psi_c = K_c \sin (\lambda_c)^{1/2} t + K_c' \cos (\lambda_c')^{1/2} t, \qquad c = 1, 2, \ldots, N \quad (2.105)$$

where K_c and K_c' are constants of integration. By analogy with the simple harmonic oscillator we conclude that

$$(\lambda_c)^{1/2} = 2\pi\nu_c \quad (2.106)$$

where ν_c is the frequency of oscillation. Our system has N possible oscillatory states, called the *normal modes* of the system; the ν_c are *the normal mode frequencies* and the ψ_c are the *normal mode coordinates.*

The normal mode coordinates are related to the displacement coordinates by

$$\eta = S\psi \quad (2.107)$$

and therefore they represent an alternative set of generalized coordinates in terms of which our system can be described. In many applications the physical system is represented by normal mode coordinates because as we have seen the equations of motion take a particularly simple form when written in terms of the normal modes. From (2.107) it follows that an arbitrary displacement of the system is simply a linear combination of the normal mode states. It is possible, with an appropriate choice of initial conditions, to excite just one of the normal modes. This will be illustrated in the examples in Section 2.3.

We may summarize the technique for handling systems oscillating with small displacements about some equilibrium configuration as follows. The potential and kinetic energy are Taylor expanded and the matrices T_{ab} and V_{ab} are evaluated at the equilibrium configuration. The normal mode frequencies are obtained from the eigenvalues of the dynamical matrix according to equation (2.106). The S matrix is constructed by placing the eigenvector corresponding to the cth eigenvalue in the cth column. The actual displacements of the system as a function of time are then obtained from (2.105) and (2.107).

C. Charged Particle Motion

In Einstein's relativity the motion of a charged particle is governed by (2.55). The $\mu = 0$ equation is

$$mc\ddot{x}^0 = (e/c)\dot{x}_k F^{0k} \quad (2.108)$$

The spatial components of (2.55) may be written as

$$m\gamma c^2\, d\boldsymbol{\beta}/dx^0 = e[\mathbf{E} + \boldsymbol{\beta}\times\mathbf{B} - (\boldsymbol{\beta}\cdot\mathbf{E})\boldsymbol{\beta}] \tag{2.109}$$

where we have used the fact that

$$\begin{aligned} \ddot{x}^k &= \ddot{x}^0\beta^k + (\dot{x}^0)^2\, d\beta^k/dx^0 \\ \boldsymbol{\beta} &= d\mathbf{x}/dx^0 \end{aligned} \tag{2.110}$$

and expressed $F^{\mu\nu}$ in terms of the electric field **E** and the magnetic field **B** according to (1.89). The $\boldsymbol{\beta}\cdot\mathbf{E}$ term comes from replacing the $\ddot{x}^0$ term appearing in (2.110) with (2.108). Thus the relativistic motion of a charged particle introduces an additional term in the Lorentz force equation. This new term, viz., $(\boldsymbol{\beta}\cdot\mathbf{E})\boldsymbol{\beta}$, may be viewed as a force whose direction is either that of $\boldsymbol{\beta}$ or $(-\boldsymbol{\beta})$ and whose magnitude is proportional to $|\boldsymbol{\beta}|^2$. Thus the effect of this term is quite small for nonrelativistic motions ($\beta \ll 1$). The other departure from the usual Lorentz force equation is the appearance of γ on the left side of (2.109).

We consider the motion of a charged particle in an electric field produced by a point charge fixed at the origin of coordinates, i.e.,

$$\mathbf{E} = q\mathbf{r}/r^3, \qquad \mathbf{B} = 0 \tag{2.111}$$

where q is the charge of the particle attached to the origin and **r** locates the moving charge relative to the origin.

If we orientate our coordinate system such that the xy plane is defined by the initial **r** and $\boldsymbol{\beta}$ of the particle, then it is clear from (2.109) that with $\boldsymbol{\beta} = 0$ there is no force perpendicular to this plane and trajectory of the particle will remain in the xy plane, i.e.,

$$m\gamma c^2\, d\beta^1/dx^0 = e[E^1 - (\boldsymbol{\beta}\cdot\mathbf{E})\beta^1] \tag{2.112a}$$

$$m\gamma c^2\, d\beta^2/dx^0 = e[E^2 - (\boldsymbol{\beta}\cdot\mathbf{E})\beta^2] \tag{2.112b}$$

$$x^3 = \text{const} = 0 \tag{2.112c}$$

It is advantageous to transform to the polar coordinates r, θ defined by

$$x^1 = r\cos\theta, \qquad x^2 = r\sin\theta \tag{2.113}$$

Equation (2.112) becomes after some manipulation

$$m\gamma c^2(r'' - r\theta'^2) = eE - eEr'^2 \tag{2.114a}$$

$$m\gamma c^2(r\theta'' + 2r'\theta') = -eErr'\theta' \tag{2.114b}$$

where the prime indicates the derivative with respect to x^0 and we have used the fact that

$$\boldsymbol{\beta} \cdot \mathbf{E} = r'E, \qquad \beta^2 = r'^2 + r\theta'^2 \tag{2.115}$$

Equation (2.114) differs from the usual Keplerian equations for central force motion by the appearance of the velocity dependent γ and the right side of (2.114b) and the r'^2 term of (2.114a).

To proceed further we assume that γ is a constant; (2.114b) yields

$$\theta' = (K/r^2)\exp(eq/m\gamma c^2 r) \tag{2.116}$$

where K is an integration constant.

Equation (2.114a) is the equation of motion for the radial coordinate as a function of time. Here, however, we would prefer to determine the orbit of the charged particle and this can be done by noting that

$$\frac{d}{dx^0} = \theta' \frac{d}{d\theta} \tag{2.117}$$

Using (2.117), (2.116), and defining $u = 1/r$ we may write (2.114a) as

$$\frac{d^2u}{d\theta^2} + u = -\frac{\delta}{K^2} e^{-2\delta u} \tag{2.118}$$

where

$$\delta = eq/m\gamma c^2 \tag{2.119}$$

Equation (2.118) differs from the usual Keplerian inverse square law orbital motion equation only by the appearance of the exponential and γ. To solve (2.118) we expand the exponential in a power series and drop terms containing u to the third or higher power. Thus (2.118) becomes

$$\frac{d^2u}{d\theta^2} + \left(1 - \frac{2\delta^2}{K^2}\right)u + \frac{4\delta^3}{K^2} u^2 = -\frac{\delta}{K^2} \tag{2.120}$$

To solve (2.118) we assume a Fourier series solution of the form

$$u = a_0 + a_1 \cos \omega\theta + a_2 \cos 2\omega\theta + \cdots \tag{2.121}$$

and substitute into (2.120) to determine the coefficients and ω. We find for $a_2 \ll a_1$

$$a_1^2 = -(a_0 + \delta - 2a_0\,\delta^2 + 4a_0^2\,\delta^3)/2\,\delta^3 \tag{2.122a}$$

$$\omega^2 = 1 - \frac{2\delta^2}{K^2}(1 - 4a_0\,\delta) \tag{2.122b}$$

$$u = a_0 + a_1 \cos \omega\theta \tag{2.122c}$$

a_0 is just the reciprocal of the position of the particle at $\theta = \pi/2$. The equation for a conic section with the origin at one of the foci is

$$u = (1 - \varepsilon \cos \theta)/a(1 - \varepsilon^2) \tag{2.123}$$

Comparing (2.122c) and (2.123)

$$a_0 = 1/a(1 - \varepsilon^2), \qquad a_1 = -\varepsilon/a(1 - \varepsilon^2) = -\varepsilon a_0 \tag{2.124}$$

and if $\omega \neq 1$ the orbit represented by (2.122c) does not close on itself, i.e., there is precession. The particle returns to the radius it had at $\theta = 0$ when

$$\theta = 2\pi/\omega$$

Using (2.122b) and expanding the square root, we find for the precession angle

$$\text{Precession angle} = \theta - 2\pi = \delta^2(1 - 4a_0\,\delta)/K^2 \tag{2.125}$$

D. Minkowski's Equation

The simplest invariant generalization of Newton's equations in Lorentz space is the *Minkowski equation*,

$$\dot{p}^\mu \equiv dp^\mu/ds = K^\mu \tag{2.126}$$

where K_μ is the *four-force*, p^μ is the four-momentum, and s is the interval. The relationship of the covariant Minkowski equation and Newton's equations may be obtained by using (1.49) and (1.42) in (2.126). Thus

$$\frac{d}{ds}\left(mc\frac{dx^\mu}{ds}\right) = \frac{dt}{ds}\frac{d}{dt}\left(mc\frac{dt}{ds}\frac{dx^\mu}{dt}\right) = K^\mu \tag{2.127}$$

or

$$\frac{d}{dt}\left(m\gamma\frac{dx^\mu}{dt}\right) = \frac{c}{\gamma}K^\mu, \qquad \gamma = \frac{1}{(1 - \boldsymbol{\beta}^2)^{1/2}} \tag{2.128}$$

In the nonrelativistic limit as $\boldsymbol{\beta} \to 0$, (2.128) reduces to Newton's equation if we require that

$$(c/\gamma)\mathbf{K} = \mathbf{F} \tag{2.129}$$

Within the Minkowski framework we may establish our dynamics in Lorentz space by converting our experimentally determined Newtonian forces to the Minkowski force via (2.129).

E. Dynamical Equation of the Spin Angular Momentum of a Particle

In Chapter 1 we derived the covariant equation of motion for the spin of a particle, Eq. (1.103), under the assumption that the spin experienced no torque in its rest frame. We found that in the lab frame the spin is observed to precess, the Thomas precession, and that this precession is not dynamical in origin. Now we extend that derivation to the situation where a torque is exerted on the spin in the rest frame of the system. Using the notation of Chapter 1, we write for the rest frame that the equation of motion of the spin is

$$dS^{j'}/dx^{0'} = N^{j'}/c \tag{2.130}$$

where $N^{j'}$ is the torque on the spin angular momentum as measured in the rest frame.

To generalize (2.130) to a covariant equation of motion of the spin valid in any Lorentz frame, we must identify those dynamical variables which may be absent in the rest frame but present elsewhere. The four-velocity is just such a variable, since its spatial components are zero in the rest frame. Thus we generalize (2.130) in the following way:

$$dS^{\mu}/ds = \mathfrak{T}^{\mu} + Au^{\mu} \tag{2.131}$$

where $\mathfrak{T}$ is the torque four-vector and A is a scalar to be determined by the subsidiary condition that, since the spin is a spacelike vector and the velocity is a timelike vector, their scalar product is zero and a constant of the motion. Thus

$$u^{\mu}S_{\mu} = 0 \tag{2.132}$$

$$\frac{d}{ds}(u^{\mu}S_{\mu}) = \frac{du^{\mu}}{ds} S_{\mu} + u^{\mu}\frac{dS_{\mu}}{ds} \tag{2.133}$$

We first investigate the nature of the torque four-vector by specializing (2.131) to the rest frame where $u^{j'} = 0$.

$$\frac{dS^{j'}}{ds} = \frac{dx^{0'}}{ds}\,\frac{dS^{j'}}{dx^{0'}} = \mathfrak{T}^{j'} \tag{2.134}$$

Comparing (2.134) and (2.130), it follows that

$$\mathfrak{T}^{j'} = N^{j'}/c \tag{2.135}$$

where we have used the fact that

$$u^{\mu\prime} u_\mu{}' = I = u^{0\prime} u_0{}' \tag{2.136}$$

Substituting (2.126) and (2.131) into (2.133), we obtain

$$K^\mu S_\mu / mc + u^\mu \mathfrak{T}_\mu + A = 0 \tag{2.137}$$

from which it follows that

$$A = -u_\mu \mathfrak{T}^\mu - K_\mu S^\mu / mc \tag{2.138}$$

and

$$dS^\mu / ds = \mathfrak{T}^\mu - u^\mu [u^\nu \mathfrak{T}_\nu + K^\nu S_\nu / mc] \tag{2.139}$$

Equation (2.139) is the covariant equation of motion of a spin angular momentum which is subject to a torque $\mathfrak{T}^\mu$.

It is interesting to consider (2.139) when the particle is a charged particle moving under the influence of an electromagnetic field. It can be shown that the torque on a charged particle having an intrinsic magnetic moment $\boldsymbol{\mu}$ is

$$\mathbf{N} = \boldsymbol{\mu} \times \mathbf{B} = (ge/2mc) \mathbf{S} \times \mathbf{B} \tag{2.140}$$

where g is the "*gyromagnetic ratio*" whose value is characteristic of the particle. For an electron g is approximately equal to 2. Equation (2.140) may be written

$$N^j = (ge/2mc) \varepsilon^{jkl} S_k B_l = (ge/2mc) F^{jk} S_k \tag{2.141}$$

where ε^{jkl} is the Levi–Civita tensor and F^{jk} is the antisymmetric field strength tensor defined in (1.89). An obvious covariant generalization of (2.141) is

$$N^\mu = (ge/2mc) F^{\mu\nu} S_\nu \tag{2.142}$$

Comparing (2.126) and (2.55) we may write for the four-vector equivalent of the Lorentz force

$$K^\mu = (e/c) F^{\mu\nu} u_\nu \tag{2.143}$$

Inserting (2.143) and (1.142) into (2.139) we find

$$\frac{dS^\mu}{ds} = \frac{ge}{2mc^2} F^{\mu\nu} S_\nu - (g-2) \frac{e}{2mc^2} u_\nu F^{\nu\lambda} S_\lambda u^\mu \tag{2.144}$$

Equation (2.144) is the covariant equation of motion of the spin of a particle having an intrinsic spin and moving under the influence of an electromagnetic field.

It is often advantageous to have an equation which describes the time rate of change of the spin of the rest frame as measured by the laboratory observer, i.e., $dS^{j'}/dx^0$. To obtain such an equation, we differentiate the Lorentz transformation equation

$$S_\mu = L^\lambda{}_\mu S_\lambda' \tag{2.145}$$

with respect to x^0. Thus

$$\frac{dS_\mu}{dx^0} = \left(\frac{d}{dx^0} L^\lambda{}_\mu\right) S_\lambda' + L^\lambda{}_\mu \frac{dS_\lambda'}{dx^0} \tag{2.146}$$

Solving for dS_λ'/dx^0 and multiplying both sides of the equation by $L^{\sigma\mu}$, we obtain

$$\frac{dS^{\sigma'}}{dx^0} = -L^{\sigma\mu} S_\lambda'\left(\frac{d}{dx^0} L^\lambda{}_\mu\right) + \frac{1}{u^0}\left[\frac{ge}{2mc^2} L^{\sigma\mu} F_\mu{}^\nu L^\varrho{}_\nu S_\varrho' - (g-2)\frac{e}{2mc^2} u_\nu F^{\nu\lambda} L^\varrho{}_\lambda S_\varrho' u^{\sigma'}\right] \tag{2.147}$$

After engaging in some tedious algebra in which Eqs. (2.65), (2.69), and (2.89) are inserted into (2.147) and noting that $S_0' = u^{j'} = 0$, it can be shown that (2.147) may be written as

$$\frac{d\mathbf{S}'}{dx^0} = \frac{ge}{2mc^2\gamma} \mathbf{S}' \times \left[\mathbf{B} + \frac{\gamma}{1+\gamma} \mathbf{E} \times \mathbf{B}\right] + \frac{(g-2)e\gamma}{2mc^2(\gamma+1)} \mathbf{S}' \times [(\boldsymbol{\beta} \times \mathbf{B} + \mathbf{E}) \times \boldsymbol{\beta}] \tag{2.148}$$

where $\mathbf{E}$ and $\mathbf{B}$ are the electric and magnetic fields as measured in the lab frame and the $\boldsymbol{\beta}$ and γ contain the velocity of the particle as measured by the lab frame. Equation (2.148) is called the *Thomas equation.*

An important conclusion from (2.148) is that $d\mathbf{S}'/dx^0$ is zero when $\mathbf{B} = 0$ and $\mathbf{E}$ is parallel to $\boldsymbol{\beta}$ as it would be if the particle were accelerated from rest in a constant uniform electric field.

When the electric field is uniform, it follows from (2.143) that

$$\frac{d}{dt}(m\gamma\mathbf{v}) = e\mathbf{E} + \frac{e}{c}\mathbf{v} \times \mathbf{B} = \frac{e}{c}\mathbf{v} \times \mathbf{B} \tag{2.149}$$

If $\boldsymbol{\beta}$ is a constant uniform field parallel to $\mathbf{v}$, it follows that

$$\frac{d}{dt}(m\gamma\mathbf{v}) = 0$$

and, therefore, the velocity remains parallel to $\mathbf{B}$. Equation (2.148) becomes

$$\frac{d\mathbf{S}'}{dx^0} = \frac{ge}{2mc^2\gamma}\mathbf{S}'\times\mathbf{B} = \mathbf{S}'\times\left(\frac{ge\mathbf{B}}{2mc^2\gamma}\right) \tag{2.150a}$$

Equation (2.150a) is the standard form [see the discussion surrounding Eqs. (1.110)–(1.112)] for a precessing vector where the rate of precession is given by

$$\boldsymbol{\omega} = ge\mathbf{B}/2mc\gamma \tag{2.150b}$$

i.e., $\mathbf{S}'$ precesses about $\mathbf{B}$ at a rate of $geB/2mc\gamma$.

Finally, we consider the important case where $\mathbf{E}$ is zero, $\mathbf{B}$ is constant and uniform, and the particle velocity is perpendicular to $\mathbf{B}$. Then the equation of motion becomes (that γ is constant follows from the equation of motion for the zeroth component of force, K^0)

$$\frac{d}{dt}(m\gamma\mathbf{v}) = \frac{e}{c}\mathbf{v}\times\mathbf{B} \tag{2.151}$$

$$\frac{d\mathbf{v}}{dt} = \mathbf{v}\times\frac{e\mathbf{B}}{mc\gamma} \tag{2.152}$$

Thus $\mathbf{v}$ precesses about $\mathbf{B}$ at the rate $eB/mc\gamma$ and, therefore, the particle moves in a circle perpendicular to $\mathbf{B}$. It can be shown that under these conditions (2.150) becomes

$$\frac{d\mathbf{S}'}{dt} = \mathbf{S}'\times\left[1 + \frac{g-2}{2}\gamma\right]\frac{e\mathbf{B}}{mc\gamma} \tag{2.153}$$

The spin also precesses about $\mathbf{B}$. If $g = 2$ the rate of precession of $\mathbf{S}$ and $\mathbf{v}$ is the same. We have here an excellent technique for measuring the gyromagnetic ratio, i.e., the amount that $\mathbf{S}'$ and $\mathbf{v}$ become out of phase is a measure of $(g - 2)$. Most of the high precession measurements of the muon and electron magnetic moments are based on this idea.

2.3. SOLVED EXAMPLES

Example 2.1. A particle of mass m near the surface of the earth moves under the influence of gravity only (Fig. 2.3). Determine its equations of motion.

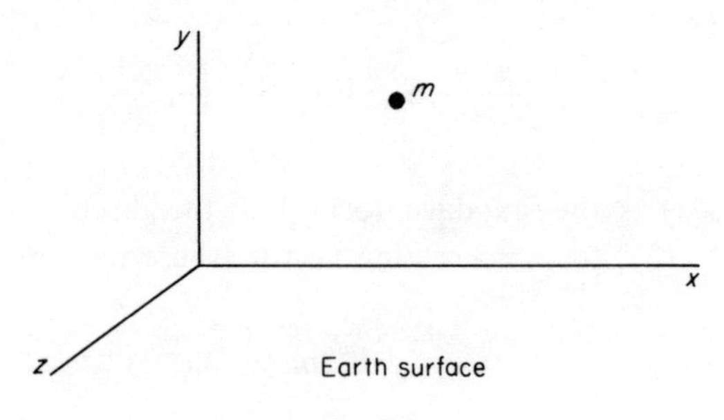

Fig. 2.3

Solution. There are three degrees of freedom. We take the xy plane to be the surface of the earth and assume a constant gravitational force, i.e., $\mathbf{F} = -mg\hat{\jmath}$.

$$V = mgy$$

and

$$T = \tfrac{1}{2}m(\dot{x}^2 + \dot{y}^2 + \dot{z}^2)$$

$$L = \tfrac{1}{2}m(\dot{x}^2 + \dot{y}^2 + \dot{z}^2) - mgy$$

$$\frac{\partial L}{\partial \dot{x}} = m\dot{x}, \qquad \frac{\partial L}{\partial \dot{y}} = m\dot{y}, \qquad \frac{\partial L}{\partial \dot{z}} = m\dot{z}$$

$$\frac{\partial L}{\partial y} = -mg, \qquad \frac{\partial L}{\partial x} = \frac{\partial L}{\partial z} = 0$$

Thus the Lagrange equations of motion are

$$m\ddot{x} = 0, \qquad m\ddot{y} = -mg, \qquad m\ddot{z} = 0$$

Example 2.2. A system of N particles moves under their mutual gravitational attraction (Fig. 2.4). Determine the equations of motion.

Solution. There are $3N$ degrees of freedom and we use the inertial coordinates to specify the configuration of the system. Thus

$$T = \tfrac{1}{2}\sum_{\alpha} m_\alpha({}_\alpha\dot{x}^i)^2, \qquad V = \tfrac{1}{2}\sum_{\alpha,\beta}{}' Gm_\alpha m_\beta/r_{\alpha\beta} \tag{2.154}$$

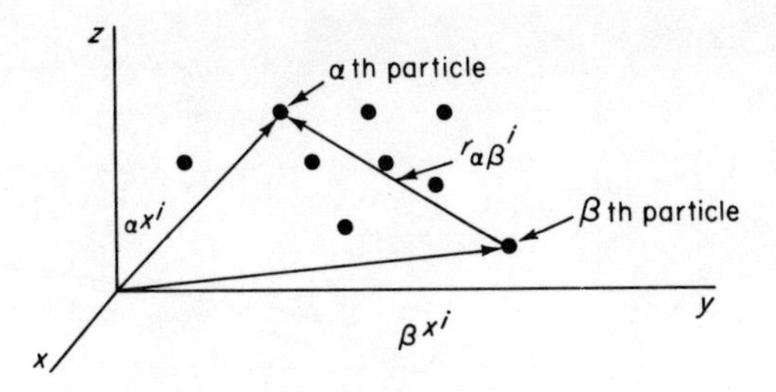

Fig. 2.4

where the prime on the summation indicates that α may not equal β.

$$L = \tfrac{1}{2}\sum_{\alpha} m_\alpha({}_\alpha\dot{x}^i)^2 - \tfrac{1}{2}\sum_{\alpha,\beta}{}' Gm_\alpha m_\beta/r_{\alpha\beta}$$
$$r_{\alpha\beta} = [({}_\alpha x^i - {}_\beta x^i)^2]^{1/2} \tag{2.155}$$

$$\partial L/\partial\, {}_\alpha x^i = \tfrac{1}{2}G\sum_{\beta,\gamma}{}' m_\beta m_\gamma({}_\beta x^i - {}_\gamma x^i)(\delta_{\beta\alpha} - \delta_{\gamma\alpha})/r_{\beta\gamma}^3$$
$$= \sum_{\beta} Gm_\alpha m_\beta({}_\alpha x^i - {}_\beta x^i)/r_{\alpha\beta}^3$$

$$\partial L/\partial\, {}_\alpha\dot{x}^i = m_\alpha\, {}_\alpha\dot{x}^i$$

The equations of motion are

$$m_\alpha\, {}_\alpha\ddot{x}^i = \sum_{\beta} Gm_\alpha m_\beta({}_\alpha x^i - {}_\beta x^i)/r_{\alpha\beta}^3 \tag{2.156}$$

Example 2.3. A particle of mass m is moving under the influence of a potential V near the surface of the earth (Fig. 2.5). Describe the equations of motion (a) relative to an inertial coordinate system at the center of the earth, and (b) relative to an observer on the surface of the earth.

Solution. (a) Relative to the inertial system at the center of the earth we may simply express the kinetic energy as

$$T = \tfrac{1}{2}m\dot{\varrho}^2 \qquad \text{or} \qquad T = \tfrac{1}{2}m(\dot{X}^2 + \dot{Y}^2 + \dot{Z}^2)$$

and the equations of motion are

$$m\ddot{\varrho}^i = -\partial V/\partial \varrho^i \tag{2.157}$$

(b) To express the kinetic energy of the mass m relative to the primed coordinate system at the surface of the earth we use (2.68). We choose

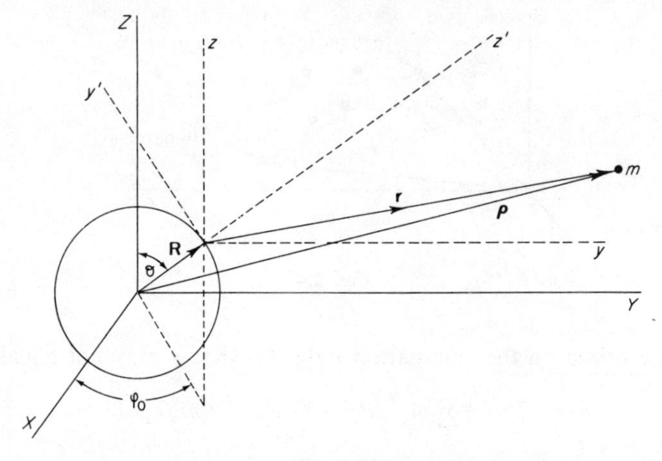

Fig. 2.5

the z' axis to run along a line drawn from the center of the earth to the location of the surface observer. Thus the primed coordinate system may be constructed as follows:

$$\hat{\mathbf{z}}' = \hat{\mathbf{R}} = \sin\theta_0 \cos\varphi_0 \hat{\mathbf{i}} + \sin\theta_0 \sin\varphi_0 \hat{\mathbf{j}} + \cos\theta_0 \hat{\mathbf{k}} \tag{2.158a}$$

$$\hat{\mathbf{x}}' = \frac{d\mathbf{R}/dt}{|\,d\mathbf{R}/dt\,|} = -\sin\varphi_0 \hat{\mathbf{i}} + \cos\varphi_0 \hat{\mathbf{j}} \tag{2.158b}$$

$$\hat{\mathbf{y}}' = \hat{\mathbf{z}}' \times \hat{\mathbf{x}}' = -\cos\theta_0 \cos\varphi_0 \hat{\mathbf{i}} - \cos\theta_0 \sin\varphi_0 \hat{\mathbf{j}} + \sin\theta_0 \hat{\mathbf{k}} \tag{2.158c}$$

To obtain the F^{ij} that appear in (2.68) we note that

$$\mathbf{r}' = x'\hat{\mathbf{x}}' + y'\hat{\mathbf{y}}' + z'\hat{\mathbf{z}}' = x\hat{\mathbf{i}} + y\hat{\mathbf{j}} + z\hat{\mathbf{k}} \tag{2.159}$$

Therefore,

$$x = -\sin\varphi_0 x' - \cos\theta_0 \cos\varphi_0 y' + \sin\theta_0 \cos\varphi_0 z' \tag{2.160a}$$

$$y = \cos\varphi_0 x' - \cos\theta_0 \sin\varphi_0 y' + \sin\theta_0 \sin\varphi_0 z' \tag{2.160b}$$

$$z = \sin\theta_0 y' + \cos\theta_0 z' \tag{2.160c}$$

Comparing (2.160) with (2.65) we obtain

$$F^{ij} = \begin{bmatrix} -\sin\varphi_0 & -\cos\theta_0 \cos\varphi_0 & \sin\theta_0 \cos\varphi_0 \\ \cos\varphi_0 & -\cos\theta_0 \sin\varphi_0 & \sin\theta_0 \sin\varphi_0 \\ 0 & \sin\theta_0 & \cos\theta_0 \end{bmatrix} \tag{2.161}$$

and

$$\dot{F}^{ij} = \dot{\varphi}_0 \begin{bmatrix} -\cos\varphi_0 & \cos\theta_0 \sin\varphi_0 & -\sin\theta_0 \sin\varphi_0 \\ -\sin\varphi_0 & -\cos\theta_0 \cos\varphi_0 & \sin\theta_0 \cos\varphi_0 \\ 0 & 0 & 0 \end{bmatrix} \tag{2.162}$$

We have constructed z' so that it runs along $\mathbf{R}$; therefore, we may write using (2.158a)

$$\mathbf{R} = R[\sin\theta_0 \cos\varphi_0 \hat{\mathbf{i}} + \sin\theta_0 \sin\varphi_0 \hat{\mathbf{j}} + \cos\theta_0 \hat{\mathbf{k}}] \tag{2.163}$$

and therefore

$$\dot{\mathbf{R}} = \dot{\varphi}_0 R[-\sin\theta_0 \sin\varphi_0 \hat{\mathbf{i}} + \sin\theta_0 \cos\varphi_0 \hat{\mathbf{j}}] \tag{2.164a}$$

$$(\dot{\mathbf{R}})^2 = (R\dot{\varphi}_0 \sin\theta_0)^2 \tag{2.164b}$$

Substituting (2.161)–(2.164) into (2.68) we obtain after some algebra (it is helpful to note that F^{ij} is an orthogonal matrix)

$$\begin{aligned} T = \tfrac{1}{2}m\{(\dot{\mathbf{r}}')^2 &+ \dot{\varphi}_0{}^2[(R\sin\theta_0 - y'\cos\theta_0 + z'\sin\theta_0)^2 + x'^2] \\ &+ 2\dot{\varphi}_0[\dot{x}'(R\sin\theta_0 - y'\cos\theta_0 + z'\sin\theta_0) + \dot{y}'x'\cos\theta_0 - \dot{z}'x'\sin\theta_0]\} \end{aligned} \tag{2.165}$$

If we assume that $\dot{\varphi}_0 = \text{const} = \omega$, the equations of motion are

$$\ddot{x}' + 2\omega(\dot{z}'\sin\theta_0 - \dot{y}'\cos\theta_0) - x'\omega^2 = -\frac{1}{m}\frac{\partial V}{\partial x'} \tag{2.166a}$$

$$\ddot{y}' + 2\omega\dot{x}'\cos\theta_0 + (R\sin\theta_0 - y'\cos\theta_0 + z'\sin\theta_0)\omega^2\cos\theta_0 = -\frac{1}{m}\frac{\partial V}{\partial y'} \tag{2.166b}$$

$$\ddot{z}' - 2\omega\dot{x}'\sin\theta_0 - (R\sin\theta_0 - y'\cos\theta_0 + z'\sin\theta_0)\omega^2\sin\theta_0 = -\frac{1}{m}\frac{\partial V}{\partial z'} \tag{2.166c}$$

We recognize the second term on the left side of these equations as the "Coriolis force" and the third term as the "centrifugal force." By comparing (2.166) and (2.157) it is clear that the Coriolis and centrifugal forces arise only because of the use of a noninertial reference system and are indeed properly referred to as "fictitious forces." Part of the power of

Lagrangian dynamics is the fact that only real forces (i.e., forces which act in an inertial coordinate system) need be considered.

Example 2.4. Find the equations of motion of a bead of mass m sliding on a frictionless wire under the influence of gravity. The wire has a parabolic shape and rotates with a constant angular velocity ω. (Fig. 2.6).

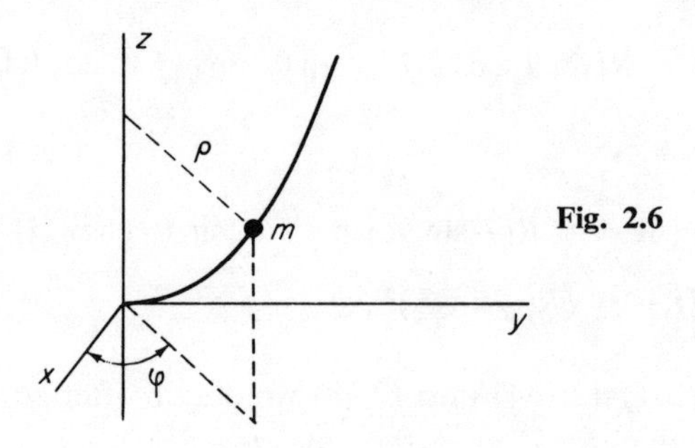

Fig. 2.6

Solution. The bead is constrained to slide on the wire. The equation of constraint is

$$z = a\varrho^2$$

This system has but one degree of freedom and we will write the equation of motion in terms of ϱ. We may write the kinetic energy as

$$T = \tfrac{1}{2}m(\dot{x}^2 + \dot{y}^2 + \dot{z}^2)$$

But

$$x = \varrho \cos \varphi, \qquad y = \varrho \sin \varphi, \qquad z = a\varrho^2, \qquad \dot{\varphi} = \omega = \text{const}$$

Therefore,

$$T = \tfrac{1}{2}m(\dot{\varrho}^2 + \varrho^2\dot{\varphi}^2 + 4a^2\varrho^2\dot{\varrho}^2)$$

$$V = mgz = mga\varrho^2$$

$$L = \tfrac{1}{2}m[\dot{\varrho}^2(1 + 4a^2\varrho^2) + \varrho^2(\omega^2 - 2ga)]$$

The equation of motion is

$$\ddot{\varrho}(1 + 4a^2\varrho^2) + 4a^2\varrho\dot{\varrho}^2 + (2ga - \omega^2)\varrho = 0$$

Instead of eliminating the z variable we may proceed by the method of Lagrange multipliers. The modified Lagrangian is

$$\bar{L} = \tfrac{1}{2}m(\dot{\varrho}^2 + \varrho^2\omega^2 + \dot{z}^2) - mgz + \lambda(z - a\varrho^2)$$

and the Euler–Lagrange equations are

$$\ddot{\varrho} - \varrho\omega^2 + 2\lambda a\varrho/m = 0, \qquad \ddot{z} + g - \lambda/m = 0, \qquad z = a\varrho^2$$

It is easy to show that the first equation reduces to the equation of motion we obtained previously and that

$$\lambda = mg + 2ma\dot{\varrho}^2 + 2ma\varrho\ddot{\varrho}$$

Example 2.5. In the system shown in Fig. 2.7 the mass m is unwinding. The rope is wound around the axle of the disk at a diameter R. (Assume that all of the rope is at this diameter and that the rope is massless, rigid, and of negligible thickness.) As the mass unwinds it is free to oscillate like a pendulum in the plane of the paper. Take the point of suspension to be at A. The disk turns in such a way that there is no slipping or sliding of the rope relative to the disk; i.e., the rope *rolls* off the disk. The disk is attached to the spring in such a way that its rotational motion is not impeded. The disk–axle system has a moment of inertia I and a mass M. Find the equations of motion of the system.

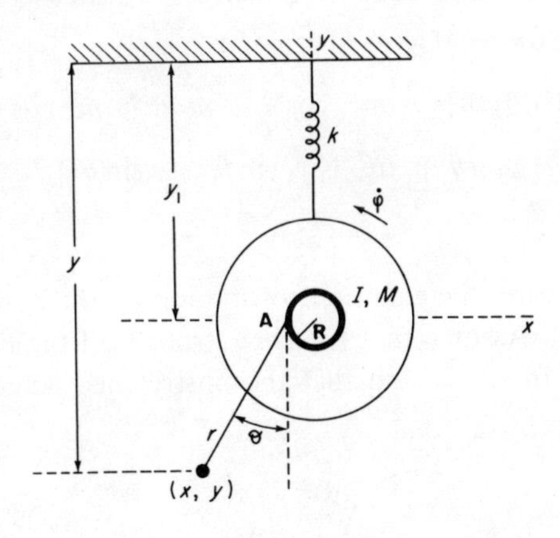

Fig. 2.7

Solution. The system has three degrees of freedom—one to locate the vertical position of the disk (y_1); one to designate the angle the pendulum makes with the vertical (θ); and one to give the length of the pendulum (r). Let the unstretched length of the spring be l_0; then the potential energy may be written as

$$V = Mgy_1 + mgy + \tfrac{1}{2}k(y_1 + l_0)^2$$

The kinetic energy is

$$T = \tfrac{1}{2}I\dot{\varphi}^2 + \tfrac{1}{2}M\dot{y}_1{}^2 + \tfrac{1}{2}m(\dot{x}^2 + \dot{y}^2)$$

where $\dot{\varphi}$ is the angular velocity of the disk. We now must express T and V in terms of three independent variables which we choose to be r, θ, and y_1. The transformation equations required here are

$$x = -(R + r\sin\theta), \qquad y = y_1 - r\cos\theta, \qquad \dot{r} = R\dot{\varphi} \quad \text{(rolling constraint)}$$

Thus the potential becomes

$$V = (M + m)gy_1 + mgr\cos\theta + \tfrac{1}{2}k(y_1 + l_0)^2$$

and the kinetic energy is

$$T = \tfrac{1}{2}(I/R^2 + m)\dot{r}^2 + \tfrac{1}{2}(M + m)\dot{y}_1{}^2 + \tfrac{1}{2}m(r^2\dot{\theta}^2 - 2\dot{y}_1[\dot{r}\cos\theta - r\dot{\theta}\sin\theta])$$

The equations of motion are

$$(m + M)\ddot{y}_1 - m(\ddot{r}\cos\theta - 2\dot{r}\dot{\theta}\sin\theta - r\ddot{\theta}\sin\theta - r\dot{\theta}^2\cos\theta)$$
$$= -(m + M)g - k(y_1 + l_0)$$
$$[(I/R^2) + m]\ddot{r} - m\ddot{y}_1\cos\theta - mr\dot{\theta}^2 + mg\cos\theta = 0$$
$$r\ddot{\theta} + 2\dot{r}\dot{\theta} + \ddot{y}_1\sin\theta = y\sin\theta$$

Example 2.6. Find the equations of motion of the pulley–spring system shown in Fig. 2.8. Assume that the rope "rolls" off the disk, that the disk has a moment of inertia I, and that the unstretched length of each spring is l_0.

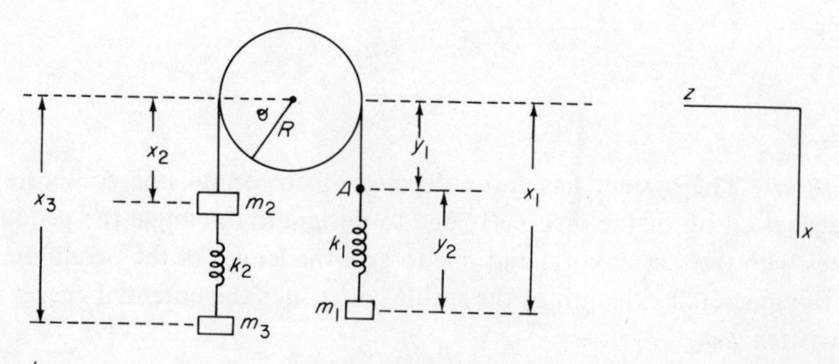

Fig. 2.8

Solution. The system has three degrees of freedom. The rope running from m_2 to A satisfies the condition

$$y_1 + \pi R + x_2 = \text{const} = K$$

and from the diagram it follows that

$$y_2 = x_1 - y_1 = x_1 + \pi R + x_2 - K$$

The fact that the rope rolls off the disk implies that

$$\dot{x}_2 = R\dot{\theta}$$

The kinetic energy is simply

$$T = \tfrac{1}{2}m_1\dot{x}_1^2 + \tfrac{1}{2}m_2\dot{x}_2^2 + \tfrac{1}{2}m_3\dot{x}_3^2 + \tfrac{1}{2}I\dot{\theta}^2$$

and the potential energy is

$$V = \tfrac{1}{2}k_1(y_2 - l_0)^2 + \tfrac{1}{2}k_2(x_3 - x_2 - l_0)^2 - (m_3x_3 + m_2x_2 + m_1x_1)g$$

Using the constraint equations we eliminate the dependent variables and express our kinetic and potential energies in terms of x_1, x_2, and x_3. Thus the Lagrangian is

$$L = \tfrac{1}{2}[m_1\dot{x}_1^2 + (m_2 + I/R^2)\dot{x}_2^2 + m_3\dot{x}_3^2] + (m_1x_1 + m_2x_2 + m_3x_3)g$$
$$- \tfrac{1}{2}k_1(x_1 + x_2 + k_1)^2 - \tfrac{1}{2}\,k_2(x_3 - x_2 - l_0)^2$$

where $K_1 = \pi R - K - l_0$. The equations of motion are

$$\ddot{x}_1 = g - (k_1/m_1)(x_1 + x_2 + K_1)$$
$$\ddot{x}_2 = [m_2g - k_1(x_1 + x_2 + K_1) + k_2(x_3 - x_2 - l_0)]/(m_2 + I/R^2)$$
$$\ddot{x}_3 = g - (k_2/m_3)(x_3 - x_2 - l_0)$$

Suppose now that we wish to determine the tension in the cord connecting m_2 and A. Using the Lagrange multiplier technique for the "taut cord" constraint and $y_2 = x_1 - y_1$, we obtain the modified Lagrangian

$$\bar{L} = \tfrac{1}{2}[m_1\dot{x}_1^2 + (m_2 + I/R^2)\dot{x}_2^2 + m_3\dot{x}_3^2] + (m_1x_1 + m_2x_2 + m_3x_3)g$$
$$- \tfrac{1}{2}k_1(x_1 - y_1 - l_0)^2 - \tfrac{1}{2}k_2(x_3 - x_2 - l_0)^2 + \lambda(y_1 + x_2 + \pi R - K)$$

The equations of motion are

$$\ddot{x}_1 = g - (k_1/m_1)(x_1 - y_1 - l_0)$$

$$\ddot{x}_2 = [m_2 g + k_2(x_3 - x_2 - l_0) + \lambda]/(m_2 + I/R^2)$$

$$\ddot{x}_3 = g - (k_2/m_3)(x_3 - x_2 - l_0)$$

$$0 = k_1(x_1 - y_1 - l_0) + \lambda$$

$$y_1 = K - \pi R - x_2$$

where from the last two equations we find $\lambda = -k_1(x_1 + x_2 + K_1)$.

Example 2.7. The cart shown in Fig. 2.9 rolls over the horizontal plane. The mass of the cart and the inclined plane is M. Each wheel may be regarded as a disk having a mass M_1 and a moment of inertia $\frac{1}{2}M_1R^2$, where R is the radius of the wheel. Find the equations of motion of the system.

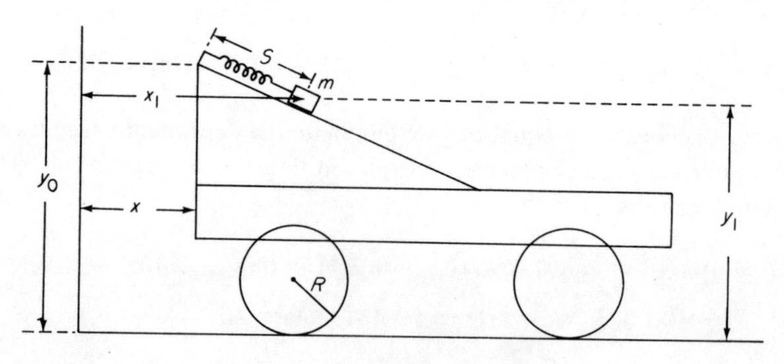

Fig. 2.9

Solution. There are two degrees of freedom—one to locate the position of the cart and one to locate the mass m on the inclined plane. The kinetic energy of a wheel may be expressed as the translational kinetic energy of the center of mass plus the kinetic energy relative to the center of mass of the particles comprising the wheel [see Eq. (2.74)].

$$T_{\text{wheel}} = \tfrac{1}{2}M_1\dot{x}^2 + \tfrac{1}{2}(\tfrac{1}{2}M_1R^2)\dot{\theta}^2$$

Since the wheels are rolling $R\dot{\theta} = \dot{x}$ and

$$T_{\text{wheel}} = \tfrac{1}{2}M_1\dot{x}^2 + \tfrac{1}{4}M_1\dot{x}^2 = \tfrac{3}{4}M_1\dot{x}^2$$

The kinetic energy of the entire system is

$$T = 3M_1\dot{x}^2 + \tfrac{1}{2}M\dot{x}^2 + \tfrac{1}{2}m(\dot{x}_1{}^2 + \dot{y}_1{}^2)$$

and

$$V = mgy_1 + \tfrac{1}{2}k(S - l_0)$$

where l_0 is the unstretched length of the spring. To express the T and V in terms of the variables S and x we note that

$$x_1 = x + S\sin\alpha, \qquad y_1 = y_0 - S\cos\alpha$$

Thus

$$L = \tfrac{1}{2}(m + M + 6M_1)\dot{x}^2 + \tfrac{1}{2}m(\dot{S}^2 + 2\dot{x}\dot{S}\sin\alpha) - mg(y_0 - S\cos\alpha)$$
$$- \tfrac{1}{2}k(S - l_0)^2$$

and the equations of motion are

$$(m + M + 6M_1)\ddot{x} + m\ddot{S}\sin\alpha = 0$$
$$\ddot{S} + \ddot{x}\sin\alpha = g\cos\alpha - (k/m)(S - l_0)$$

It follows from these equations that the mass m moves as a harmonic oscillator with a natural frequency of

$$\omega^2 = k(m + M + 6M_1)/[m(m\cos^2\alpha + M + 6M_1)]$$

Example 2.8. Determine the equations of motion of the spinning symmetrical top moving under the influence of gravity with the bottom of the top being constrained to remain at the origin (Fig. 2.10).

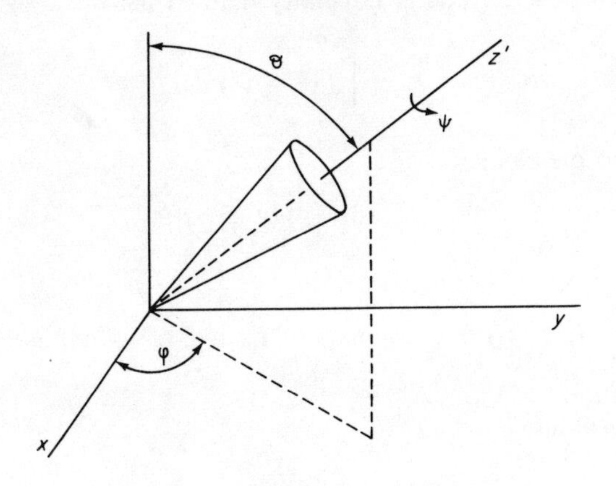

Fig. 2.10

Solution. We may adapt (2.77) to express the kinetic energy. For our case $R^i = \dot{R}^i = 0$ and (2.77) becomes

$$T = \tfrac{1}{2}\dot{F}^{ij}\dot{F}^{ik}I^{jk'}$$

where $I^{jk'}$ is the moment of inertia relative to the body axes (whose origin is at the bottom of the spinning top) and F^{ij} is the matrix which transforms from the body system to the unprimed system. φ and θ are the Euler angles locating the body axes relative to the unprimed axes and in Appendix D it is shown that

$$F = \begin{bmatrix} \cos\varphi\cos\psi & -\cos\varphi\sin\psi & \sin\varphi\sin\theta \\ \quad -\sin\varphi\cos\theta\sin\psi & \quad -\sin\varphi\cos\theta\cos\psi & \\ \sin\varphi\cos\psi & -\sin\varphi\sin\psi & -\cos\varphi\sin\theta \\ \quad +\cos\varphi\cos\theta\sin\psi & \quad +\cos\varphi\cos\theta\cos\psi & \\ \sin\theta\sin\psi & \sin\theta\cos\psi & \cos\theta \end{bmatrix}$$

It is not difficult to show that the body axes are the principal axes of the top, i.e., $I^{jk} = 0$ for $j \neq k$. Furthermore, because of the symmetry of the top $I^{11} = I^{22}$. Thus the kinetic energy is just

$$T = \tfrac{1}{2}[(\dot{F}^{i1})^2 + (\dot{F}^{i2})^2]I^{11'} + \tfrac{1}{2}I^{33'}(\dot{F}^{i3})^2$$

or more explicitly

$$T = \tfrac{1}{2}I^{33'}(\dot{\theta}^2 + \dot{\varphi}^2\sin^2\theta) + \tfrac{1}{2}I^{11'}(2\dot{\psi}^2 + 4\dot{\varphi}\dot{\psi}\cos\theta + \dot{\varphi}^2\cos^2\theta + \dot{\varphi}^2 + \dot{\theta}^2)$$

This expression takes a somewhat simpler form if we note that the moment of inertia about the z' axis is normally defined as

$$I_z \equiv \int (x^2 + y^2)\,dm$$

Thus for our problem

$$I_z = I^{11'} + I^{22'} = 2I^{11'}$$

Also

$$I_x \equiv \int (z^2 + y^2)\,dm, \qquad I_y \equiv \int (z^2 + x^2)\,dm$$

or for our problem

$$I_x = I^{33'} + I^{22'} = I^{33'} + I^{11'} = I_y$$

and

$$I^{33} = I_x - I^{11'} = I_x - \tfrac{1}{2}I_z$$

Replacing $I^{33'}$ and $I^{11'}$ with I_z and I_x, we obtain for the kinetic energy

$$T = \tfrac{1}{2}I_x(\dot{\theta}^2 + \dot{\varphi}^2 \sin^2 \theta) + \tfrac{1}{2}I_z(\dot{\psi} + \dot{\varphi} \cos \theta)^2$$

The potential energy is

$$V = mgz = mgR \cos \theta$$

where R is the distance from the origin to the center of mass. The equations of motion are

$$I_x\ddot{\theta} - (I_x - I_z)\dot{\varphi}^2 \sin \theta \cos \theta + I_z\dot{\varphi}\dot{\psi} \sin \theta = mgR \sin \theta$$

$$I_x\ddot{\varphi} + I_z\ddot{\psi} \cos \theta + 2(I_x - I_z)\dot{\varphi}\dot{\theta} \sin \theta \cos \theta - I_z\dot{\theta}\dot{\psi} \sin \theta = 0$$

$$\ddot{\psi} + \ddot{\varphi} \cos \theta - \dot{\varphi}\dot{\theta} \sin \theta = 0$$

It follows from the fact that the Lagrangian is cyclic in φ and ψ that the corresponding momentum is conserved. Thus the angular momentum about the unprimed z axis and the angular momentum about the body z axis are conserved, i.e.,

$$I_z(\dot{\psi} + \dot{\varphi} \cos \theta) = \text{const} = k$$

$$I_x\dot{\varphi} \sin^2 \theta + k \cos \theta = \text{const} = k_1$$

Example 2.9. Determine the normal mode frequencies and the normal mode displacements of the system shown in Fig. 2.11. The spring is unstretched when $\theta_1 = \theta_2 = 0$.

Solution. The system has two degrees of freedom and we shall express the Lagrangian in terms of θ_1 and θ_2. The kinetic energy and potential

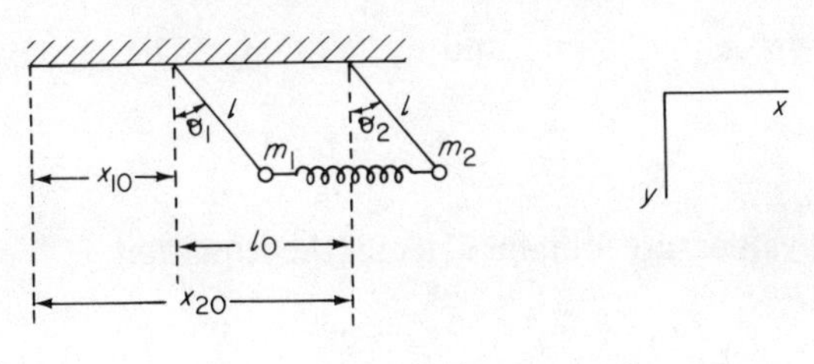

Fig. 2.11

energy are

$$T = \tfrac{1}{2}m_1(\dot{x}_1^2 + \dot{y}_1^2) + \tfrac{1}{2}m_2(\dot{x}_2^2 + \dot{y}_2^2)$$

$$V = \tfrac{1}{2}k(x_2 - x_1 - l_0)^2 - m_1 g y_1 - m_2 g y_2$$

The transformation equations relating x_1, x_2, y_1, y_2 with θ_1 and θ_2 are

$$x_1 = x_{10} + l \sin\theta_1, \qquad y_1 = l\cos\theta_1$$

$$x_2 = x_{20} + l \sin\theta_2, \qquad y_2 = l\cos\theta_2$$

Thus the Lagrangian is

$$L = \tfrac{1}{2}m_1 l^2\dot{\theta}_1^2 + \tfrac{1}{2}m_2 l^2\dot{\theta}_2^2 - \tfrac{1}{2}kl^2(\sin\theta_2 - \sin\theta_1)^2 + gl(m_1\cos\theta_1 + m_2\cos\theta_2)$$

We first determine the equilibrium positions.

$$\partial L/\partial\theta_1 = 0 = -kl^2(\sin\theta_2 - \sin\theta_1)\cos\theta_1 + glm_1\sin\theta_1$$

$$\partial L/\partial\theta_2 = 0 = kl^2(\sin\theta_2 - \sin\theta_1)\cos\theta_2 + glm_2\sin\theta_2$$

These equations are satisfied if $\theta_1 = \theta_2 = 0$ (Stable equilibrium). Thus we write for small displacements from equilibrium

$$\theta_1 = \theta_1' + \eta_1 = \eta_1, \qquad \dot{\theta}_1 = \dot{\eta}_1, \qquad \theta_2 = \theta_2' + \eta_2 = \eta_2, \qquad \dot{\theta}_2 = \dot{\eta}_2$$

T_{ab} defined by (2.91) is

$$T = l^2\begin{bmatrix} m_1 & 0 \\ 0 & m_2 \end{bmatrix}$$

and V_{ab} defined by (2.86) is

$$V = \begin{bmatrix} kl^2 + glm_1 & -kl^2 \\ -kl^2 & kl^2 + glm_2 \end{bmatrix}$$

The dynamical matrix is

$$A = T^{-1}V = \begin{bmatrix} \omega_1^2 + \omega_0^2 & -\omega_1^2 \\ -\omega_2^2 & \omega_2^2 + \omega_0^2 \end{bmatrix}$$

where $\omega_0^2 = g/l$, $\omega_1^2 = k/m_1$, and $\omega_2^2 = k/m_2$. The eigenvalue equation for A is

$$A\Phi = \lambda\Phi$$

and the eigenvalues are obtained from the equation

$$\det\begin{bmatrix} \omega_1^2 + \omega_0^2 - \lambda & -\omega_1^2 \\ -\omega_2^2 & \omega_2^2 + \omega_0^2 - \lambda \end{bmatrix} = 0$$

or

$$(\omega_1^2 + \omega_0^2 - \lambda)(\omega_2^2 + \omega_0^2 - \lambda) - \omega_1^2\omega_2^2 = 0$$

The solutions of this second degree algebraic equation are

$$\lambda_1 = \omega_1^2 + \omega_2^2 + \omega_0^2, \qquad \lambda_2 = \omega_0^2$$

To determine the eigenvectors we write

$$A\Phi_1 = \lambda_1\Phi_1, \qquad \Phi_1 = \begin{bmatrix} \phi_{11} \\ \phi_{12} \end{bmatrix}$$

This matrix equation yields one independent equation

$$(\omega_1^2 + \omega_0^2)\phi_{11} - \omega_1^2\phi_{12} = \lambda_1\phi_{11}$$

Substituting the value of λ_1 and letting $\phi_{11} = 1$ (eigenvectors are determined only up to an arbitrary constant so we may choose one of the components at will), we find

$$\Phi_1 = \begin{bmatrix} 1 \\ -m_1/m_2 \end{bmatrix}$$

Proceeding in a similar fashion for ψ_2 we obtain

$$A\Phi_3 = \lambda_2\Phi, \qquad \Phi_2 = \begin{bmatrix} 1 \\ 1 \end{bmatrix}$$

The S matrix is

$$S = \begin{bmatrix} 1 & 1 \\ -m_1/m_2 & 1 \end{bmatrix}$$

Equation (2.105) becomes

$$\psi_1 = A_1 \sin (\lambda_1)^{1/2}t + B_1 \cos (\lambda_1)^{1/2}t$$
$$\psi_2 = A_2 \sin (\lambda_2)^{1/2}t + B_2 \cos (\lambda_2)^{1/2}t$$

and

$$\eta_a = S_{ab}\psi_b$$

Therefore,

$$\eta_1 = A_1 \sin (\lambda_1)^{1/2}t + B_1 \cos (\lambda_1)^{1/2}t + A_2 \sin (\lambda_2)^{1/2}t + B_2 \cos (\lambda_1)^{1/2}t$$
$$\eta_2 = -(m_1/m_1)[A_1 \sin (\lambda_1)^{1/2}t + B_1 \cos (\lambda_1)^{1/2}t] + A_2 \sin (\lambda_2)^{1/2}t + B_2 \cos (\lambda_2)^{1/2}t$$

For normal mode #1 to be excited it is necessary that $A_2 = B_2 = 0$ and at $t = 0$

$$\eta_1/\eta_2 = -m_1/m_2 \quad \Rightarrow$$

For normal mode #2 $A_1 = B_1 = 0$ and at $t = 0$

$$\eta_1/\eta_2 = 1 \quad \Rightarrow$$

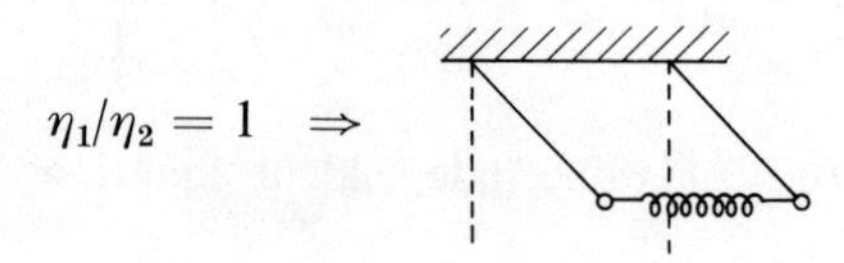

Example 2.10. The Lagrangian describing the motion of a particle of mass m in an external scalar field is given by

$$L = mc - k\varphi(x^\mu)$$

where k is the coupling constant. Find the equations of motion within the framework of Einstein's relativity.

Solution. To use Eq (2.50) we require

$$\partial L/\partial \dot{x}^\mu = 0, \qquad \partial L/\partial x^\mu = -k\varphi_{,\mu}$$

and (2.50) becomes

$$k\varphi_{,\mu} + \frac{d}{ds}\,[\dot{x}_\mu(mc - k\varphi)] = 0$$

or

$$\ddot{x}_\mu(mc - k\varphi) - k\dot{x}_\mu\dot{x}^\nu\varphi_{,\nu} + k\varphi_{,\mu} = 0$$

PROBLEMS

2.1 Find the equations of motion of the double pendulum shown in Fig. P.1. Mass m_1 and mass m_2 are connected by a massless spring. Assume that the spring can displace only along a line connecting m_1 and m_2.

Let r measure the distance between m_1 and m_2 and l_0 is the unstretched length of the spring.

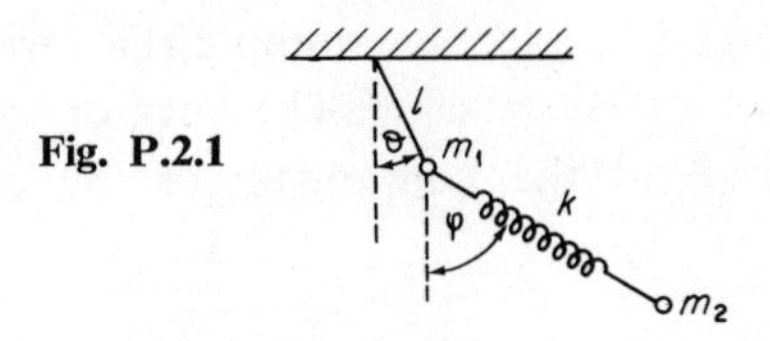

Fig. P.2.1

2.2 Determine the equations of motion of the system shown in Fig. P.2. Assuming a small displacement what is the period of oscillation of the pendulum? The point of connection A is constrained to move horizontally only.

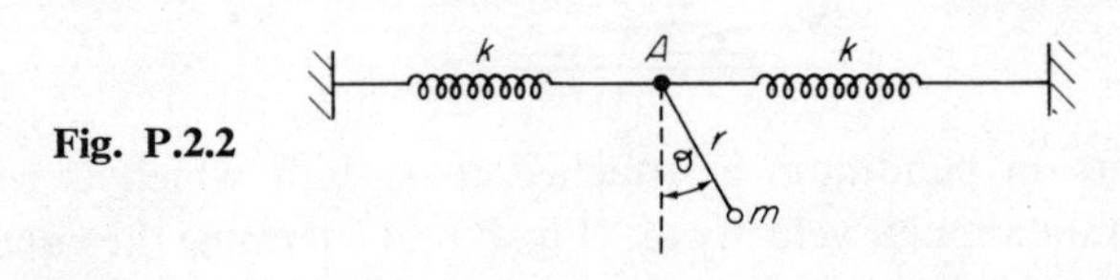

Fig. P.2.2

2.3 A solid uniform disk of mass M and radius R has a small mass m attached to its face at a distance r from its center (Fig. P.3). The disk is free to roll without sliding along a horizontal straight line. Find the equations of motion and the period of oscillation if the disk is displaced only slightly from equilibrium.

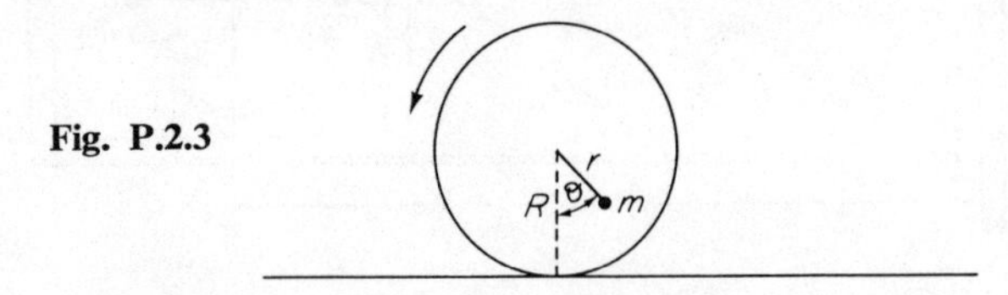

Fig. P.2.3

2.4 The pulley shown in Fig. P.4 is massless and m_1 and m_2 are free to slide in smooth horizontal tubes. The tubes and shaft have moment of inertia I about the vertical axis. Find the equations of motion.

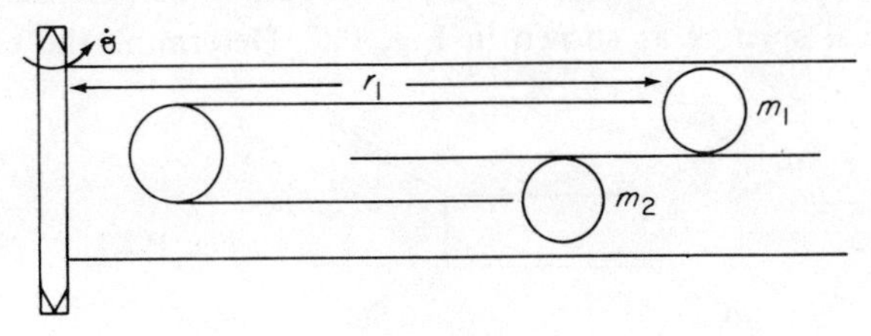

Fig. P.2.4

2.5 A rotating platform P_1 is mounted on a second rotating platform P_2 which is in turn mounted on an elevator E (Fig. P.5). P_1 and P_2 revolve at angular speeds of ω_1 and ω_2, respectively. The simple pendulum swings in a plane containing the vertical axis of rotation of P_1 and its point of suspension. The elevator has a constant vertical acceleration a. Find the Lagrangian of the pendulum system.

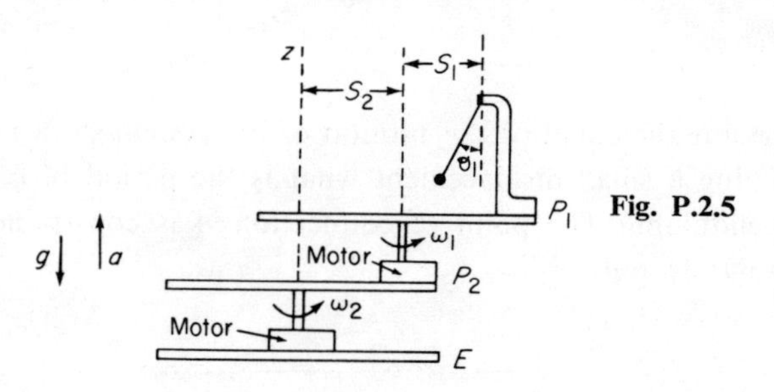

Fig. P.2.5

2.6 A torsion pendulum is attached to a shaft which is rotating with constant angular velocity $\boldsymbol{\omega}_1$ (Fig. P.6). Determine the equations of the pendulum. I_1 is the moment of inertia of the system (except for the masses m_1 and m_2) about the shaft. Assume a Hooke's law potential for the pendulum.

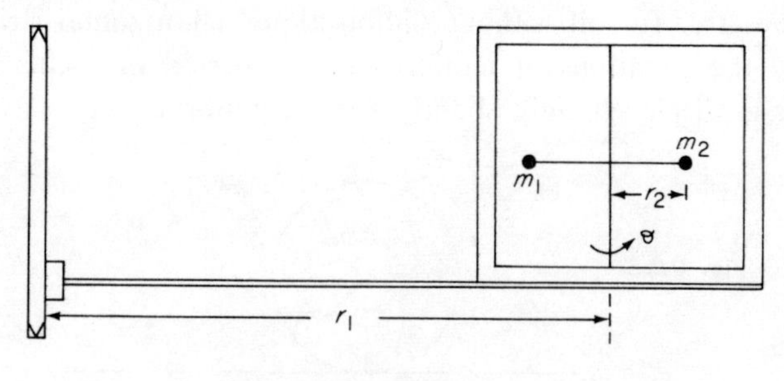

Fig. P.2.6

2.7 A homogeneous rectangular plate can move on a frictionless horizontal plane. Initially, the plate is at rest under the action of four identical springs as shown in Fig. P.7. Determine the equations of

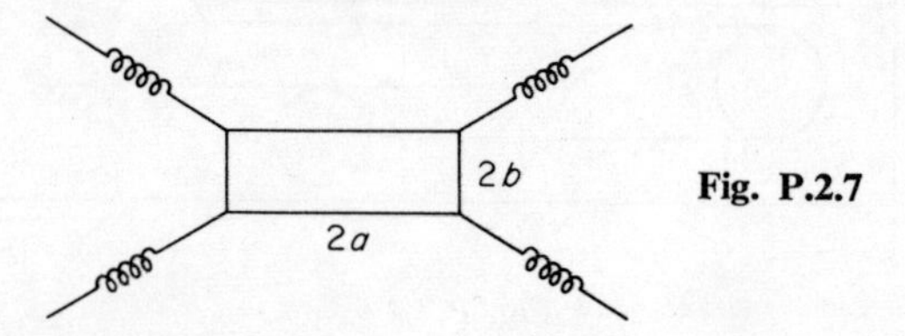

Fig. P.2.7

motion and investigate the motion for small displacements from equilibrium.

2.8 A hollow homogeneous sphere slides on a frictionless, fixed, horizontal plane. At the same time, a point mass slides without friction on the inside of the sphere. Determine the motion of the system.

2.9 A circular disk turns about a rod AB which is connected by a frictionless ball-joint to the end of another rod OA (Fig. P.9). The latter is connected by a frictionless ball-joint to a fixed point O. Determine the motion of the system.

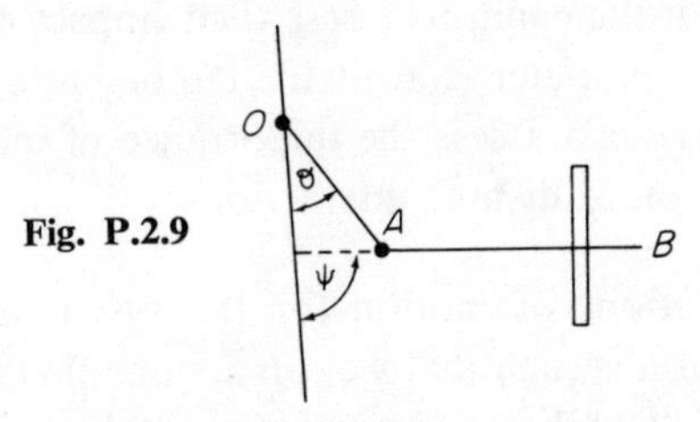

Fig. P.2.9

2.10 Assuming that the earth is a homogeneous, oblate ellipsoid of revolution, and that it rotates about an axis which makes an infinitesmal angle with the earth's axis of symmetry, calculate the angular velocity of precession.

2.11 A projectile is fired due south from a launching pad at 45° latitude. The velocity of projection is 100 ft/sec and the angle of elevation at takeoff is 45° above the horizon plane. Find the point of impact of the projectile.

2.12 Determine the angular deviation of a plumb line from the true vertical at a point on the earth's surface at a latitude λ. Where is the deviation a maximum? a minimum?

2.13 Determine the equations of motion of a symmetrical top one point of which is fixed at the surface of the earth at latitude λ. Discuss the nature of its motion.

2.14 Determine the effect of the earth's rotation on small oscillations of a pendulum at the surface of the earth at a latitude λ. (Foucault's pendulum.)

2.15 Discuss the motion of a gyroscope at the surface of the earth and explain how it might be used as a compass.

2.16 Considerable insight into the nature of the vibratory motions of a crystalline solid can be obtained by investigating the normal modes of a linear chain. The spring constants corresponding to various molecules can be obtained from Ref. 3. An alternative procedure is to consider an ionic bonding where the potential is a combination of an electrostatic potential and a smaller repulsive potential (see Harrison [2] for details). It is interesting to Taylor expand this potential, calculate the normal mode spectrum, and compare this spectrum with that obtained by using the spring constant data on Ref. 3. The problem can be made even more meaningful by substituting various "impurities" in the chain to assess their impact on the normal mode behavior. If a computer is available ten or more masses can be considered and one can assess the importance of including next nearest neighbor and more distant interactions.

2.17 Find the equations of motion for the system shown in Fig. P.17 when the motion is such that we can assume that the springs elongate only in the vertical direction (no horizontal component). Determine the normal mode frequencies and the corresponding normal mode displacements. The pins at the left are frictionless.

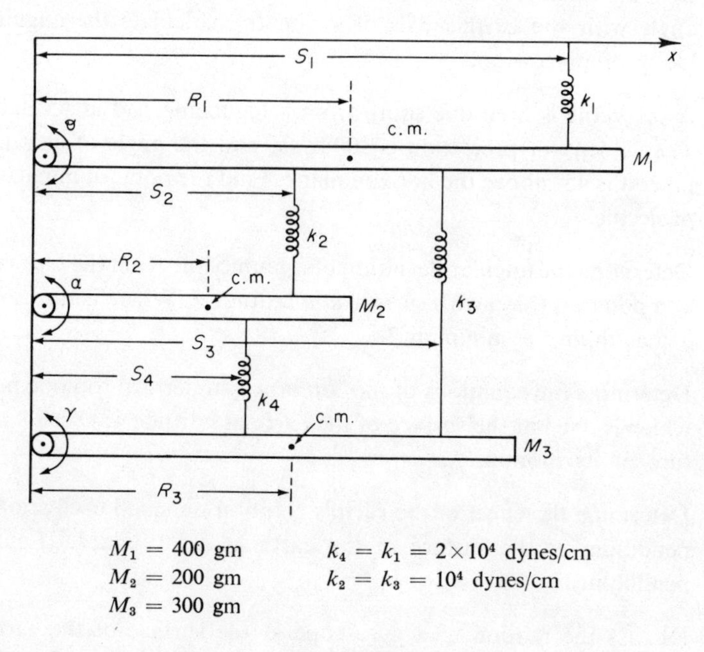

$M_1 = 400$ gm $\qquad k_4 = k_1 = 2\times 10^4$ dynes/cm
$M_2 = 200$ gm $\qquad k_2 = k_3 = 10^4$ dynes/cm
$M_3 = 300$ gm

Fig. P.2.17

2.18 A system of springs and pulleys is shown in Fig. P.18. Determine the normal modes of vibration; i.e., the oscillation frequencies and the nature of the motion of the system when it is in a normal mode.

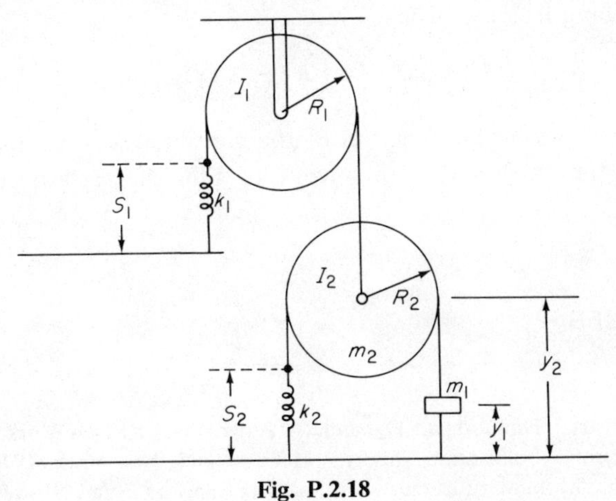

Fig. P.2.18

2.19 Discuss the nature of the motion of a charged particle moving in a uniform static magnetic field.

2.20 Determine the relativistic precession of the planet Mercury by assuming that the sun is a fixed attracting center and by replacing the electric charge appearing in (2.55) with the mass of the planet Mercury and defining $F^{\mu\nu}$ as

$$F^{\mu\nu} = \begin{bmatrix} 0 & G_1 & G_2 & G_3 \\ -G_1 & 0 & 0 & 0 \\ -G_2 & 0 & 0 & 0 \\ -G_3 & 0 & 0 & 0 \end{bmatrix}, \quad \text{where} \quad \mathbf{G} = \frac{GM\mathbf{r}}{r^3}$$

mass of the earth $= 5.983 \times 10^{27}$ gm $= m_e$
mass of the sun $(M) = 329390 m_e$
mass of Mercury $(m_M) = 0.056 m_o$
gravitational constant $(G) = 6.67 \times 10^{-8}$ cm^3 gm^{-1} sec^{-2}
semimajor axis of Mercury's orbit $= 57.9 \times 10^6$ km
Mercury's orbital eccentricity $= 0.20563$
How does your result compare with the 42.6″ per century that is observed (and predicted by the general theory of relativity)?

2.21 Derive Eq. (2.148) from Eq. (2.147).

2.22 Derive (2.153) from (2.148) under the conditions: $\mathbf{E} = 0$, $\mathbf{B}$ is uniform and constant, $\mathbf{v}$ is perpendicular to $\mathbf{B}$.

2.23 A particle moves under the influence of a scalar field. The Lagrangian describing its motion is

$$L = mc - \varphi(x^\mu)$$

where φ is a scalar function of the coordinates. Find the equations of motion within the framework of Einstein's relativity.

REFERENCES

1. C. W. Kilmister, "Hamiltonian Dynamics." Amer. Elsevier, New York, 1965.
2. W. A. Harrison, "Solid State Theory." McGraw Hill, New York, 1970.
3. G. Herzberg, "Molecular Structure and Molecular Spectra." Van Nostrand-Reinhold, Princeton, New Jersey, 1945.

BIBLIOGRAPHY

A. O. Barut, "Electrodynamics and Classical Theory of Fields and Particles." Macmillan, New York, 1964.

T. C. Bradbury, "Theoretical Mechanics." Wiley, New York, 1968.

R. W. Brehme, The relativistic Lagrangian, *Amer. J. Phys.* **39**, 275 (1971).

C. Caratheodory, "Calculus of Variations and Partial Differential Equations of the First Order." Holden-Day, San Francisco, California, 1965.

H. Corben and P. Stehle, "Classical Mechanics." Wiley, New York, 1960.

H. Goldstein, "Classical Mechanics." Addison-Wesley, Reading, Massachusetts, 1950.

B. R. Gossick, "Hamilton's Principle and Physical Systems." Academic Press, New York, 1967.

S. W. Groesberg, "Advanced Mechanics." Wiley, New York, 1968.

A. Katz, "Classical Mechanics, Quantum Mechanics, Field Theory." Academic Press, New York, 1965.

C. W. Kilmister, "Lagrangian Dynamics." Plenum, New York, 1967.

G. L. Kotkin and V. G. Serbo, "Collection of Problems in Classical Mechanics." Pergamon, Oxford, 1971.

C. Lanczos, "Variational Principles of Mechanics." Univ. of Toronto Press, Toronto, 1966.

J. B. Marion, "Classical Dynamics." Academic Press, New York, 1965.

G. Rosen, "Formulations of Classical and Quantum Dynamical Theory." Academic Press, New York, 1969.

E. J. Saletan and A. H. Cromer, A variational principle for nonholonomic systems. *Amer. J. Phys.* **38**, 892 (1970).

E. J. Saletan and A. H. Cromer, "Theoretical Mechanics." Wiley, New York, 1971.

D. Ter Haar, "Elements of Hamiltonian Dynamics." Pergamon, Oxford, 1971.

E. T. Whittaker, "Analytical Dynamics of Particles." Cambridge Univ. Press, London and New York, 1937.

E. B. Wilson, J. C. Decius, and P. C. Cross, "Molecular Vibrations." McGraw-Hill, New York, 1955.

W. Yourgrau and S. Mandelstam, "Variational Principles in Dynamics and Quantum Theory." Saunders, Philadelphia, Pennsylvania, 1968.

3

CONSERVATION LAWS

3.1. RELATIONSHIP OF CONSERVATION LAWS AND SYMMETRY TRANSFORMATIONS

The importance of conservation laws is amply demonstrated in Newtonian vectorial physics. These laws are no less important in analytical mechanics. We have already seen that the momentum conjugate to a cyclic variable is conserved. However, within the Lagrange formalism the conservation laws are most suitably discussed within a tour de force known as Noether's theorem, which we now develop.

A. Symmetry Transformations

Among those transformations to which we may subject the variables describing our system, there exists a particularly important set, viz., those which leave the equations of motion invariant in form [1]. Such transformations are called *symmetry transformations*. The relativity transformations of Chapter 1 are examples of symmetry transformations.

The transformations may be expressed by

$$q_a' = q_a'(q_1, q_2, \ldots, q_N, \tau) \tag{3.1}$$

$$\tau' = \tau'(\tau) \tag{3.2}$$

If the transformations are continuous, (3.1) and (3.2) may be replaced by

$$q_a' = q_a + \delta q_a \tag{3.3}$$

$$\tau' = \tau + \delta\tau \tag{3.4}$$

where δq_a and $\delta\tau$ are infinitesmal transformations and repeated application of (3.3) and (3.4) generate the finite transformations (3.1) and (3.2). The action in terms of the primed variables is

$$I' = \int_{\tau_1'}^{\tau_2'} L'(q_a', \dot{q}_a', \tau')\, d\tau' \tag{3.5}$$

We now maintain the numerical invariance of the action by requiring that

$$\int L'(q_a', \dot{q}_a', \tau')\, d\tau' = \int L(q_a, \dot{q}_a, \tau)\, d\tau \tag{3.6}$$

This may be regarded as a definition of L'.

Now if the transformations (3.1) and (3.2) are to be symmetry transformations, it is necessary that the Euler–Lagrange equations have the same form in the primed coordinates as they had in the unprimed coordinates. This in turn imposes certain restrictions on the Lagrangian of (3.5) from which the equations are derived. It was shown in Section 2.2 that the Lagrangian determining a specific set of equations is not unique. Indeed it was demonstrated that the equations of motion are unchanged if the Lagrangian is modified as follows:

$$L \rightarrow L + \frac{d}{d\tau} G(q_a, \tau) \tag{3.7}$$

where G is an arbitrary function of the $\{q_a, \tau\}$, but is independent of the $\dot{q}_a$. Equations (3.7) provides a sufficient condition that the equations of motion preserve their form; it is also a necessary condition [2].

We thus require that the functional form of L' arising from a symmetry transformation differs from the original functional form of L by

$$L'(q_a', \dot{q}_a', \tau') = L(q_a', \dot{q}_a', \tau') + \frac{d}{d\tau'} G(q_a', \tau') \tag{3.8}$$

Let us represent the change in the Action resulting from the symmetry transformation by δI_1. From (3.6) it follows immediately that

$$\delta I_1 = \int L'(q_a', \dot{q}_a', \tau')\, d\tau' - \int L(q_a, \dot{q}_a, \tau)\, d\tau = 0 \tag{3.9}$$

Using (3.8), (3.9) becomes

$$\delta I_1 = \int L(q_a', \dot{q}_a', \tau')\, d\tau' - \int L(q_a, \dot{q}_a, \tau)\, d\tau + \int \frac{dG(q_a', \tau')}{d\tau'}\, d\tau' \tag{3.10}$$

Replacing the primed variables according to (3.3) and (3.4) and assuming that the infinitesmal transformation can produce only an infinitesmal change in the functional form of the Lagrangian, (3.1) becomes

$$\delta I_1 = \int L(q_a + \delta q_a, \dot{q}_a + \delta \dot{q}_a, \tau + \delta\tau)\left(1 + \frac{d\,\delta\tau}{d\tau}\right) d\tau - \int L(q_a, \dot{q}_a, \tau)\, d\tau$$

$$+ \int \frac{d\tau}{d\tau'} \frac{d}{d\tau}\, \delta G(q_a + \delta q_a, \tau + \delta\tau)\, d\tau \tag{3.11}$$

where the δG has replaced G simply to indicate explicitly the infinitesmal character of the term altering the Lagrangian. The first two terms of (3.11) have the same form as δI of (2.16). It is well to keep in mind the difference between the δq_a as they appear in (2.16) and (3.11). In (2.16) the δq_a represent an infinitesmal displacement of the system's path in configuration space from its actual path; whereas, in (3.11) the δq_a are infinitesmal displacements of the system as defined by the set of symmetry transformations. Recognition of the formal similarity is useful in that now we can proceed as we did in Chapter 2 and obtain from (2.18)

$$\delta I_1 = \int_{\tau_1}^{\tau_2} \left[\frac{\partial L}{\partial q_a}\, \delta q_a + \frac{\partial L}{\partial \dot{q}_a}\, \delta \dot{q}_a + \frac{\partial L}{\delta \tau}\, \delta\tau + L \frac{d\,\delta\tau}{d\tau}\right] d\tau$$

$$+ \int_{\tau_1}^{\tau_2} \frac{d}{d\tau}\, \delta G(q_a, \tau)\, d\tau = 0 \tag{3.12}$$

where the last integral of (3.12) is the result of Taylor expanding the integrand of (3.11) and retaining only the lowest-order infinitesmals.

It is possible to extract from (3.12) a *test* to determine whether a given transformation is a symmetry transformation. To be a symmetry transformation it is necessary that

$$\left[\delta q_a \frac{\partial}{\partial q_a} + \delta \dot{q}_a \frac{\partial}{\partial \dot{q}_a} + \delta\tau \frac{\partial}{\partial \tau} + \frac{d\,\delta\tau}{d\tau}\right] L = -\frac{d}{d\tau}\, \delta G(q_a, \tau) \tag{3.13}$$

the test consists in demonstrating that the terms on the left side of (3.13) can be cast into a form similar to that of the right side, viz., the total derivative with respect to the independent variable. If the left side of (3.13) vanishes identically, it follows immediately from (3.8) that the Lagrangian is *form invariant* under the transformation.

B. Noether's Theorem

Equation (3.12) may be cast into a most useful form by proceeding as we did in going from (2.18) to (2.22). The result is

$$\int_{\tau_1}^{\tau_2} \frac{d}{d\tau}\left[\delta q_a \frac{\partial L}{\partial \dot{q}_a} + \left(L - \dot{q}_a \frac{\partial L}{\partial \dot{q}_a}\right)\delta\tau + \delta G\right] d\tau + \int_{\tau_1}^{\tau_2} \left(\frac{\partial L}{\partial q_a} - \frac{d}{d\tau}\frac{\partial L}{\partial \dot{q}_a}\right)(\delta q_a - \dot{q}_a\,\delta\tau)\,d\tau = 0 \tag{3.14}$$

Equation (3.14) results only from the requirement that the infinitesmal continuous transformation be a symmetry transformation. If now we require that the Euler–Lagrange equations (2.26) are valid, the second integral of (3.14) is zero and there remains

$$\int_{\tau_1}^{\tau_2} \frac{d}{d\tau}\left[\delta q_a \frac{\partial L}{\partial \dot{q}_a} + \left(L - \dot{q}_a \frac{\partial L}{\partial \dot{q}_a}\right)\delta\tau + \delta G\right] d\tau = 0 \tag{3.15}$$

From (3.15) it follows that

$$\frac{d}{d\tau}\left[\delta q_a \frac{\partial L}{\partial \dot{q}_a} + \left(L - \dot{q}_a \frac{\partial L}{\partial \dot{q}_a}\right)\delta\tau + \delta G\right] = 0 \tag{3.16}$$

Equation (3.16) is the usual form for a *conservation law* with the term in the brackets being the conserved quantity; i.e.,

$$\left[\delta q_a \frac{\partial L}{\partial \dot{q}_a} + \left(L - \dot{q}_a \frac{\partial L}{\partial \dot{q}_a}\right)\delta\tau + \delta G\right] = \text{const} \tag{3.17}$$

The important point is this. If (3.3) and (3.4) constitute a symmetry transformations [as can be tested by (3.13)] then the conservation law expressed by (3.17) is a consequence. In other words, for every continuous infinitesmal symmetry transformation there is a corresponding conservation law. This connection constitutes Noether's theorem [1, 3].

The coefficient of the δq_a term is just the conjugate momentum p_a. Consider now the coefficient of $\delta\tau$ and call it H.

$$-H = L - \dot{q}_a\,\partial L/\partial \dot{q}_a = L - \dot{q}_a p_a \tag{3.18}$$

If we restrict ourselves to Galilean relativity, $L = T - V$. If $V \neq V(\dot{q}_a)$

$$\partial L/\partial \dot{q}_a = \partial T/\partial \dot{q}_a \tag{3.19}$$

Now the general expression for the kinetic energy is presented in (2.62)

$$T = d_{ab}\dot{q}_a\dot{q}_b + d_a\dot{q}_a + d \tag{2.62}$$

where $d_{ab}d_a$, and d are defined in (2.63). For our purpose here it is sufficient to note that these coefficients are not functions of the velocity and therefore (3.19) becomes

$$\partial L/\partial \dot{q}_a = 2d_{ab}\dot{q}_b + d_a \tag{3.20}$$

and

$$\dot{q}_a\,\partial L/\partial \dot{q}_a = 2\,d_{ab}\dot{q}_a\dot{q}_b + d_a\dot{q}_a \tag{3.21}$$

From (2.63) it is clear that if the transformation equations are not explicit functions of the time, d_a and d are zero, and (3.21) becomes

$$\dot{q}_a\,\partial L/\partial \dot{q}_a = 2\,d_{ab}\dot{q}_a\dot{q}_b = 2T \tag{3.22}$$

It follows that (3.18) becomes

$$H = -(T - V) + 2T = T + V = E \tag{3.23}$$

In other words, the coefficient of $\delta\tau$ in (3.17) is the total energy provided that we are within the realm of Galilean relativity and the transformation equations connecting the inertial Cartesian coordinates and the generalized coordinates do not explicitly depend on the time. We also required that the potential energy be independent of the velocities and time. Thus (3.17) may be written

$$[\delta q_a p_a - H\,\delta\tau + \delta G] = \text{const} \tag{3.24}$$

According to our discussion in Chapter 1, the Galilean transformations should leave the equations of motion invariant and therefore we expect them to be symmetry transformations. We now pursue this point.

C. Galilean Transformations

The Galilean transformations are defined in (1.24). Rather than apply Noether's theorem to the entire group, we focus our attention on characteristic subsets. First consider a pure time translation (i.e., displacement of the time origin)

$$t' - t = \delta t = t_0 = \text{const}, \qquad \delta x_\alpha{}^i = \delta \dot{x}_\alpha{}^i = 0 \tag{3.25}$$

where $x_\alpha{}^i$ is the ith component of the vector locating the αth particle. Applying the test (3.13) we obtain

$$t_0 \frac{\partial L}{\partial t} = -\frac{d}{dt}\,\delta G(x_\alpha{}^i, t) \tag{3.26}$$

Thus time translation is a symmetry transformation under which the Lagrangian is form invariant provided that the Lagrangian does not depend explicitly on the time, i.e., $L \neq L(t)$. From (3.24) we see that the conservation law resulting from this transformation is

$$H = \text{const} \tag{3.27}$$

We have established a most important connection between energy and time, viz., that energy conservation follows from the invariance of the laws of motion under a displacement of the system in time. In other words, if time is homogeneous, the laws of physics described now are the same as the laws of physics at any previous or future time; and, therefore, the conservation of energy depends on the homogeneity of time.

Consider now a pure spatial displacement of the origin in the x^1 direction; i.e.,

$$x_\alpha^{i\prime} - x_\alpha{}^i = \delta x_\alpha{}^i = C^i, \qquad \delta t = 0, \qquad \delta \dot{x}_\alpha{}^i = 0 \tag{3.28}$$

where C^i is a constant vector. Substituting (3.28) into the test (3.13), we obtain

$$\sum_{\alpha=1}^{N} C^i \frac{\partial L}{\partial x_\alpha{}^i} = -\frac{d}{dt}\,\delta G(x_\alpha{}^i, t) \tag{3.29}$$

It follows immediately that this is a symmetry transformation leaving the Lagrangian form invariant if

$$\sum_{\alpha=1}^{N} \partial L/\partial x_\alpha{}^i = Q^i = 0 \tag{3.30}$$

where Q^i is the total force in the x^i direction acting on the system of N particles.

A sufficient condition for the validity of (3.30) is Newton's third law which states that the forces between two particles always appear in action–reaction pairs. A single particle must be a free particle system.

A system for which (3.30) is valid is an isolated system and is Galilean invariant. For the sake of simplicity or convenience, it is often advantageous to deal with nonisolated system, i.e., systems for which some of the sources of forces on the particles of interest are not included in the dynamics of the system (e.g., often the motion of the earth is ignored in discussing the motion of objects over short distances near the surface of the earth, and only its gravitational force on the objects is included in the dynamical description). When the system is not an isolated system, it is entirely possible that it will not be Galilean invariant, and its symmetry properties must be investigated.

The conservation law associated with this symmetry transformation is obtained from (3.17), viz.,

$$\sum_\alpha C^i \,\partial L/\partial \dot{x}_\alpha{}^i = \text{const}, \qquad \sum_{\alpha=1}^{N} p_\alpha{}^i = \text{const} \tag{3.31}$$

Thus, the total momentum is conserved provided that the total force is zero, or alternatively the total momentum is conserved if space is homogeneous.

Consider now an infinitesimal rotation of the system described by

$$x_\alpha^{i'} = \omega^{ij} x_\alpha{}^j = \delta^{ij} x_\alpha{}^j + \Omega^{ij} x_\alpha{}^j \tag{3.32}$$

or

$$x_\alpha^{i'} - x_\alpha{}^i = \delta x_\alpha{}^i = \Omega^{ij} x_\alpha{}^j, \qquad \delta t = 0 \tag{3.33}$$

where δ^{ij} is the Kronecker delta, ω^{ij} is the rotation operator, and Ω^{ij} is the antisymmetric part of the rotation operator. For example, the rotation by an angle $\delta\theta$ about the x^3 axis is described by

$$\Omega^{ij} = -\Omega^{ji} = \begin{bmatrix} 0 & -\delta\theta & 0 \\ \delta\theta & 0 & 0 \\ 0 & 0 & 0 \end{bmatrix} \tag{3.34}$$

Using (2.14), we find that

$$\delta \dot{x}_\alpha{}^i = \frac{d}{dt}\,\delta x_\alpha{}^i = \Omega^{ij} \dot{x}_\alpha{}^j \tag{3.35}$$

Using (3.33) and (3.35) in the test (3.13), we obtain

$$\Omega^{ij} \sum_\alpha \left[x_\alpha{}^j \frac{\partial L}{\partial x_\alpha{}^i} + \dot{x}_\alpha{}^j \frac{\partial L}{\partial \dot{x}_\alpha{}^i} \right] = -\frac{d}{dt}\,\delta G(x_\alpha{}^i, t) \tag{3.36}$$

To simplify the discussion, we assume that the $x_\alpha{}^i$ coordinates are all independent and that the potential does not depend explicitly on the velocities. Then

$$\partial L/\partial \dot{x}_\alpha{}^i = m_\alpha \dot{x}_\alpha{}^i$$

and the second term on the left side of (3.36) becomes

$$\sum_{\alpha} m_{\alpha} \Omega^{ij} \dot{x}_{\alpha}^{j} \dot{x}_{\alpha}^{i} = 0 \tag{3.37}$$

Equation (3.37) follows from the fact that Ω^{ij} is antisymmetric in i and j, whereas $\dot{x}_{\alpha}^{j}\dot{x}_{\alpha}^{i}$ is symmetric in i and j. Thus (3.36) reduces to

$$\sum_{\alpha} \Omega^{ij} x_{\alpha}^{j} \frac{\partial L}{\partial x_{\alpha}^{i}} = \sum_{\alpha} \Omega^{ij} x_{\alpha}^{j} Q_{\alpha}^{i} = -\frac{d}{dt}\, \delta G(x_{\alpha}^{i}, t) \tag{3.38}$$

where Q_{α}^{i} is the force on the αth particle. But the $\Omega^{ij} x_{\alpha}^{j} Q_{\alpha}^{i}$ are the components of the total torque acting on the αth particle. Therefore the rotation is a symmetry transformation leaving the Lagrangian form invariant if the total torque is zero.

The corresponding conservation law is

$$\sum_{\alpha} \Omega^{ij} x_{\alpha}^{j} p_{\alpha}^{i} = \text{const} \tag{3.39}$$

Thus for rotational symmetry about the x^3 axis we find from (3.34) and (3.39) that

$$\sum_{\alpha} \Omega^{ij} x_{\alpha}^{j} p_{\alpha}^{i} = \text{const} \tag{3.40}$$

But we recognize the term appearing in (3.40) as the third component of the angular momentum of the system. Thus we have established the connection between the rotational symmetry of a system and the conservation of its angular momentum. Thus if space is isotropic (i.e., the laws of motion of the system are the same if the entire system is rotated about some axis to a new orientation in space) then it follows that the total angular momentum of the system is conserved.

To complete our discussion of the Galilean group we have only to consider the transformations that connect inertial observers moving with constant velocity relative to each other. In this case

$$\delta x_{\alpha}^{i} = v^{i} t \tag{3.41a}$$

$$\delta \dot{x}_{\alpha}^{i} = v^{i} \tag{3.41b}$$

$$\delta t = 0 \tag{3.41c}$$

We consider v^{i} to be an infinitesimal so as to preserve the infinitesimal character of δx_{α}^{i} and $\delta \dot{x}_{\alpha}^{i}$. (Of course any inertial frame can be reached by successively applying the above transformations.) The symmetry test yields

$$v^i \sum_\alpha \left(t \frac{\partial L}{\partial x_\alpha{}^i} + \frac{\partial L}{\partial \dot{x}_\alpha{}^i} \right) = - \frac{d}{dt} \delta G \tag{3.42}$$

Now if the system is invariant under spatial translations (e.g., an isolated system), $\sum_\alpha \partial L / \partial x_\alpha{}^i = 0$ and (3.42) becomes

$$\mathbf{v} \cdot \mathbf{P} = - \frac{d}{dt} \delta G \tag{3.43}$$

where $\mathbf{P}$ is the total momentum of the system and a constant. It is clear that in general only transformations to inertial frames moving at right angles to the total momentum of the system are form invariant transformations.

Of particular interest is the single particle (i.e., free particle) system for which

$$L = \tfrac{1}{2} m (\dot{\mathbf{r}})^2, \qquad \mathbf{P} = m \dot{\mathbf{r}} \tag{3.44}$$

Then (3.43) becomes

$$m(\mathbf{v} \cdot \mathbf{r}) = - \frac{d}{dt} \delta G$$

from which it follows that

$$\delta G = -m \mathbf{v} \cdot \mathbf{r} \tag{3.45}$$

The corresponding conserved quantity is from (3.17).

$$m \mathbf{r} - \mathbf{P} t = \mathbf{K} = m(\mathbf{r} - \dot{\mathbf{r}} t) \tag{3.46}$$

Thus it follows from invariance of the single particle system under spatial and velocity transformations that a free particle in Galilean relativity has uniform motion, and this is the connection between Galilean relativity and the principle of inertia.

D. Momentum under a General Galilean Transformation

It is often important to know how essential physical quantities such as momentum and energy transform under a general Galilean transformation. To this end we note that in the new frame the momentum is written

$$p_\alpha^{i'} = \partial L' / \partial \dot{x}_\alpha^{i'} \tag{3.47}$$

where $p_\alpha^{i'}$ is the ith component of the αth particle as measured by the primed frame observer. Using (3.7), we may write

$$p_\alpha^{i'} = \frac{\partial \dot{x}_\beta{}^k}{\partial \dot{x}_\alpha^{i'}} \frac{\partial}{\partial \dot{x}_\beta{}^k} \left(L + \frac{d}{dt}\, \delta G \right) \tag{3.48}$$

The most general Galilean transformation is

$$\delta x_\alpha{}^i = C^i + \Omega^{ij} x_\alpha{}^j + v^i t, \qquad \delta t = C \tag{3.49}$$

and from this it follows that

$$\delta \dot{x}_\alpha{}^i = \Omega^{ij} \dot{x}_\alpha{}^j + v^i \tag{3.50}$$

Therefore,

$$\dot{x}_\beta^{k'} = \dot{x}_\beta{}^k + \Omega^{kj} \dot{x}_\beta{}^j + v^k \tag{3.51}$$

Taking the derivative of both sides of (3.51) with respect to $\dot{x}_\alpha^{i'}$, we find

$$\delta^{ki}\, \delta^{\beta\alpha} = (\delta^{kj} + \Omega^{kj})\, \partial \dot{x}_\beta{}^j / \partial \dot{x}_\alpha^{i'} \tag{3.52}$$

Multiplying both sides of (3.52) by $(\delta^{lk} - \Omega^{lk})$ and neglecting infinitesimals of the second order, we obtain

$$\partial \dot{x}_\beta{}^k / \partial \dot{x}_\alpha^{i'} = (\delta^{ki} - \Omega^{ki})\, \delta^{\beta\alpha} \tag{3.53}$$

Assuming that the system conserves energy and momentum, it follows from (3.26), (3.29), (3.38), and (3.42) that

$$\frac{d}{dt}\, \delta G = -v^l \sum_\gamma \frac{\partial L}{\partial \dot{x}_\gamma{}^l} \tag{3.54}$$

Inserting (3.54) and (3.53) into (3.48) we obtain

$$p_\alpha^{i'} = p_\alpha{}^i + \Omega^{ik} p_\alpha{}^k + v^k \sum_\beta \frac{\partial^2 L}{\partial \dot{x}_\beta{}^k\, \partial \dot{x}_\alpha{}^i} + v^l \Omega^{ik} \sum_\beta \frac{\partial^2 L}{\partial \dot{x}_\beta{}^l\, \partial \dot{x}_\alpha{}^k} \tag{3.55}$$

The last term in (3.55) is an infinitesimal of the second order (v^l and Ω^{ik} are infinitesimals) and will be dropped. If the potential energy is independent of the velocities, the Lagrangian becomes for Cartesian coordinates

$$L = \tfrac{1}{2} \sum_\gamma m_\gamma \dot{\mathbf{x}}_\gamma{}^2 \tag{3.56}$$

Inserting (3.56) into (3.55) we obtain

$$p_\alpha^{i'} = p_\alpha{}^i + \Omega^{ik} p_\alpha{}^k + m_\alpha v^i \tag{3.57}$$

or

$$\mathbf{p}_\alpha{}' = R\mathbf{p}_\alpha + m_\alpha \mathbf{v} \tag{3.58}$$

where R is the rotation operator. The transformation properties of the energy are left to the problems at the end of the chapter.

3.2. CONSERVATION LAWS ASSOCIATED WITH LORENTZ TRANSFORMATIONS

In Section 3.1 we established the formalism of conservation laws primarily within the framework of Galilean relativity. This formalism can easily be modified to yield equivalent results for Einstein's relativity.

Recall that in deriving the Euler–Lagrange equations of motion compatible with Einstein's relativity it was necessary to account for the constraint

$${}_a\dot{x}^\mu \, {}_a\dot{x}_\mu = 1$$

This constraint means that not all of the variables are independent. The method of Lagrange multipliers was introduced to remedy the situation. We proceed in a similar fashion here and, as in Chapter 2, we limit our discussion to one particle systems.

Introducing the modified Lagrangian

$$\bar{L} = L + \tfrac{1}{2}\lambda(S)(\dot{x}^\mu \dot{x}_\mu - 1) \tag{3.59}$$

and replacing τ with the proper time S, we proceed as in Section 3.1 and obtain for the symmetry test of Eq. (3.13)

$$\left[\delta x^\mu \frac{\partial}{\partial x^\mu} + \delta\lambda \frac{\partial}{\partial \lambda} + \delta\dot{x}^\mu \frac{\partial}{\partial \dot{x}^\mu} + \delta s \frac{\partial}{\partial s} + \frac{d\,\delta s}{ds}\right]\bar{L} = -\frac{d}{ds}\,\delta G(x^\mu, \lambda, s) \tag{3.60}$$

Similarly, Noether's theorem becomes

$$\left[\delta x^\mu \frac{\partial \bar{L}}{\partial \dot{x}^\mu} + \left(\bar{L} - \dot{x}^\mu \frac{\partial \bar{L}}{\partial \dot{x}^\mu}\right)\delta s + \delta G\right] = \text{const} \tag{3.61}$$

where in (3.60) and (3.61) we have used the fact that $\dot{\lambda}$ does not appear explicitly in $\bar{L}$.

In Chapter 2 we found in Eq. (2.49) that

$$\lambda = L - \dot{x}^\mu \, \partial L/\partial \dot{x}^\mu \tag{2.49}$$

where we are now assuming that $\partial L/\partial S = 0$.

Using (2.49) and the constraint equation in (3.61) we find that the coefficient of δs is identically zero, i.e.,

$$\bar{L} - \dot{x}^\mu \, \partial \bar{L}/\partial \dot{x}^\mu = 0 \tag{3.62}$$

And if we define the coefficient of δx^μ as the four-momentum, we obtain

$$\mathcal{P}^\mu \equiv \partial \bar{L}/\partial \dot{x}^\mu = \dot{x}^\mu L \tag{3.63}$$

A. Free Particle

According to Chapter 2 the Lagrangian of a free particle is

$$\bar{L} = mc + \tfrac{1}{2}\lambda(\dot{x}^\mu \dot{x}_\mu - 1) \tag{3.64}$$

Let us first consider the translation of the system in Lorentz space, i.e.,

$$\delta x^\mu = a^\mu \tag{3.65}$$

Then since $\partial \bar{L}/\partial x^\mu = 0$, it follows immediately from (3.60) that (3.65) is a symmetry transformation, and it follows from (3.61) that

$$\mathcal{P}^\mu = mc\dot{x}^\mu = \text{const} \tag{3.66}$$

i.e., the conservation of the four-momentum follows from the invariance of the system under a space–time translation. This result agrees with that obtained in Section 3.1.

The remaining transformations of the Poincaré group, viz., the Lorentz transformations $x^{\mu\prime} = L^\mu{}_\nu \, x^\nu$, may be investigated quite simply by recognizing that the continuous transformations of the Lorentz group represent rotations in Lorentz space. Therefore, we write

$$L^\mu{}_\nu = g^\mu{}_\nu + \omega^\mu{}_\nu \tag{3.67}$$

where $g^\mu{}_\nu$ is the mixed form of the metric tensor. Thus

$$x^{\mu\prime} = L^\mu{}_\nu x^\nu = g^\mu{}_\nu x^\nu + \omega^\mu{}_\nu x^\nu \tag{3.68}$$

and

$$\delta x^\mu = x^{\mu\prime} - x^\mu = \omega^\mu{}_\nu x^\nu \tag{3.69}$$

where $\omega_{\mu\nu} = -\omega_{\nu\mu}$ is the infinitesimal rotation operator that plays the same role in Lorentz space as the Ω^{ij} used in the previous section plays in Euclidean space. If we require that $\delta s = 0$, it follows from (3.69) that

$$\delta \dot{x}^\mu = \omega^\mu{}_\nu \dot{x}^\nu \tag{3.70}$$

Inserting (3.70), (3.69), and (3.64) into the symmetry test we find

$$\lambda \omega_{\mu\nu} \dot{x}^\mu \dot{x}^\nu = -\frac{d}{ds}\,\delta G \tag{3.71}$$

But

$$\omega_{\mu\nu} \dot{x}^\mu \dot{x}^\nu = 0$$

since $\omega_{\mu\nu}$ is antisymmetric in $\mu\nu$ and $\dot{x}^\mu \dot{x}^\nu$ is symmetric in $\mu\nu$. Therefore the Lorentz transformations are symmetry transformations for the free particle and the corresponding conservation law is

$$\omega^\mu{}_\nu x^\nu \mathcal{P}_\mu = \text{const} \tag{3.72}$$

It is clear from (3.71) that if the rotations are independent (i.e., ω_{12} is independent of ω_{23}), then these rotations are individually symmetry transformations and (3.72) becomes

$$\omega_{\mu\nu}(x^\mu \mathcal{P}^\nu - \mathcal{P}^\nu x^\mu) = \omega_{\mu\nu} M^{\mu\nu} = \text{const}, \qquad \text{no sum on } \mu,\nu \tag{3.73}$$

where $M^{\mu\nu}$ is the *angular momentum tensor*. Thus invariance of our system to rotations in Lorentz space leads to the conservation of angular momentum and this is analogous to the result obtained for Euclidean space in Section 3.1.

B. Particle in an Electromagnetic Field

The dynamics of a particle in an electromagnetic field was discussed in Chapter 2. The Lagrangian for this system is

$$\bar{L} = mc + (e/c)A^\mu \dot{x}_\mu + \tfrac{1}{2}\lambda(\dot{x}^\mu \dot{x}_\mu - 1) \tag{3.74}$$

It is not difficult to show that in general the transformations such as (3.65), (3.69), and (3.70) are not symmetry transformations and thus momentum and angular momentum are not conserved. This follows mathematically from the fact that A^μ is a known function of the coordinates; this is physically reasonable since we no longer have an isolated system, i.e., there is some outside agency, the source of A^μ, which supplies energy to the particle. Only a formalism which included the particle and field sources could be expected to be invariant under Lorentz transformations and thereby momentum and angular momentum conserving.

If we again define momentum as the coefficient of δx^μ in (3.61), we obtain

$$\mathcal{P}^\mu = \partial \bar{L}/\partial \dot{x}^\mu = (e/c)A^\mu + \lambda \dot{x}^\mu \tag{3.75}$$

and for our case

$$\lambda = mc \tag{3.76}$$

and therefore

$$\mathcal{P}^\mu = mc\dot{x}^\mu + (e/c)A^\mu \tag{3.77}$$

3.3. SOLVED EXAMPLES

Example 3.1. Discuss the symmetry properties and conservation laws of a system of N particles moving under their mutual gravitational attraction.

Solution. The Lagrangian and Euler–Lagrange equations for this system were obtained in Example 2.2.

$$L = \tfrac{1}{2} \sum_\alpha m_\alpha (\dot{x}_\alpha{}^i)^2 - \tfrac{1}{2} \sum_{\alpha,\beta}{}' G m_\alpha m_\beta / r_{\alpha\beta} \tag{2.155}$$

Since L does not depend explicitly on the time we conclude from (3.26) and (3.27) that time translation is a symmetry transformation and that H, the total energy of the system, is conserved.

To test space translation we calculate

$$\begin{aligned} \frac{\partial L}{\partial x_\gamma{}^1} &= - \sum_{\alpha \neq \gamma} \frac{x_\alpha{}^1 - x_\gamma{}^1}{r_{\alpha\gamma}^3} G m_\alpha m_\gamma \\ \sum_\gamma \frac{\partial L}{\partial x_\gamma{}^1} &= - \sum_\gamma \sum_{\alpha \neq \gamma} \frac{x_\alpha{}^1 - x_\gamma{}^1}{r_{\alpha\gamma}^3} G m_\alpha m_\gamma = 0 \end{aligned} \tag{3.78}$$

That the double sum is zero follows immediately from the fact that $m_\alpha m_\gamma / r^3_{\alpha\gamma}$ is symmetric in $\alpha\gamma$, whereas $x_\alpha{}^1 - x_\gamma{}^1$ is antisymmetric in $\alpha\gamma$. From (3.78) and (3.30) we conclude that space translations are symmetry transformations leaving the Lagrangian form invariant. The total momentum in a given direction is conserved, e.g.,

$$\mathcal{P}^1 = \sum_\alpha \partial L / \partial \dot{x}_\alpha{}^1 = \sum_\alpha m_\alpha \dot{x}_\alpha{}^1 = \text{const}$$

Thus we conclude that the system represented by (2.155) has a high degree of symmetry and satisfies all of the familar conservation laws. This is primarily because the system is isolated. We shall see in the subsequent example what effect an outside agency can have on the symmetry of a system.

Example 3.2. Discuss the symmetry properties and conservation laws of a bead sliding down a frictionless parabolic wire which is rotating with constant velocity.

Solution. The Lagrangian and equation of motion of this system has been obtained in Example 2.4.

$$L = \tfrac{1}{2} m[\dot{\varrho}^2(1 + 4a^2\varrho^2) + \varrho^2(\omega^2 - 2ga)] \tag{3.79}$$

Since L does not depend explicitly on the time, we conclude that time translation is a symmetry transformation leaving the Lagrangian of the system form invariant and that the total energy of the system is conserved.

To test for space translation we insert $\partial L / \partial \varrho$ into (3.13) and obtain

$$m\varrho(4a^2\dot{\varrho}^2 + \omega^2) = -\frac{d}{dt}\,\delta G$$

Since δG cannot be a function of $\dot{\varrho}$, we conclude that the right side cannot be put into the form of a total time derivative and, therefore, space translation is not a symmetry transformation; thus the momentum $\partial L / \partial \dot{\varrho}$ is not conserved. This is primarily due to the fact that the system is not isolated since not only gravity but also a constraint is acting on the bead.

It is not difficult to show that rotation about the z axis is not a symmetry transformation.

PROBLEMS

3.1 A mass m is attached to a spring having a spring constant k. The system lies along the x axis. The mass is driven by the force $\mathbf{F} = A \sin \omega t \, \hat{\imath}$. Determine the symmetry properties and conservation laws for this system.

3.2 Determine the symmetry properties and conservation laws of the system described in Problems 2.2, 2.3, and 2.6.

3.3 Determine the transformation properties of the energy defined in (3.18) under a general Galilean transformation.

3.4 Show that for a single particle the quantity $U = E - \mathbf{p}^2/2m$ is a Galilean invariant. Show that U is the energy in the rest frame of the particle. (E is the total energy of the particle.)

3.5 Extend the observations of the previous problem to a system of N particles.

3.6 Show that the transformation $\delta x = \varepsilon \sin \omega t$ is a symmetry transformation of the harmonic oscillator ($L = \frac{1}{2}m\dot{x}^2 - \frac{1}{2}kx^2$) if $\omega^2 = k/m$. Determine the corresponding constant of the motion.

REFERENCES

1. E. L. Hill, *Rev. Mod. Phys.* **23**, 253 (1951).
2. R. Courant and D. Hilbert, "Methods of Mathematical Physics." Volume I. Wiley, New York, 1953.
3. E. Noether, *Nachr. Akad. Wiss. Goettingen Math. Phys. Kl.* p. 235 (1918).

BIBLIOGRAPHY

J. M. Levy-Leblond, *Amer. J. Phys.* **39**, 502 (1971).
C. Palmieri and B. Vitale, *Nuovo Cimento A* **66**, 299 (1970).

4

THE CANONICAL FORMALISM

4.1. BASIC THEORY

In Chapters 2 and 3 we developed Lagrangian dynamics which described the motion of a physical system in terms of the positions and velocities of the particles comprising that system. The equations of motion were derived from a "state function," the Lagrangian, which was a function of the positions, velocities, and time. In this chapter we develop an alternative dynamical description by introducing a state function, the Hamiltonian, which is a function of the coordinates and conjugate momenta. There are several ways of developing Hamiltonian dynamics. We prefer once again to develop the formalism from an extremum principle.

A. Hamilton's Equations of Motion

Let the state of a physical system be specified by the set of coordinates $\{q_a\}$ and the set of momenta $\{p_a\}$, where $a = 1, 2, \ldots, N$ and N is the number of degrees of freedom of the system. Thus the state of a system may be represented by a point in a $2N$-dimensional space, called *phase space*, spanned by the coordinates and momenta. We assume that there exists a "state function," the Hamiltonian,

$$H = H(q_a, p_a, \tau) \tag{4.1}$$

As in thermodynamics the state functions of dynamics differ from one another by the variables of which they are a function. Following the procedure common to thermodynamics we relate the state functions by a Legendre transformation [1].

$$H(q_a, p_a, \tau) = p_a\dot{q}_a - L(q_a, \dot{q}_a, \tau) \tag{4.2}$$

By comparing (4.2) and (3.18) we note that if the transformation $\delta\tau = \text{const}$ is a symmetry transformation of our system, our new state function H has the advantage of being a conserved quantity which is under appropriate conditions the total energy of the system.

Using (4.2) we may write the action (2.2) as

$$I = \int_{\tau_1}^{\tau_2} [p_a \dot{q}_a - H(q_a, p_a, \tau)]\, d\tau \tag{4.3}$$

Then Hamilton's principle states that the initial state $q_a(\tau_1)$, $p_a(\tau_1)$ evolves into the state $q_a(\tau_2)$, $p_a(\tau_2)$ in such a way that the action is an extremum, i.e.,

$$\delta I = \delta \int_{\tau_1}^{\tau_2} [p_a \dot{q}_a - H(q_a, p_a, \tau)]\, d\tau = 0 \tag{4.4}$$

It is important to note that in the Hamiltonian formalism the coordinates and momenta are independent of one another. Thus in performing the variation indicated in (4.4), δq_a and δp_a are independent.

Proceeding as in Chapter 2, (4.4) becomes to first order in the variations

$$\delta I = \int_{\tau_1}^{\tau_2} \left[\dot{q}_a\, \delta p_a + p_a \frac{d}{d\tau}\, \delta q_a - \frac{\partial H}{\partial q_a}\, \delta q_a - \frac{\partial H}{\partial p_a}\, \delta p_a \right] d\tau = 0 \tag{4.5}$$

In deriving (4.5) we have used (2.14) and the fact that in Hamilton's principle $\delta\tau = 0$; i.e., the variations are performed at a given value of τ. Now

$$\int_{\tau_1}^{\tau_2} p_a \frac{d}{d\tau}\, \delta q_a\, d\tau = \int_{\tau_1}^{\tau_2} \frac{d}{d\tau} (p_a\, \delta q_a)\, d\tau - \int_{\tau_1}^{\tau_2} \dot{p}_a\, \delta q_a\, d\tau \tag{4.6}$$

The first integral on the right side of (4.6) is zero since in Hamilton's principle $\delta q_a(\tau_1) = \delta q_a(\tau_2) = 0$. Thus (4.5) may be written

$$\delta I = \int_{\tau_1}^{\tau_2} \left[\left(\dot{q}_a - \frac{\partial H}{\partial p_a} \right) \delta p_a - \left(\dot{p}_a + \frac{\partial H}{\partial q_a} \right) \delta q_a \right] d\tau = 0 \tag{4.7}$$

Since δp_a and δq_a are arbitrary and independent, it is necessary that

$$\dot{q}_a = \partial H / \partial p_a, \qquad a = 1, 2, \ldots, N \tag{4.8a}$$

$$\dot{p}_a = -\partial H / \partial q_a, \qquad a = 1, 2, \ldots, N \tag{4.8b}$$

Equations (4.8a) and (4.8b) are the *Hamilton equations of motion*; they are often referred to as the *canonical equations of motion*. We note that they differ from the Euler–Lagrange equations of motion in that they constitute a system of $2N$ first-order ordinary differential equations; whereas the Euler–Lagrange equations are a set of N second-order ordinary differential equations.

If the Lagrangian of a system is known, the Hamiltonian may be determined from (4.2); however, it must be remembered that H is a function of the coordinates and momenta, and therefore the velocities appearing on the right side of (4.2) must be replaced with what they are equal to in terms of the coordinates and momenta. To do this we solve the equations

$$p_a = \partial L / \partial \dot{q}_a \tag{4.9}$$

for the velocities and substitute these results back into (4.2).

For Galilean relativity we may obtain the Hamiltonian directly by recalling that from (3.23) the Hamiltonian is just

$$H = T + V \tag{4.10}$$

provided that the conditions stated in the paragraph following (3.23) are satisfied.

B. Poisson Brackets

The equation of motion of any function $F = F(q_a, p_a, \tau)$ is

$$\dot{F} = \frac{dF}{d\tau} = \frac{\partial F}{\partial q_a}\dot{q}_a + \frac{\partial F}{\partial p_a}\dot{p}_a + \frac{\partial F}{d\tau} \tag{4.11}$$

Using (4.8), we may write (4.11) as

$$\dot{F} = \frac{\partial F}{\partial q_a}\frac{\partial H}{\partial p_a} - \frac{\partial F}{\partial p_a}\frac{\partial H}{\partial q_a} + \frac{\partial F}{\partial \tau} = [F, H] + \frac{\partial F}{\partial \tau} \tag{4.12}$$

where $[F, H]$ is called the *Poisson bracket* of F with H and (4.12) is said to be the equation of motion of F in Poisson bracket form. Thus the Poisson bracket of any two functions f and g is defined as

$$[f, g] \equiv \frac{\partial f}{\partial q_a}\frac{\partial g}{\partial p_a} - \frac{\partial f}{\partial p_a}\frac{\partial g}{\partial q_a} = -[g, f] \tag{4.13}$$

In particular the Poisson brackets of the canonical variables are

$$[q_b, q_c] = \frac{\partial q_b}{\partial q_a}\frac{\partial p_c}{\partial p_a} - \frac{\partial q_b}{\partial p_a}\frac{\partial p_c}{\partial q_a} = \delta_{ba}\,\delta_{ca} = \delta_{bc}$$
$$[q_b, q_a] = [p_b, p_c] = 0 \tag{4.14}$$

The Poisson bracket relations of (4.14) are called the *fundamental brackets*.

The canonical equations may be written as

$$\dot{q}_a = [q_a, H], \qquad \dot{p}_a = [p_a, H] \tag{4.15}$$

From (4.12) we see that a function F is a constant of the motion, i.e., $\dot{F} = 0$ if it is not an explicit function of τ and if its Poisson bracket with the Hamiltonian is zero.

C. Canonical Transformations

Suppose now that we wish to describe our system in terms of a new set of coordinates and momenta (Q_a, P_a) which are related to the original set (q_a, p_a) by the equations

$$Q_a = Q_a(q_b, p_b, \tau) \tag{4.16a}$$

$$P_a = P_a(q_b, p_b, \tau) \tag{4.16b}$$

The set of transformations (4.16a) and (4.16b) are said to be canonical if the equations of motion expressed in terms of (Q_a, P_a) are of the same form as they were for (q_a, p_a); i.e., the new equations of motion must be

$$\dot{Q}_a = \partial H'/\partial P_a \tag{4.17a}$$

$$\dot{P}_a = -\partial H'/\partial Q_a \tag{4.17b}$$

where H' is the Hamiltonian expressed in terms of the new coordinates and momenta (Q_a, P_a).

From our derivation of (4.8) it follows that (4.17) is assured if

$$\delta \int_{\tau_1}^{\tau_2} [\dot{Q}_a P_a - H'(Q_b, P_b, \tau)]\, d\tau = 0 \tag{4.18}$$

Comparing (4.18) and (4.4), we see that

$$\int_{\tau_1}^{\tau_2} [p_a\dot{q}_a - H(q_b, p_b, \tau)]\, d\tau - \int_{\tau_1}^{\tau_2} [\dot{Q}_a\dot{P}_a - H'(Q_b, P_b, \tau)]\, d\tau$$
$$= \int_{\tau_1}^{\tau_2} \frac{dW}{d\tau}\, d\tau \tag{4.19}$$

where W is an arbitrary function of the coordinates, momenta, and τ. Since the variation of momenta and coordinates at τ_1 and τ_2 is zero, it follows that

$$\delta \int_{\tau_1}^{\tau_2} \frac{dW}{d\tau}\, d\tau = 0 \tag{4.20}$$

W is called the *generator* of the canonical transformation.

The requirement that the variables appearing in (4.19) be connected by the transformations (4.16) reduces the number of independent quantities in (4.19) from $4N + 1$ to $2N + 1$. The arbitrary function W may then be regarded as a function of whatever set of $2N$ coordinates and momenta we prefer to use as independent. We distinguish the following combinations:

$$\begin{aligned} W_1 &= W_1(q_a, Q_a, \tau), \qquad & W_2 &= W_2(q_a, P_a, \tau), \\ W_3 &= W_3(p_a, Q_a, \tau), \qquad & W_4 &= W_4(p_a, P_a, \tau) \end{aligned} \tag{4.21}$$

We now pursue the consequences of choosing one of the forms appearing in (4.21). Consider first the choice of the q_a and Q_a as independent. Equation (4.19) becomes

$$\int_{\tau_1}^{\tau_2} \Big[p_a\dot{q}_a - P_a\dot{Q}_a + H'(Q_b, P_b, \tau) - H(q_b, p_b, \tau)$$
$$- \frac{\partial W_1}{\partial q_a}\dot{q}_a - \frac{\partial W_1}{\partial Q_a}\dot{Q}_a - \frac{\partial W}{\partial \tau}\Big]\, d\tau = 0 \tag{4.22}$$

In order that (4.22) may be valid for all values of τ_1 and τ_2, it is necessary that the integrand be zero and this in turn requires that the coefficients of the independent quantities $\dot{q}_a$ and $\dot{Q}_a$ vanish separately, i.e.,

$$p_a = \partial W_1/\partial q_a \tag{4.23a}$$

$$P_a = -\partial W_1/\partial Q_a \tag{4.23b}$$

$$H' = H + \partial W_1/\partial T \tag{4.23c}$$

The N equations (4.18a) do not contain the P_a and therefore they may be used to obtain the NQ_a's in terms of the p_a and q_a. Using these results in

(4.23b) we can also express the P_a's in terms of the p_a and q_a. Thus if we know W_1, we may derive the canonical transformations (4.16a) and (4.16b) from (4.23a) and (4.23b). This is why W is called the *generator* of the transformation.

A similar set of equations may be derived when our choice of independent quantities is q_a and P_a. It is somewhat simpler to obtain these equations by use of the Legendre transformation, i.e.,

$$W_2(q_a, P_a, \tau) = W_1(q_a, Q_a, \tau) + P_a Q_a \tag{4.24}$$

Differentiating (4.24) with respect to the independent variables, we obtain

$$\frac{\partial W_2}{\partial q_a}\dot{q}_a + \frac{\partial W_2}{\partial P_a}\dot{P}_a + \frac{\partial W_2}{\partial \tau} = \frac{\partial W_1}{\partial q_a}\dot{q}_a + \frac{\partial W_1}{\partial Q_a}\dot{Q}_a + \frac{\partial W_1}{\partial \tau} + P_a\dot{Q}_a \tag{4.25}$$

Using (4.23) in (4.25) we find

$$\frac{\partial W_2}{\partial q_a}\dot{q}_a + \frac{\partial W_2}{\partial P_a}\dot{P}_a + \frac{\partial W_2}{\partial \tau} = p_a\dot{q}_a - P_a\dot{Q}_a + H' - H + P_a\dot{Q}_a + \dot{P}_a Q_a \tag{4.26}$$

Equating coefficients of the independent quantities $\dot{q}_a$ and $\dot{P}_a$, we obtain

$$p_a = \partial W_2/\partial q_a \tag{4.27a}$$

$$Q_a = \partial W_2/\partial P_a \tag{4.27b}$$

$$H' = H + \partial W_2/\partial \tau \tag{4.27c}$$

Once again the canonical transformations may be obtained by solving (4.27a) for the P_a in terms of the q_a, p_a and substituting these results into (4.27b) to obtain Q_a in terms of q_a and p_a.

Similarly it can be shown that

$$q_a = -\partial W_3/\partial p_a \tag{4.28a}$$

$$P_a = -\partial W_3/\partial Q_a \tag{4.28b}$$

$$H' = H + \partial W_3/\partial \tau \tag{4.28c}$$

and

$$q_a = -\partial W_4/\partial p_a \tag{4.29a}$$

$$Q_a = \partial W_4/\partial P_a \tag{4.29b}$$

$$H' = H + \partial W_4/\partial \tau \tag{4.29c}$$

To illustrate the use of a generator and to prepare for a subsequent discussion, consider the following generating function:

$$W_2 = q_a P_a \tag{4.30}$$

Using (4.30) in (4.27), we obtain

$$p_a = P_a \tag{4.31a}$$

$$Q_a = q_a \tag{4.31b}$$

$$H' = H \tag{4.31c}$$

The canonical transformations defined by (4.31) are just the identity *transformations*, and therefore $q_a P_a$ is the generator of an identity transformation.

It is instructive to generalize (4.30) by replacing q_a with an arbitrary function of the coordinates and the independent variable

$$W_2 = f_a(q_b, \tau) P_a \tag{4.32}$$

From (4.27) we obtain

$$p_a = f_a(q_b, \tau) \tag{4.33a}$$

$$Q_a = f_a(q_b, \tau) \tag{4.33b}$$

$$H' = H + P_a \, \partial f_a / \partial \tau \tag{4.33c}$$

Now (4.33b) is the kind of coordinate transformation with which we dealt in previous chapters, and we therefore conclude that these transformations, often called "point transformations," are also canonical transformations. In particular, the relativity transformations are canonical.

Consider now the generating function

$$W_1 = q_a Q_a \tag{4.34}$$

From (4.23) it follows that

$$p_a = Q_a, \qquad P_a = -q_a, \qquad H' = H \tag{4.35}$$

We see that the generator of (4.34) simply transforms the coordinates into momenta and vice versa. Thus in the canonical formalism coordinates and momenta are indeed on an equal footing as the "independent variables describing the motion of my system" since no actual distinction can be made between the momentum-type variable and the coordinate-type variable; different symbols are but a matter of convenience.

We now illustrate a more ambitious use of the canonical transformation. Consider a particle of mass m moving under the influence of a harmonic oscillator potential $V = \frac{1}{2}kq^2$. The Hamiltonian for such a system is

$$H = p^2/2m + kq^2/2 = p^2/2m + m\omega^2 q^2/2 \tag{4.36}$$

where

$$\omega^2 = k/m$$

and the equations of motion are

$$\dot{q} = p/m, \qquad \dot{p} = -m\omega^2 q \tag{4.37}$$

Rather than try to solve this system of equations we note that if the Hamiltonian were cyclic in q, the momentum would be a constant of the motion and only the differential equation in q would have to be solved. Thus we seek a canonical transformation which will provide a new set of coordinates for which the Hamiltonian will be cyclic in Q. Consider the generator

$$W_1 = \tfrac{1}{2}m\omega q^2 \cot Q \tag{4.38}$$

From (4.23) it follows that

$$p = \frac{\partial W_1}{\partial q} = m\omega q \cot Q, \qquad P = -\frac{\partial W}{\partial Q} = \frac{m\omega q^2}{2\sin^2 Q} \tag{4.39}$$

or from (4.39)

$$q = (2P/m\omega)^{1/2} \sin Q, \qquad p = (2m\omega P)^{1/2} \cos Q \tag{4.40}$$

and using (4.39) in (4.36)

$$H' = H = \omega P \tag{4.41}$$

Therefore, since H is cyclic in Q

$$P = \text{const} = E/\omega \tag{4.42}$$

and

$$\dot{Q} = \omega, \qquad Q = \omega t + \alpha \tag{4.43}$$

Transforming back to q we find from (4.40) and (4.43)

$$q = (2E/m\omega^2)^{1/2} \sin(\omega t + \alpha) \tag{4.44}$$

By choosing an appropriate system of coordinates we simplified considerably the differential equations that had to be solved. The generator yielding the preferable set of coordinates was pulled out of thin air. A systematic technique for obtaining such generators will be presented later.

D. Conservation Laws and Symmetry Properties

Consider the following infinitesimal canonical transformations:

$$Q_a = q_a + \delta q_a \tag{4.45a}$$

$$P_a = p_a + \delta p_a \tag{4.45b}$$

Since the transformations are infinitesimal we expect that the generator of the transformations will differ from the generator of the identity transformation by an infinitesimal, i.e.,

$$W_2 = q_a P_a + \varepsilon G(q_b, P_b) \tag{4.46}$$

where ε is an infinitesimal parameter. Substituting (4.46) into (4.27), we obtain

$$p_a = P_a + \varepsilon\, \partial G/\partial q_a \tag{4.47a}$$

$$Q_a = q_a + \varepsilon\, \partial G/\partial P_a \tag{4.47b}$$

$$H' = H \tag{4.47c}$$

Noting that

$$\frac{\partial G}{\partial P_a} = \frac{\partial p_b}{\partial P_a}\frac{\partial G}{\partial p_b} = \left(\delta_{ab} + \varepsilon \frac{\partial^2 G}{\partial P_a\, \partial q_b}\right)\frac{\partial G}{\partial p_b}$$

we may rewrite (4.47a) and (4.47b) as

$$P_a - p_a = \delta p_a = -\varepsilon\, \partial G/\partial q_a \tag{4.48a}$$

$$Q_a - q_a = \delta q_a = \varepsilon\, \partial G(q, P)/\partial p_a = \varepsilon\, \partial G(q, p)/\partial p_a \tag{4.48b}$$

In (4.48a) we have retained only those terms which were of the first order in the infinitesimal parameter ε. We replaced $G(p, P)$ with $G(q, p)$ by replacing P with $p + \delta p$ and Taylor expanding the function G.

Now let us specialize these results to the transformation having the following generator and infinitesimal parameter:

$$G = H(q, p), \qquad \varepsilon = \delta\tau \tag{4.49}$$

Substituting (4.49) into (4.48b), we obtain

$$\delta q_a = d\tau\, \partial H/\partial p_a = d\tau\, \dot{q}_a = dq_a \tag{4.50a}$$

Also

$$\delta p_a = -d\tau\, \partial H/\partial q_a = d\tau\, \dot{p}_a = dp_a \tag{4.50b}$$

From (4.50a) and (4.50b) we see that the canonical transformation generated by (4.49) changes the coordinates and momenta at τ to their values at $\tau + d\tau$. Therefore the values of q_a and p_a at τ can be obtained from their initial values at τ_0 by a canonical transformation generated by the Hamiltonian.

Consider now the change induced in a function $u = u(q_a, p_a)$ by an infinitesimal displacement of the location of the system in phase space from $\{q_a, p_a\}$ to $\{q_a + \delta q_a, p_a + \delta p_a\}$. In the usual fashion we define

$$\delta u \equiv u(q_a + \delta q_a, p_a + \delta p_a) - u(q_a, p_a) \tag{4.51}$$

Performing a Taylor expansion of the first term on the right side of (4.51) about the initial point $\{q_a, p_a\}$ and neglecting infinitesimals of the second and higher orders, we find

$$\delta u = \frac{\partial u}{\partial q_a}\,\delta q_a + \frac{\partial u}{\partial p_a}\,\delta p_a \tag{4.52}$$

Using (4.48), (4.52) becomes

$$\delta u = \varepsilon\left(\frac{\partial u}{\partial q_a}\frac{\partial G}{\partial p_a} - \frac{\partial u}{\partial p_a}\frac{\partial G}{\partial q_a}\right) = \varepsilon[u, G] \tag{4.53}$$

If the function u is the Hamiltonian, (4.53) is

$$\delta H = \varepsilon[H, G] \tag{4.54}$$

If G is a constant of the motion then its Poisson bracket with the Hamiltonian is zero, and we conclude from (4.54) that the Hamiltonian is invariant under the infinitesimal transformation generated by G, i.e.,

$$\delta H = 0 \tag{4.55}$$

It follows immediately that we may characterize constants of the motion as generating functions of those infinitesimal canonical transformations which leave the Hamiltonian invariant, and one could in principle determine all the constants of motion by finding canonical transformations leaving H invariant. Once again we have established that intimate connection between conservation laws and the properties of the system under certain canonical transformations. We illustrate these results by considering the canonical transformation which performs a spatial displacement of the entire system by an infinitesimal amount, i.e.,

$$\delta q_a = \varepsilon \tag{4.56}$$

From (4.48b) it follows that the generator of such a transformation is

$$G = \sum_a p_a \tag{4.57}$$

The Poisson bracket of G with H is

$$[H, G] = \sum_{a=1}^{N} \partial H/\partial q_a \tag{4.58}$$

Therefore, if the Hamiltonian is cyclic in the coordinates q_a, the total momentum (i.e., G) is conserved. This result agrees with that obtained from Noether's theorem.

E. Hamilton-Jacobi Theory

In our discussion of the harmonic oscillator we demonstrated that a canonical transformation could be used to simplify the solution of the dynamical equations. Hamilton–Jacobi theory provides a systematic method for determining generating functions which lead to equations of motion that can be solved easily. But now we will be more ambitious than we were in the harmonic oscillator problem, where we sought a canonical transformation which made our Hamiltonian cyclic in Q. We now seek a canonical transformation to a new set of canonical coordinates for which our Hamiltonian will be cyclic in both the coordinates and momenta. If we are successful, it follows immediately that the momenta and coordinates are constants, and if we identify the constants as the values of our original coordinates and momenta at the initial time t_0, i.e., $Q_a = q_{0_a}$ and $P_a = p_{0_a}$, then our canonical transformations back to the original coordinates have the form

$$q_a = q_a(q_{0_b}, p_{0_b}, t) \tag{4.59a}$$

$$p_a = p_a(q_{0_b}, p_{0_b}, t) \tag{4.59b}$$

But these are just the solutions to our dynamical equations of motion.

Thus to accomplish our objective we require

$$\partial H'/\partial P_a = \dot{Q}_a = 0, \qquad -\partial H'/\partial Q_a = \dot{P}_a = 0 \tag{4.60}$$

Now (4.60) will be satisfied if

$$H' = 0 \tag{4.61}$$

But the relationship between H and H' is

$$H' = H + \partial W/\partial t \tag{4.62}$$

Let us choose a generator of the W_2 type, i.e., $W_2 = W_2(q_a, P_a, t)$, then with (4.61), (4.62) becomes

$$H(q_a, p_a, t) + \partial W_2/\partial t = 0 \tag{4.63}$$

But with this generator the p_a are not independent but are given by (4.27a), i.e.,

$$H(q_a, \partial W_2/\partial q_a, t) + \partial W_2/\partial t = 0 \tag{4.64}$$

This is the *Hamilton–Jacobi equation*. It is more commonly written with W_2 replaced by S and we will follow that procedure. Thus we write (4.64) as

$$H(q_a, \partial S/\partial q_a, t) + \partial S/\partial t = 0, \qquad a = 1, 2, \ldots, N \tag{4.65}$$

The generator S is called *Hamilton's Principle Function.*

In (4.65) it is understood that wherever the momentum appears in the Hamiltonian of the system it is to be replaced with $\partial S/\partial q_a$, and therefore (4.65) is a first-order partial differential equation for S in terms of the $(N + 1)$-independent variables q_a and t. Thus the solution will contain $(N + 1)$ arbitrary constants of integration, $\alpha_1, \alpha_2, \ldots, \alpha_N, \alpha$. Since S appears in (4.65) only through its derivatives, one of the constants of integration must be additive. Thus the solution to (4.65) has the form

$$S = S'(q_a, \alpha_a, t) + \alpha \tag{4.66}$$

Now comparing (4.66) with the usual form of the generator W_2, it is convenient to identify the α_a with the momenta P_a (more generally we could let the α_a be a function of the P_a). Thus Eqs. (4.27a) and (4.27b) become

$$p_a = \partial S'(q_b, P_b, t)/\partial q_a \tag{4.67a}$$

$$Q_a = \partial S'(q_b, P_b, t)/\partial P_a \tag{4.67b}$$

To obtain the solutions of our dynamical problem in the form of Eqs. (4.59), we proceed as follows. In (4.67a) we set $t = t_0$, $q_b = q_{b_0}$, $p_a = p_{a_0}$, and solve for P_b in terms of t_0, q_{b_0}, p_{b_0}

$$P_b = P_b(q_{a_0}, p_{a_0}, t_0) \tag{4.68}$$

Then in (4.67a) we set $t = t_0$, $q_b = q_{b_0}$, replace P_b with q_{b_0}, p_{b_0}, t_0, according to (4.68) and solve for Q_a in terms of q_{a_0}, p_{a_0}, and t_0

$$Q_b = Q_b(q_{a_0}, p_{a_0}, t_0) \tag{4.69}$$

With Q_a and P_a now expressed in terms of the initial values, we solve (4.67b) for q_a and obtain using (4.68) and (4.69)

$$q_a = q_a(q_{0_b}, p_{0_b}, t_0, t) \tag{4.70}$$

Substituting (4.68), (4.69), and (4.70) into (4.67a), we obtain

$$p_a = p_a(q_{0_b}, p_{0_b}, t_0, t) \tag{4.71}$$

Equations (4.70) and (4.71) are the solutions to our dynamical problem. Thus classical dynamics has essentially been reduced to the solution of a first-order partial differential equation. Elegant as this formalism is, its viability as a method of solving dynamical problems depends on whether the differential equation can be solved. Since the Hamiltonian contains the kinetic energy which is always quadratic in the momentum, Eq. (4.65) will contain terms of the form $(\partial S/\partial q_a)^2$. Thus the Hamilton–Jacobi equation is a *nonlinear* first-order partial differential equation. The mathematical technique, the Jacobi theory, for proving the existence of solutions to first-order nonlinear equations is to show that the partial differential equation can be converted into a system of $2N$ ordinary differential equations. These are of course just Hamilton's equations. Thus the general mathematical technique for solving (4.65) is to convert (4.65) into the corresponding Hamilton equations of motion. As far as we are concerned this process is circular and we might just as well ignore Hamilton–Jacobi theory, unless we can find some other technique for extracting a solution from (4.65). In some interesting cases (4.65) can be solved directly by a separation of variables technique.

When the Hamiltonian is not an explicit function of the time, S' may be separated into

$$S'(q_a, \alpha_a, t) = S_1(q_a, \alpha_a) - \alpha_1 t \tag{4.72}$$

and (4.65) becomes

$$H(q_a, \partial S_1/\partial q_a) = \alpha_1 \tag{4.73}$$

and

$$S = S_1(q_a, \alpha_a) - \alpha_1 t + \alpha \tag{4.74}$$

The Hamilton–Jacobi technique is now illustrated by applying it to the motion of a one-dimensional harmonic oscillator.

The Hamiltonian for a single particle moving in a simple harmonic oscillator potential is presented in (4.36)

$$H = (p^2/2m) + \tfrac{1}{2}m\omega^2 q^2$$

The corresponding Hamilton–Jacobi equation is

$$(1/2m)(\partial S/\partial q)^2 + \tfrac{1}{2}m\omega^2 q^2 + (\partial S/\partial t) = 0 \tag{4.75}$$

To solve (4.75) we employ the separation of variables

$$S(q, \alpha, t) = S_1(q, \alpha) - \alpha t \tag{4.76}$$

where α is a constant of integration which we identify with P. Substituting (4.76) into (4.75), we obtain

$$(1/2m)(\partial S_1/\partial q)^2 + \tfrac{1}{2}m\omega^2 q^2 - P = 0 \tag{4.77}$$

or

$$dS_1/dq = [2mP - m^2\omega^2 q^2]^{1/2} \tag{4.78}$$

Equation (4.78) may be written

$$S_1 = \int (2mP - m^2\omega^2 q^2)^{1/2}\, dq \tag{4.79}$$

and

$$S = m\omega \int [(2P/m\omega^2) - q^2]^{1/2}\, dq - Pt \tag{4.80}$$

We can save some effort by realizing that we never actually use S but only derivatives of S to obtain the dynamical solutions. Thus Eqs. (4.67) become

$$p = \frac{\partial S}{\partial q} = m\omega\left(\frac{2P}{m\omega^2} - q^2\right)^{1/2} \tag{4.81a}$$

$$Q = \frac{\partial S}{\partial P} = \frac{1}{\omega}\int\left(\frac{2P}{m\omega^2} - q^2\right)^{-1/2} dq - t = -\left[t + \frac{1}{\omega}\cos^{-1}\frac{q\omega}{(2P/m)^{1/2}}\right] \tag{4.81b}$$

If at $t = 0$, $p_0 = 0$, and $q_0 = A$, (4.81a) yields

$$P = \tfrac{1}{2}m\omega^2 A^2 \tag{4.82a}$$

and (4.81b) yields as a possible solution

$$Q = 0 \tag{4.82b}$$

Now using (4.82a) and (4.82b) in (4.81b) and solving for q we find

$$q = A \cos \omega t \tag{4.83}$$

which is the usual solution for the harmonic oscillator. Equation (4.81a) yields

$$p = -m\omega A \sin \omega t \tag{4.84}$$

Furthermore, since the Hamiltonian is the total energy of the system we have

$$H + \partial S/\partial t = 0 = E + \partial S/\partial t$$

or from (4.76) and (4.82a)

$$E = -\partial S/\partial t = P = (m\omega^2/2)A^2 \tag{4.85}$$

We conclude our discussion of the Hamilton–Jacobi theory by noting that in general

$$dS/dt = (\partial S/\partial q_a)\dot{q}_a + \partial S/\partial t \tag{4.86}$$

But using (4.65) and (4.67a), (4.86) may be written

$$dS/dt = p_a\dot{q}_a - H = L \tag{4.87}$$

or

$$S = \int L\, dt + \text{const} \tag{4.88}$$

Thus Hamilton's principle function is just the indefinite time integral of the Lagrange function. Interesting as the relationship is it cannot be used to obtain S since the Lagrangian integration can be performed only after its arguments have been expressed as functions of time and this can be done only after the problem has been solved.

F. Hamiltonian Dynamics in Lorentz Space

In Chapter 2 it was stated that there were serious difficulties associated with an effort to construct a many particle Lagrangian dynamics compatible with E.S.R. (Einstein's Special Theory of Relativity). A similar situation prevails for Hamilton's dynamics. As in Chapter 2 we will confine our discussion to a single particle system, leaving a more general discussion for the next chapter.

The development of a Lagrangian dynamics compatible with E.S.R. was achieved by choosing the proper time as the independent variable and incorporating the constraint $\dot{x}^{\mu}\dot{x}_{\mu} = 1$ by means of the Lagrange multiplier technique. To establish a corresponding Hamiltonian theory we are confronted with several additional problems. Presumably we wish to construct a dynamics based on a "Hamiltonian" that is a function of the coordinates and momenta. This may be accomplished by employing a Legendre transformation as in Eq. (4.2), but in anticipation of the subsequent variational problem, we define the Hamiltonian, $\bar{H}$ in terms of the modified Lagrangian $\bar{L}$ which includes the constraint on the four velocities.

$$\bar{H}(\bar{p}^{\mu}, {}_{\lambda}\bar{p}, x^{\mu}, \lambda) = \bar{p}^{\mu}\dot{x}_{\mu} + {}_{\lambda}\bar{p}\dot{\lambda} - \bar{L}(x^{\mu}, \dot{x}^{\mu}, \lambda) \tag{4.89}$$

where λ is the Lagrange multiplier, ${}_{\lambda}\bar{p}$ the momentum conjugate to λ, and

$$\bar{p}^{\mu} = \partial\bar{L}/\partial\dot{x}_{\mu}$$

From (4.84) we obtain the action

$$I = \int_{s_1}^{s_2} [\bar{p}^{\mu}\dot{x}_{\mu} + {}_{\lambda}\bar{p}\dot{\lambda} - \bar{H}(x^{\mu}, p^{\mu}, \lambda, {}_{\lambda}p)]\, ds \tag{4.90}$$

Hamilton's principle, i.e., $\delta I = 0$, yields the following equations of motion:

$$\dot{x}^{\mu} = \partial\bar{H}/\partial\bar{p}_{\mu}, \qquad \dot{\lambda} = \partial\bar{H}/\partial_{\lambda}\bar{p} \tag{4.91a}$$

$$\dot{\bar{p}}^{\mu} = \partial\bar{H}/\partial x_{\mu}, \qquad {}_{\lambda}\dot{\bar{p}} = -\partial\bar{H}/\partial\lambda \tag{4.91b}$$

Using the relativistic free particle Lagrangian (2.51) and the fact that ${}_{\lambda}p = 0$, (4.89) becomes

$$\bar{H} = \bar{p}^{\mu}\bar{p}_{\mu}/2\lambda - mc + \tfrac{1}{2}\lambda \tag{4.92}$$

and the equations of motion are

$$\dot{\bar{p}}^{\mu} = 0, \qquad \dot{x}^{\mu} = \bar{p}^{\mu}/\lambda, \qquad p_{\lambda} = 0 = \partial\bar{H}/\partial\lambda = 1 - \bar{p}^{\mu}\bar{p}_{\mu}/\lambda^2 \tag{4.93}$$

Thus we may choose $\lambda = mc$ to obtain the usual relativistic free particle equations. As in relativistic Lagrangian dynamics we may consider the motion of a particle under the influence of a field whose sources are ignored. A particle in an electromagnetic field is considered in Section 4.3.

The covariant single particle canonical formalism is developed in the same manner as in Galilean invariant systems. Our Hamiltonian transforms as a scalar and may be considered as a generator of a displacement of the particle along its world line. Unlike its Galilean counterpart, it is not the total energy of the system, in fact, from (3.54) we see that it is numerically equal to zero. As discussed in Chapter 3, the total energy of the system is the zeroth component of the four-momentum, and since we defined the Hamiltonian as a scalar it should come as no surprise that it is not the total energy of the system. Barut [2] may be consulted for other definitions of the relativistic Hamiltonian.

4.2. DEVELOPMENT AND EXTENSION OF THE BASIC THEORY

A. Dirac Extension of Hamiltonian Dynamics

In the derivation of the canonical equations of motion it was assumed that the coordinates and momenta were all independent. However, suppose that there exist equations of constraint of the form

$$\varphi_\alpha(q_a, p_a) = 0, \qquad \alpha = 1, 2, \ldots, \Gamma \tag{4.94}$$

The number of independent variables is reduced from $2N$ to $2N - \Gamma$ and the canonical Eqs. (4.8) no longer follow from (4.7). This is exactly the problem that was confronted when discussing Lorentz invariant systems.

The Lagrange multiplier technique was developed to handle variational problems to constraints. Applying this technique to our problem, we multiply the constraints φ_α by the multipliers λ_α and add the resulting equations to the integrand of (4.4)

$$\delta I = \delta \int_{\tau_1}^{\tau_2} [p_a \dot{q}_a - H(q_a, p_a, \tau) - \sum_\alpha \lambda_\alpha \varphi_\alpha(q_a, p_a, \tau)]\, d\tau = 0 \tag{4.95}$$

The variation of the action is to be performed as though the $\{p_a, q_a, \lambda_\alpha\}$ are independent variables. Proceeding in the usual fashion, we obtain to first order in the variations

$$\int_{\tau_1}^{\tau_2} \Big[\Big(\dot{q}_a - \frac{\partial H}{\partial p_a} - \sum_\alpha \lambda_\alpha \frac{\partial \varphi_\alpha}{\partial p_a} \Big) \delta p_a - \Big(\dot{p}_a + \frac{\partial H}{\partial q_a} + \sum_\alpha \lambda_\alpha \frac{\partial \varphi_\alpha}{\partial q_a} \Big) \delta q_a$$
$$+ \sum_\alpha \varphi_\alpha \, \delta\lambda_\alpha - \Big(\frac{\partial H}{\partial \tau} + \sum_\alpha \lambda_\alpha \frac{\partial \varphi_\alpha}{\partial \tau} \Big) \delta\tau - (H + \sum_\alpha \lambda_\alpha \varphi_\alpha) \frac{d\,\delta\tau}{d\tau} \Big] d\tau = 0 \tag{4.96}$$

Now in conformity with Hamilton's principle we require that $\delta\tau = 0$ and that δp_a, δq_a, and $\delta\lambda_\alpha$ be independent. Thus for (4.96) to be satisfied it is necessary that

$$\dot{q}_a = \frac{\partial H}{\partial p_a} + \sum_\alpha \lambda_\alpha \frac{\partial \varphi_\alpha}{\partial p_a} \tag{4.97a}$$

$$\dot{p}_a = -\frac{\partial H}{\partial q_a} - \sum_\alpha \lambda_\alpha \frac{\partial \varphi_\alpha}{\partial q_a} \tag{4.97b}$$

$$\varphi_\alpha = 0 \tag{4.97c}$$

or in Poisson bracket notation

$$\dot{q}_a = [q_a, H] + \sum_\alpha \lambda_\alpha [q_a, \varphi_\alpha] \tag{4.98a}$$

$$\dot{p}_a = [p_a, H] + \sum_\alpha \lambda_\alpha [p_a, \varphi_\alpha] \tag{4.98b}$$

The constraints φ_α are called *primary constraints.*

We now define the *total Hamiltonian* H_{T} as

$$H_{\mathrm{T}} = H + \sum_\alpha \lambda_\alpha \varphi_\alpha \tag{4.99}$$

Recalling that $\lambda_\alpha \neq \lambda(q_a, p_a)$ and using (4.99) and the definition of the Poisson bracket, we may write (4.98) as

$$\dot{q}_a = [q_a, H_{\mathrm{T}}] \tag{4.100a}$$

$$\dot{p}_a = [p_a, H_{\mathrm{T}}] \tag{4.100b}$$

$$\varphi_\alpha = 0 \tag{4.100c}$$

Similarly, if $F = F(q_a, p_a)$, its equation of motion may be written as

$$\dot{F} = \frac{\partial F}{\partial q_a} \dot{q}_a + \frac{\partial F}{\partial p_a} \dot{p}_a = \frac{\partial F}{\partial q_a}\frac{\partial H_{\mathrm{T}}}{\partial p_a} - \frac{\partial F}{\partial p_a}\frac{\partial H_{\mathrm{T}}}{\partial q_a}$$
$$= [F, H_{\mathrm{T}}] \tag{4.101}$$

For the equations to be consistent it is necessary that

$$\dot{\varphi}_\alpha = 0 = [\varphi_\alpha, H_{\mathrm{T}}] \tag{4.102}$$

It should be noted that φ_α is set equal to zero only after the Poisson bracket has been evaluated. Equations (4.102) impose a set of consistency conditions the nature of which we now investigate. Equation (4.53) may be written

$$[\varphi_\alpha, H] + \sum_\alpha \lambda_\beta [\varphi_\alpha, \varphi_\beta] = 0 \tag{4.103}$$

Thus (4.103) represents Γ consistency relations. We distinguish three ways in which (4.103) may be satisfied.

Case A: It may be identically satisfied, i.e., $0 = 0$, with the help of the primary constraints (4.94).

Case B: It may reduce to an equation which is independent of the Lagrange multipliers and thus involves only the p_a and q_a. Such an equation must not go to zero because of the primary constraints or also it would be equivalent to Case A. Thus it is of the form

$$\chi_\gamma(q_a, p_a) = 0 \tag{4.104}$$

Case C: It is satisfied by imposing conditions on the Lagrange multipliers.

The consistency requirements of type B result in an additional set of constraints, viz., (4.104), called secondary constraints. The secondary constraints differ from primary constraints in that they are consequences of the equations of motion whereas the primary constraints follow directly from the defining equations of momentum. Now the secondary constraints impose yet another consistency condition,

$$\chi = 0 = [\chi, H] + \sum_\alpha \lambda_\alpha [\chi, \varphi_\alpha] \tag{4.105}$$

Equation (4.105) may also be classified as of type A, B, or C. We can continue this procedure until all of the consistency conditions have been exhausted, i.e., until we obtain no additional constraints. Thus in the end we are left with a number of secondary constraints and a number of conditions on the Lagrange multipliers.

We can consolidate our notation somewhat if we designate both primary and secondary constraints by

$$\varphi_\eta = 0, \qquad \eta = 1, 2, \ldots, M, M+1, \ldots, M+N \tag{4.106}$$

where M is the number of primary constraints and N is the number of secondary constraints.

Consider now consistency conditions of the type C. Then the equations

$$[\varphi_\eta, H] + \sum_\alpha \lambda_\alpha[\varphi_\eta, \varphi_\alpha] = 0, \qquad \alpha = 1, 2, \ldots, M$$

may be regarded as a set of equations for λ_α, i.e., we seek solutions of the form

$$\lambda_\alpha = U_\alpha(q_a, p_a) \tag{4.107}$$

If such solutions would not exist it would indicate that the Lagrange equations are inconsistent. Solutions of the form (4.102) are not unique in that we may add to them any solution $V_\alpha(q_a, p_a)$ of the homogeneous equations associated with (4.103), i.e.,

$$V_\alpha(q_a, p_a) = 0 \tag{4.108}$$

Thus the most general solution is

$$\lambda_\alpha = U_\alpha + \sum_\beta v_\beta V_{\beta\alpha} \tag{4.109}$$

and

$$H_{\mathrm{T}} = H + \sum_\alpha U_\alpha\varphi_\alpha + \sum_{\alpha,\beta} v_\beta V_{\beta\alpha}\varphi_\alpha \tag{4.110}$$

where the v_β are arbitrary coefficients which may be functions of τ. Thus when type C consistency conditions arise we have arbitrary functions of τ occurring in the general solution of the equation of motion with given initial conditions. As a result, the dynamical variables describing the evolution of the system from its configuration at τ_1 to its configuration at τ_2 is not completely determined by the configuration at τ_1. Dirac [3] investigates the ramifications of type C conditions in electrodynamics and general relativity.

B. Action-Angle Variables

Within the Hamilton–Jacobi formalism there is a powerful technique for extracting certain information about periodic systems without actually obtaining the complete dynamical solutions. It is common to identify two types of periodic motion, *libration* and *rotation*. In libration the coordinate and its conjugate momentum are periodic functions of τ with the same frequency; whereas in rotation the coordinate is not itself periodic but the conjugate momentum repeats itself everytime the coordinate goes through some interval q_0. The situation is illustrated in Fig. 4.1 for a system with one degree of freedom; Fig. 4.1c shows that the same system, in this case a simple pendulum, may exhibit both libration and rotation. For systems having more than one degree of freedom, the orbit of the system in the multidimensional phase space will generally be rather complicated, but if the Hamilton–Jacobi equation is separable in the canonical variables (q_a, p_a) the concept of libration and rotation can be applied to the projection of the orbit on each (q_a, p_a) plane. Since the Hamilton–Jacobi equation is separable the projected motions are independent of each other and therefore their orbits may be readily examined.

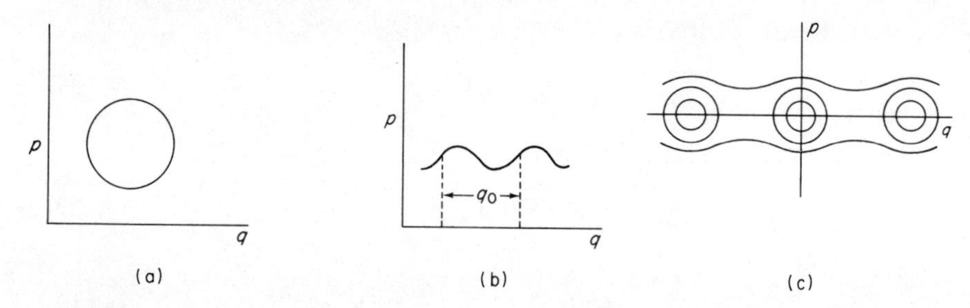

Fig. 4.1. The orbits in phase space corresponding to possible one-dimensional periodic motions: (a) libration, (b) rotation, and (c) simple pendulum exhibiting both libration and rotation.

We now introduce a new pair of canonical variables which are conventionally designated by J_a and w_a rather than P_a and Q_a. The J_a are the new momentum variables and are called the *action variables*; the w_a are the new coordinates and are called the *angle variables*. The action variables will replace the constants α_a [see Eq. (4.74)] and are defined as

$$J_a = \int p_a \, dq_a, \qquad \text{no sum on } a, \quad a = 1, 2, \ldots, N \tag{4.111}$$

where the integration is performed over one complete period of q_a. It should be noted in passing that if one of the q_a's is cyclic, the corresponding momentum is constant and (4.111) yields

$$J_a = p_a \oint dq_a = 2\pi p_a \tag{4.112}$$

where the scale for q_a has been chosen such that its period is 2π.

It follows from (4.67a) and the identification of the α_a with the P_a that

$$J_a = \oint \frac{\partial S_1(q_b, \alpha_b)}{\partial q_a} dq_a \tag{4.113}$$

The separability of the Hamilton–Jacobi equation in the variable pairs (q_a, p_a) assures the independence of the pairs and therefore the J_a form a set of N independent functions of the α_b's and may be used as a set of new constant momenta. Thus

$$S_1 = S_1(q_a, J_a) \tag{4.114}$$

and from (4.73) it follows that

$$H = \alpha_1 = H(J_a) \tag{4.115}$$

From (4.67b) it follows that the angle variables are obtained from the transformation equations

$$w_a = \partial S_1 / \partial J_a \tag{4.116}$$

The canonical equations of motion for the angle variables are

$$\dot{w}_a = \partial H(J_b)/\partial J_a = \nu_a(J_b) \tag{4.117}$$

where the ν_a are constant functions of the action variables. Equation (4.117) yields immediately

$$w_a = \nu_a t + \beta_a \tag{4.118}$$

Consider now the change that occurs in the action variable Δw_a when one of the coordinates q_b goes through a complete cycle, i.e.,

$$\Delta w_a = \oint \delta_b w_a \tag{4.119}$$

where $\delta_b w_a$ is the infinitesimal change in w_a due to an infinitesimal change in q_b:

$$\delta_b w_a = \frac{\partial w_a}{\partial q_b} dq_b, \qquad \text{no sum on } b \tag{4.120}$$

Therefore,

$$\Delta w_a = \oint \frac{\partial w_a}{\partial q_b} dq_b = \oint \frac{\partial^2 S_1}{\partial q_b \, \partial J_a} dq_b = \frac{\partial}{\partial J_a} \oint p_b dq_b = \delta_{ab} \tag{4.121}$$

Thus, if τ_a is the period associated with the coordinate q_a, it follows from

(4.118) and (4.121) that

$$\Delta w_a = \nu_a \tau_a = 1 \tag{4.122}$$

We conclude, therefore, that

$$\nu_a = 1/\tau_a \tag{4.123}$$

i.e., that the constants ν_a are the frequencies of the motion. The great advantage of the action-angle formalism is that it permits an evaluation of the frequencies of the motion without finding a complete solution for the motion of the system. The procedure is to express the Hamiltonian in terms of the action variables, defined by (4.111), and use (4.117) to determine the frequencies.

To illustrate the action-angle variable technique we consider again the one-dimensional harmonic oscillator. There is only one action variable given by

$$J = \oint p\, dq = \oint \frac{\partial S_1}{\partial q}\, dq = \oint [2m\alpha_1 - m^2\omega^2 q^2]^{1/2}\, dq \tag{4.124}$$

where we have used Eq. (4.78) which presents $\partial W/\partial q$ in terms of α_1 and q. To evaluate (4.124) we use the substitution

$$q = (2\alpha_1/m\omega^2)^{1/2} \sin\theta$$

and integrate. We obtain

$$J = 2\pi\alpha_1/\omega \tag{4.125}$$

The Hamiltonian is

$$\begin{aligned} H = (p^2/2m) + \tfrac{1}{2}m\omega^2 q^2 &= (1/2m)[2m\alpha_1 - m^2\omega^2 q^2] + \tfrac{1}{2}m\omega^2 q^2 \\ &= \alpha_1 = J\omega/2\pi \end{aligned} \tag{4.126}$$

The frequency of oscillation is

$$\nu = \partial H/\partial J = \omega/2\pi$$

which is the usual formula for the harmonic oscillator.

C. Adiabatic Invariance

A system experiences an *adiabatic* change if its motion is characterized by a parameter λ, which varies very slowly with time as a result of some activity external to the system. Because of the external influence the energy is not conserved, but it changes at a rate which is proportional to the rate at which λ changes. Under adiabatic conditions, there are certain quantities which remain as constants of the motion; they are called *adiabatic invariants*. It is advantageous to describe adiabatically varying systems in terms of action–angle variables because the action is an adiabatic invariant.

The point of departure is the realization that the Hamiltonian is now an explicit function of the time through the dependence of λ on time. The *adiabatic approximation* consists in deriving and solving the dynamical equations as if λ were a constant. However, the solution is recognized as holding only at a particular time t_i; at some later time t_f, the same *form* for the solution is used, but the values of the parameters appearing in the equation (e.g., "constants of integration," λ, etc.) are different than the corresponding values at t_i. These new values for the various parameters can often be obtained from the adiabatic invariants of the system.

To illustrate the technique we consider a harmonic oscillator whose natural frequency is changing adiabatically. Although the Hamiltonian is now written as

$$H(p, q, t) = (p^2/2m) + \tfrac{1}{2}m[\omega(t)]^2q^2 \tag{4.127}$$

the solutions describing the motion of the constant ω oscillator are assumed to be valid, viz., (4.83) and (4.84). We can write for t_i and t_f

$$q(t_i) = A_i \cos \omega_i t, \qquad p = -m\omega_i A_i \sin \omega_i t \tag{4.128}$$

$$q(t_f) = A_f \cos \omega_f t, \qquad p = -m\omega_f A_f \sin \omega_f t \tag{4.129}$$

At this point the solutions in (4.128) are not very useful because although A_i may be determined by inserting the "initial conditions" at $t = t_i$, A_f has no such conditions to permit its evaluation at $t = t_f$. The relationship between A_f and A_i can be determined from the action variable J, which is an adiabatic invariant (this will be demonstrated later). Recall that

$$J_i = \oint p_i \, dq_i \qquad \text{(no sum on } i) \tag{4.130}$$

Inserting (4.128) into (4.129), we find

$$J_i = \pi m \omega_i A_i^2 \tag{4.131}$$

Similarly,

$$J_f = \pi m \omega_f A_f^2 \tag{4.132}$$

Now if $J_f = J_i$ (i.e., the action is an adiabatic invariant), it follows that

$$\omega_f A_f^2 = \omega_i A_i^2$$

or

$$A_f^2 = (\omega_i/\omega_f) A_i^2 \tag{4.133}$$

Equation (4.133) tells us how the amplitude of oscillation changes under the influence of an adiabatic change of the natural frequency of a harmonic oscillator; i.e., as the frequency increases the amplitude decreases. From (4.85) it follows that the energy of the oscillator changes according to

$$E_f = \tfrac{1}{2} m \omega_f^2 A_f^2 = \tfrac{1}{2} m \omega_i \omega_f A_i^2 = (\omega_f/\omega_i) E_i \tag{4.134}$$

i.e., the energy increases as the natural frequency increases.

We conclude our discussion of adiabatic invariance by showing that the action may be considered an adiabatic invariant. We begin by noting that the generator and the Hamiltonian of (4.62) are now functions of the time, and, therefore, so is the new Hamiltonian. Hence (4.63) must now be written

$$H' = H + \frac{\partial W_2}{\partial t} = H(J_a, \lambda) + \frac{\partial W_2(w_a, J_a, \lambda)}{\partial \lambda} \dot{\lambda} \tag{4.135}$$

where we are immediately expressing H and W_2 in terms of the action-angle variables. The Hamilton equations of motion for the action-angle variables are

$$\dot{w}_a = \frac{\partial H'}{\partial J_a} = \frac{\partial H}{\partial J_a} + \frac{\partial^2 W_2}{\partial J_a \, \partial \lambda} \dot{\lambda} = \nu_a(J_b, \lambda) + \frac{\partial^2 W_2}{\partial J_a \, \partial \lambda} \dot{\lambda} \tag{4.136}$$

$$\dot{J}_a = -\frac{\partial H'}{\partial w_a} = -\frac{\partial^2 W_2}{\partial w_a \, \partial \lambda} \dot{\lambda} \tag{4.137}$$

where the ν_a are the frequencies introduced in (4.117), but now they are functions of the time through λ. Although it is clear from (4.137) that the action variables are not constants of the motion, we now show that under appropriate conditions (the adiabatic conditions) J_a varies so slowly that it is a good approximation to consider them as constant.

From (4.137) it follows that the change in J_a in the time τ is

$$|\Delta J_a|_\tau = \left| \int_0^\tau \dot{\lambda} \frac{\partial^2 W_2}{\partial w_a \, \partial \lambda} dt \right| = \left| \int_0^{w_a(\tau)} \dot{\lambda} \frac{\partial^2 W_2}{\partial w_a \, \partial \lambda} \frac{dw_a}{\dot{w}_a} \right| \qquad \text{(no sum on } a\text{)} \tag{4.138}$$

But now suppose that $\lambda(t)$ is a slowly varying function of the time, i.e.,

$$|\dot{\lambda}| < \varepsilon \tag{4.139}$$

where ε is a small positive number, then (4.136) yields

$$|\dot{w}_a| \leq |\nu_a| + \left|\frac{\partial^2 W_2}{\partial J_a\,\partial\lambda}\right| \varepsilon \tag{4.140}$$

Taking the inverse of both sides of (4.140), we find

$$|\dot{w}_a|^{-1} \leq |\nu_a|^{-1} + O(\varepsilon) \tag{4.141}$$

where $O(\varepsilon)$ represents terms in ε to the first and higher powers. Substituting (4.141) in (4.138), we find

$$|\Delta J_a|_\tau < \varepsilon \left|\frac{1}{\nu_a}\int_0^{w_a(\tau)} \frac{\partial^2 W_2}{\partial w_a\,\partial\lambda}\,dw_a + 0(\varepsilon)\right| \qquad \text{(no sum on } a) \tag{4.142}$$

It can be shown that the integral is bounded, i.e.,

$$\int_0^{w_a(\tau)} \frac{\partial^2 W_2}{\partial w_a\,\partial\lambda}\,dw_a \leq B \neq B(\tau) \qquad \text{(no sum on } a) \tag{4.143}$$

where B is a positive number independent of τ. Then (4.142) becomes

$$|\Delta J_a|_\tau < \varepsilon B/\nu_a + O(\varepsilon) \tag{4.144}$$

Now (4.144) states that to first order in ε, the change in the action J_a is independent of the time τ and can be made as small as we please by choosing ε to be sufficiently small. But ε is just the rate of change of λ; so we have found that as long as we change λ sufficiently slowly, J_a may be considered a constant (an adiabatic invariant) even if the total change in λ over a large time τ is large because J_a depends on ε but is independent of τ.

That the integral of (4.142) is bounded and independent of τ, i.e., that (4.143) is valid, is clear if first we expand $\partial W_2/\partial\lambda$ in a Fourier series.

$$\frac{\partial W_2}{\partial\lambda} = \sum_a \sum_{n=-\infty}^{\infty} \exp(i2\pi n_a w_a) \tag{4.145}$$

Then the derivative of (4.145) with respect to w_a eliminates the constant term in the series, and every other term of the series averages to zero over a cycle of the angle variable

$$\int_0^1 \frac{\partial^2 W_2}{\partial w_a \, \partial \lambda} \, dw_a = 0 \qquad \text{since} \quad \int_0^1 \exp(i2\pi n_2 w_a) \, dw_a = 0 \qquad \text{for} \quad n \neq 0 \tag{4.146}$$

Equation (4.143) follows immediately from (4.146).

D. Canonical Perturbation Theory

Throughout the text we have generally terminated our discussion of a physical system when we acquired the equations of motion. Of course in applications the solutions are essential. For most systems of sufficient complexity to warrant the utilization of the analytical methods developed in this book, the dynamical equations are too difficult to solve exactly. Frequently, the equations are solved by numerical techniques with the aid of a computer. The high-speed, large memory, and interactive modes of modern computers permit not only excellent solutions but also considerable insight into the relationship between the solutions and the parameters that characterize the system. A vast literature has developed on this subject, and the references [4–10] may be consulted for an introduction to the subject.

Often far greater insight may be obtained by employing analytical techniques that provide approximate solutions to the dynamical equations. An enormous lore on approximation schemes has developed, and the book by Hagihara [4] is recommended for an extensive presentation of the standard methods employed in celestial mechanics and further references. We content ourselves here with a presentation of a perturbation theory that is used in conjunction with Hamilton–Jacobi theory.

Perturbation theory is widely applied to systems which can be approximated by a system whose dynamical equations can be solved exactly. The terms in the dynamical equations (or the Hamiltonian) which represent the difference between the equations of the approximate system and the equations of the actual system are called the perturbation of the system. Thus for the one dimensional anharmonic oscillator the actual system is represented by the Hamiltonian

$$H = (p^2/2m) + \tfrac{1}{2}m\omega^2 q^2 + \lambda q^3 \tag{4.147}$$

and the equations of motion are

$$\dot{q} = \frac{\partial H}{\partial p} = \frac{p}{m}, \qquad \dot{p} = -\frac{\partial H}{\partial q} = -m\omega^2 q - 3\lambda q^2 \tag{4.148}$$

Now if λ is sufficiently small the system may be represented by the harmonic oscillator whose solutions are known exactly. The term λq^3 is the perturbation of the Hamiltonian of the harmonic oscillator. The simplified Hamiltonian

$$H_0 = (p^2/2m) + \tfrac{1}{2}m\omega^2 q^2 \tag{4.149}$$

is called the unperturbed Hamiltonian. We now develop a method which permits us to proceed from the solution of the unperturbed system to a solution which better represents the motion of the perturbed system. For simplicity we restrict the development to a separable bound system having one degree of freedom; the extension to systems having many degrees of freedom is straightforward.

Let the actual Hamiltonian of the system be given by $H(q, p, \lambda)$, where λ is a parameter characterizing the strength of the perturbation. When $\lambda = 0$ the Hamiltonian represents the unperturbed system and quantities associated will have a zero subscript, e.g., $H_0 = H(q, p, \lambda = 0)$. Suppose now that the canonical transformation from the variables (q, p) to the action-angle variables is known for the unperturbed system. Then the equations of motion for the unperturbed system are

$$\dot{w}_0 = \partial H_0'(J_0)/\partial J_0 = \nu_0(J_0), \qquad \dot{J}_0 = 0 \tag{4.150}$$

Assuming that the actual system can be represented by action-angle variables (i.e., it is bounded), then according to (4.115) we may write for the perturbed system

$$H(q, p, \lambda) = H''(J, \lambda) = H'(w_0, J_0, \lambda) \tag{4.151}$$

The single primed Hamiltonian is the Hamiltonian obtained from the original Hamiltonian by the canonical transformation from (q, p) to (w_0, J_0); whereas the double primed Hamiltonian is obtained by the canonical transformation from (q, p) to (w, J). Since both (w_0, J_0) and (w, J) are obtained from (q, p) by (different) canonical transformations, there must exist a canonical transformation connecting (w_0, J_0) and (w, J). Our goal then is to find the generator of this transformation.

The generator should reduce to the generator of an identity transformation when $\lambda = 0$. We have seen that the generator of an identity transformation is given by (4.30), viz.,

$$W_2(w_0, J, \lambda = 0) = w_0 J \tag{4.152}$$

It therefore seems advantageous to choose a type-two generator for the

$\lambda \neq 0$ case. We assume that the generator is analytic in λ and perform a Taylor expansion of the generator about $\lambda = 0$. Thus

$$W_2(w_0, J, \lambda) = W_{20}(w_0, J) + \lambda W_{21}(w_0, J) + \lambda^2 W_{22}(w_0, J) + \cdots \tag{4.153}$$

where W_{20} is given by (4.152).

According to (4.27), the transformations generated by (4.153) are

$$J_0 = \frac{\partial W_2}{\partial w_0} = J + \lambda \frac{\partial W_{21}}{\partial w_0} + \lambda^2 \frac{\partial W_{22}}{\partial w_0} + \cdots \tag{4.154}$$

$$w = \frac{\partial W_2}{\partial J} = w_0 + \lambda \frac{\partial W_{12}}{\partial J} + \lambda^2 \frac{\partial W_{22}}{\partial J} + \cdots \tag{4.155}$$

Similarly, we expand $H'(w_0, J_0, \lambda)$

$$H'(w_0, J_0, \lambda) = H_0'(J_0) + \lambda H_1'(w_0, J_0) + \lambda^2 H_2'(w_0, J_0) + \cdots \tag{4.156}$$

where the coefficients of the expansion H_0', H_1', ... are known since H' is known. Likewise

$$H''(J, \lambda) = H_0''(J) + \lambda H_1''(J) + \lambda^2 H_2''(J) + \cdots \tag{4.157}$$

While the coefficients H_0'', H_1'', H_2'', ... are not known, they may be obtained from (4.156) by equating like powers of λ provided we first express H_0', H_1', ... in terms of J rather than J_0. This is accomplished by expanding $H'(w_0, J_0, \lambda)$ about $J_0 = J$ and inserting (4.154). Thus

$$\begin{aligned}
H'(w_0, J_0, \lambda)& \\
&= H'(w_0, J, \lambda) + \left.\frac{\partial H'}{\partial J_0}\right|_{J_0=J} (J_0 - J) + \frac{1}{2} \left.\frac{\partial^2 H'}{\partial J_0^2}\right|_{J_0=J} (J_0 - J)^2 + \cdots \\
&= H'(w_0, J, \lambda) + \left.\frac{\partial H'}{\partial J_0}\right|_{J_0=J} \left[\lambda \frac{\partial W_{21}}{\partial w_0} + \lambda^2\left(\frac{\partial W_{22}}{\partial w_0}\right) + \cdots\right] \\
&\quad + \frac{1}{2} \left.\frac{\partial^2 H'}{\partial J_0^2}\right|_{J_0=J} \left[\lambda^2\left(\frac{\partial W_{21}}{\partial w_0}\right)^2 + 2\lambda^3\left(\frac{\partial W_{21}}{\partial w_0} \frac{\partial W_{22}}{\partial w_0}\right) + \cdots\right] \\
&= H'(w_0, J, \lambda) + \lambda \left.\frac{\partial H'}{\partial J_0}\right|_{J_0=J} \frac{\partial W_{21}}{\partial w_0} \\
&\quad + \lambda^2\left[\left.\frac{\partial H'}{\partial J_0}\right|_{J_0=J} \frac{\partial W_{22}}{\partial w_0} + \frac{1}{2} \left.\frac{\partial^2 H'}{\partial J_0^2}\right|_{J_0=J}\left(\frac{\partial W_{21}}{\partial w_0}\right)^2\right] + \cdots
\end{aligned} \tag{4.158}$$

Inserting (4.156) evaluated at $J_0 = J$, we find

$$H'(w_0, J_0, \lambda) = H_0'(J) + \lambda H_1'(w_0, J) + \lambda^2 H_2'(w_0, J) + \cdots$$
$$+ \lambda \frac{\partial W_{21}}{\partial w_0}\left[\frac{\partial H_0'}{\partial J_0} + \lambda \frac{\partial H_1'}{\partial J_0} + \cdots\right]_{J_0=J}$$
$$+ \lambda^2 \left\{ \frac{\partial W_{22}}{\partial w_0}\left[\frac{\partial H_0'}{\partial J_0} + \lambda \frac{\partial H_1'}{\partial J_0} + \cdots\right]_{J_0=J}\right.$$
$$\left. + \frac{1}{2}\left(\frac{\partial W_{21}}{\partial w_0}\right)^2 \left[\frac{\partial^2 H_0'}{\partial J_0^2} + \lambda \frac{\partial^2 H_1'}{\partial J_0^2} + \cdots\right]_{J_0=J}\right\} + \cdots \tag{4.159}$$

Because of (4.151) we can equate the coefficients of like powers of λ in (4.157) and (4.159), thereby obtaining

$$H_0''(J) = H_0'(J) \tag{4.160}$$

$$H_1''(J) = H_1'(w_0, J) + \left.\frac{\partial H_0'}{\partial J_0}\right|_{J_0=J} \frac{\partial W_{21}}{\partial w_0}$$
$$= H_1'(w_0, J) + \nu_0(J) \frac{\partial W_{21}}{\partial w_0} \tag{4.161}$$

$$H_2''(J) = H_2'(w_0, J) + \left.\frac{\partial H_1'}{\partial J_0}\right|_{J_0=J} \frac{\partial W_{21}}{\partial w_0} + \left.\frac{\partial H_0'}{\partial J_0}\right|_{J_0=J} \frac{\partial W_{22}}{\partial w_0}$$
$$+ \frac{1}{2} \left.\frac{\partial^2 H_0'}{\partial J_0^2}\right|_{J_0=J} \left(\frac{\partial W_{21}}{\partial w_0}\right)^2 \tag{4.162}$$
$$\vdots$$

Recall now that (4.121) demonstrated that the angle variable changes by one when the coordinate q goes through one cycle. Or to put it another way, q and p go through one cycle when w_0 changes by one. Therefore, q and p are periodic functions of w_0 with a period of one. But since w, J, and J_0 are functions of q and p, we conclude that $w - w_0$, J, J_0 are periodic functions of w_0 with a period of one. Thus it follows from (4.155) that $\partial W_{2j}/\partial J$ is periodic in w_0 and from (4.154) that $\partial W_{2j}/\partial w_0$ is periodic in w_0. Therefore, we expand the generator in a Fourier series in w_0.

$$W_{2j}(w_0, J) = \sum_{n=-\infty}^{\infty} W_{2jn}(J) \exp(2\pi i n w_0) \; j \neq 0 \tag{4.163}$$

It follows from (4.163) that

$$\int_0^1 \frac{\partial W_{2j}}{\partial w_0}\, dw_0 = 0$$

because the constant term in the Fourier expansion of $\partial W_{2j}/\partial w_0$ is absent and the complex exponential terms integrate to zero over one period.

Multiplying (4.161) by dw_0 and integrating from 0 to 1, we obtain (remember that H_1'' is not a function of w_0)

$$H_1''(J) = \langle H_1' \rangle \tag{4.164}$$

where the symbol $\langle\ \rangle$ represents

$$\langle H_1' \rangle = \int_0^1 H_1' \, dw_0$$

Substituting (4.164) back into (4.161) we find that

$$\frac{\partial W_{21}}{\partial w_0} = \frac{\langle H_1' \rangle - H_1'}{\nu_0} \tag{4.165}$$

Similarly, from (4.161) and (4.165) we find

$$\begin{aligned}\frac{\partial W_{22}}{\partial w_0} = \frac{1}{\nu_0}(\langle H_2' \rangle - H_2') &+ \frac{1}{\nu_0^2}\left\{\left\langle \frac{\partial H_1'}{\partial J_0}\bigg|_{J_0=J}\right\rangle \langle H_1' \rangle\right. \\ &- \left\langle \frac{\partial H_1'}{\partial J_0} H_1' \bigg|_{J_0=J}\right\rangle - \frac{\partial H_1'}{\partial J_0}\bigg|_{J_0=J} \langle H_1' \rangle + \left.\frac{\partial H_1'}{\partial J_0}\bigg|_{J_0=J} H_1'\right\} \\ &+ \frac{1}{2\nu_i^3} \frac{\partial^2 H_0'}{\partial J_0^2}\bigg|_{J_0=J} \{\langle H_1'^2 \rangle - 2\langle H_1' \rangle^2 + 2\langle H_1' \rangle H_1' - H_1'^2\}\end{aligned} \tag{4.166}$$

From Eqs. (4.165) and (4.166) and similar equations the coefficients of the generator expansion (4.153) may be determined, and then the perturbed action-angle variables may be found in terms of the unperturbed action-angle variables by using (4.154) and (4.155).

The perturbed Hamiltonian (i.e., the perturbed energy) is obtained by substituting (4.165), (4.166), and (4.161) into (4.157).

$$\begin{aligned}H''(J, \lambda) = H_0'(J) &+ \lambda \langle H_1' \rangle \\ &+ \lambda^2 \left\{ \langle H_2' \rangle + \frac{1}{\nu_0}\left(\left\langle \frac{\partial H_1'}{\partial J_0} \right\rangle \langle H_1' \rangle - \left\langle \frac{\partial H_1'}{\partial J_0} H_1' \right\rangle \right)_{J_0=J} \right. \\ &+ \left. \frac{1}{2\nu_0^2} \frac{\partial^2 H_0'}{\partial J_0^2}\bigg|_{J_0=J} (\langle H_1'^2 \rangle - \langle H_1' \rangle^2 \right\} + \cdots\end{aligned} \tag{4.167}$$

This equation is then used to obtain the frequency of the perturbed system, i.e.,

$$\nu = \partial H''/\partial J \tag{4.168}$$

The extension of this method to periodic systems having more than one degree of freedom is straightforward provided that the system is not degenerate (see Ter Haar [1]). Canonical perturbation theory is exhibited in Example 4.14.

4.3. SOLVED EXAMPLES

Example 4.1. Determine the Hamilton equations of motion of a mass m moving under the influence of gravity near the surface of the earth.

Solution. The Lagrangian for this system has been obtained in Example 2.1 and is

$$L = \tfrac{1}{2}m(\dot{x}^2 + \dot{y}^2 + \dot{z}^2) - mgy$$

The momenta are

$$p_x = \frac{\partial L}{\partial \dot{x}} = m\dot{x}, \qquad p_y = \frac{\partial L}{\partial \dot{y}} = m\dot{y}, \qquad p_z = \frac{\partial L}{\partial \dot{z}} = m\dot{z}$$

Thus the Hamiltonian is obtained from

$$\begin{aligned} H &= p_x\dot{x} + p_y\dot{y} + p_z\dot{z} - \tfrac{1}{2}m(\dot{x}^2 + \dot{y}^2 + \dot{z}^2) + mgy \\ &= (1/2m)(p_x^2 + p_y^2 + p_z^2) + mgy \end{aligned} \tag{4.169}$$

In this case (4.169) could have been written down immediately by recourse to (4.10); i.e., the Hamiltonian is the total energy of the system. The equations of motion are

$$\dot{x} = \frac{\partial H}{\partial p_x} = \frac{p_x}{m}, \qquad \dot{p}_x = -\frac{\partial H}{\partial x} = 0$$

$$\dot{y} = \frac{\partial H}{\partial p_y} = \frac{p_y}{m}, \qquad \dot{p}_y = -\frac{\partial H}{\partial y} = -mg$$

$$\dot{z} = \frac{\partial H}{\partial p_z} = \frac{p_z}{m}, \qquad \dot{p}_z = -\frac{\partial H}{\partial z} = 0$$

Example 4.2. A particle of mass m is moving under the influence of a potential V near the surface of the earth. Find the Hamilton equations of motion valid for an observer on the surface of the rotating earth.

Solution. The kinetic energy is presented in Example 2.19. The conjugate momenta are

$$p_x' = \frac{\partial L}{\partial \dot{x}'} = m[\dot{x}' + \dot{\varphi}_0(R \sin \theta_0 - y' \cos \theta_0 + z' \sin \theta_0)]$$

$$p_y' = \frac{\partial L}{\partial \dot{y}'} = m[\dot{y}' + \dot{\varphi}_0 x' \cos \theta_0]$$

$$p_z' = \frac{\partial L}{\partial \dot{z}'} = m[\dot{z}' - \dot{\varphi}_0 x' \sin \theta_0]$$

Using these equations we eliminate the velocities from the defining expression for the Hamiltonian to obtain

$$H' = (p_x'^2 + p_y'^2 + p_z'^2)/2m - \dot{\varphi}_0[p_x'(R \sin \theta_0 - y' \cos \theta_0 + z' \sin \theta_0) + x'(p_y' \cos \theta_0 - p_z' \sin \theta_0)] + V \tag{4.170}$$

The equations of motion are

$$\dot{x}' = \frac{p_x'}{m} - \dot{\varphi}_0(R \sin \theta_0 - y' \cos \theta_0 + z' \sin \theta_0),$$

$$\dot{p}_x = \dot{\varphi}_0(p_y' \cos \theta_0 - p_z' \sin \theta_0) - \frac{\partial V}{\partial x'}$$

$$\dot{y}' = \frac{p_y'}{m} - \dot{\varphi}_0 x' \cos \theta_0 .$$

$$\dot{p}_y = -\dot{\varphi}_0 p_x \cos \theta_0 - \frac{\partial V}{\partial y'}$$

$$\dot{z}' = \frac{p_z'}{m} + \dot{\varphi}_0 x' \sin \theta_0$$

$$\dot{p}_z = \dot{\varphi}_0 p_x \sin \theta_0 - \frac{\partial V}{\partial z'}$$

Does our expression for H' agree with that obtained by writing $H' = T + V$? Why?

Example 4.3. Find the Hamilton equations of motion of a bead of mass m sliding on a frictionless wire under the influence of gravity. The wire has a parabolic shape and rotates with a constant angular velocity ω.

Solution A. The diagram and Lagrangian of this system is presented in Example 3.4. The constraints reduce the system to one degree of freedom ϱ whose conjugate momentum is

$$p = \frac{\partial L}{\partial \dot{\varrho}} = m(1 + 4a^2\varrho^2)\dot{\varrho}$$

The Hamiltonian of the system is

$$H = p\dot{\varrho} - L = p^2/[2m(1 + 4a^2\varrho^2)] - \tfrac{1}{2}m\varrho^2(\omega^2 - 2ga)$$

and the equations of motion are

$$\dot{\varrho} = \frac{\partial H}{\partial p} = \frac{p}{m(1 + 4a^2\varrho^2)}$$

$$\dot{p} = -\frac{\partial H}{\partial \varrho} = \frac{4a^2\varrho p^2}{m(1 + 4a^2\varrho^2)^2} + m(\omega^2 - 2ga)\varrho$$

Solution B. This problem may also be solved by the Dirac method. The parabolic constraint is expressed by

$$\varphi = z - a\varrho^2$$

and the Lagrangian is (now z and ϱ are the independent variables)

$$L = \tfrac{1}{2}m(\dot{\varrho}^2 + \varrho^2\omega^2 + \dot{z}_0) - mgz$$

It follows that H_T is

$$H_T = (p_z{}^2/2m) + (p_\varrho{}^2/2m) - \tfrac{1}{2}m\varrho^2\omega^2 + mgz + \lambda(z - a\varrho^2)$$

where

$$p_z = \partial L/\partial \dot{z} = m\dot{z}, \qquad p_\varrho = \partial L/\partial \dot{\varrho} = m\dot{\varrho}$$

Let us first consider the primary constraint:

$$\dot{\varphi} = [\varphi, H_T] = [\varphi, H] + \lambda[\varphi, \varphi] = p_z/m - (2a\varrho/m)p_\varrho = 0$$

Thus the primary constraint is of Class B type leading to the secondary constraint:

$$\chi = p_z - 2a\varrho p_\varrho$$

The secondary constraint must satisfy

$$\begin{aligned}\dot{\chi} &= [\chi, H_T] = [\chi, H] + \lambda[\chi, \varphi] \\ &= -(mg + 2ma\omega^2\varrho^2 + (2a/m)p_\varrho^{\,2}) - (1 + 4a^2\varrho^2)\lambda = 0\end{aligned}$$

Thus the secondary constraint can be satisfied only if

$$\lambda = \frac{-(mg + 2ma\omega^2\varrho^2 + (2a/m)p_\varrho^{\,2})}{1 + 4a^2\varrho^2}$$

and therefore it is a type C constraint. It is important to note that $\lambda = 0$ is the solution to the homogeneous equation since $(1 + 4a^2\varrho^2)$ is not zero and from this it follows that arbitrary functions will not appear in the dynamical equations. The equations of motion are

$$\dot{p}_\varrho = \frac{m\varrho\omega^2 - 2ma\varrho g - (4a^2/m)\varrho p_\varrho^{\,2})}{1 + 4a^2\varrho^2}$$

$$\dot{p}_z = \frac{(2a/m)p_\varrho^{\,2} - ma\omega^2\varrho^2 - 4a^2\varrho^2 mg}{1 + 4a^2\varrho^2}$$

Solution B dynamical equations differ from solution A equations because $p_\varrho \neq p$. It is not difficult, however, to show that both solutions lead to the $\ddot{\varrho}$ solution obtained in Example 2.4.

Example 4.4. Determine the Hamilton equations of motion of a spinning symmetrical top moving under the influence of gravity with the bottom of the top constrained to remain at the origin.

Solution. The diagram, kinetic energy, and potential energy for this problem are presented in Example 2.8. The conjugate momenta are

$$p_\theta = \frac{\partial L}{\partial \dot{\theta}} = I_x\dot{\theta}, \qquad p_\psi = \frac{\partial L}{\partial \dot{\psi}} = I_z(\dot{\psi} + \dot{\varphi}\cos\theta)$$

$$p_\varphi = \frac{\partial L}{\partial \dot{\varphi}} = I_x\dot{\varphi}\sin^2\theta + I_z(\dot{\psi} + \dot{\varphi}\cos\theta)\cos\theta$$

Using these equations to eliminate the velocities from the defining equation of the Hamiltonian we find

$$H = \frac{1}{2}\frac{p_\theta^{\,2}}{I} + \frac{1}{2}\frac{p_\psi^{\,2}}{I_z} + \frac{1}{2I_x}\left(\frac{p_\varphi - p_\psi\cos\theta}{\sin\theta}\right)^2 + mgR\cos\theta$$

The dynamical equations are

$$\dot{p}_\varphi = -\frac{\partial H}{\partial \varphi} = 0, \qquad \dot{p}_\psi = -\frac{\partial H}{\partial \psi} = 0$$

$$p_\theta = -\frac{\partial H}{\partial \theta}$$
$$= mgR \sin\theta - \frac{1}{I_x \sin^3\theta} [p_\varphi p_\psi(1 + \cos^2\theta) - (p_\varphi{}^2 + p_\psi{}^2)\cos\theta]$$

As in Example 2.8 we find that the momenta p_φ and p_ψ are conserved since the Hamiltonian is cyclic in φ and ψ. It is not difficult to show that the $\dot{p}_\theta$ equation reduces to the $\ddot{\theta}$ equation obtained in Example 2.8.

Example 4.5. Verify that the transformation

$$P = q \cot p, \qquad Q = \ln(q^{-1} \sin p)$$

is a canonical transformation.

Solution. To be a canonical transformation it is necessary that if

$$\dot{P} = -\partial H/\partial Q, \qquad \dot{Q} = \partial H/\partial P$$

then

$$\dot{p} = -\partial H/\partial q, \qquad \dot{q} = \partial H/\partial q$$

Thus

$$\dot{P} = \frac{\partial P}{\partial q}\dot{q} + \frac{\partial P}{\partial p}\dot{p} = -\frac{\partial H}{\partial Q} = -\frac{\partial q}{\partial Q}\frac{\partial H}{\partial q} - \frac{\partial p}{\partial Q}\frac{\partial H}{\partial p}$$

$$\dot{Q} = \frac{\partial Q}{\partial q}\dot{q} + \frac{\partial Q}{\partial p}\dot{p} = \frac{\partial H}{\partial P} = \frac{\partial q}{\partial P}\frac{\partial H}{\partial q} + \frac{\partial p}{\partial P}\frac{\partial H}{\partial p}$$

It follows from the transformation equations that

$$\frac{\partial P}{\partial q} = \cot p, \qquad \frac{\partial P}{\partial p} = -q\csc^2 p, \qquad \frac{\partial Q}{\partial q} = -q^{-1}, \qquad \frac{\partial Q}{\partial p} = \cot p$$

$$\frac{\partial q}{\partial Q} = -q\csc^2 p, \qquad \frac{\partial p}{\partial Q} = \cot p, \qquad \frac{\partial q}{\partial P} = -\cot p, \qquad \frac{\partial p}{\partial P} = -q^{-1}$$

The equations for $\dot{P}$ and $\dot{Q}$ become

$$\dot{q}\cot p - q\dot{p}\csc^2 p = q\csc^2 p\,\frac{\partial H}{\partial q} + \cot p\,\frac{\partial H}{\partial p}$$

$$-q^{-1}\dot{q} + \dot{p}\cot p = -\cot p\,\frac{\partial H}{\partial q} - q^{-1}\,\frac{\partial H}{\partial p}$$

Multiplying the first equation by q^{-1} and the second by $\cot p$ and adding, we obtain

$$-\dot{p} = \partial H/\partial q$$

and substituting this into the first equation it follows that

$$\dot{q} = \partial H/\partial p$$

Thus the transformation from the canonical coordinates (P, Q) to (p, q) is canonical.

Example 4.6. Determine the canonical transformations from the canonical coordinates (p, q) to the coordinates (P, Q) generated by the generator

$$W_3 = -(e^Q - 1)^2 \tan p$$

Solution. We employ Eqs. (4.28a) and (4.28b) to obtain the canonical transformations,

$$q = -\partial W_3/\partial p = (e^Q - 1)^2 \sec^2 p$$

$$P = -\partial W_3/\partial Q = 2(e^Q - 1)e^Q \tan p$$

From the first equation we find

$$Q = \ln(1 + q^{1/2}\cos p)$$

and from the second equation

$$P = 2(1 + q^{1/2}\cos p)\sin p$$

Example 4.7. Determine the Poisson brackets of the angular momentum of a particle $\mathbf{L} = \mathbf{r} \times \mathbf{p}$ and its linear momentum $\mathbf{p}$.

Solution. The angular momentum may be written

$$L_i = \varepsilon_{ijk} r_j p_k$$

where ε_{ijk} is the Levi–Civita tensor. Thus

$$[p_l, L_i] = \frac{\partial p_l}{\partial r_m}\frac{\partial L_i}{\partial p_m} - \frac{\partial p_l}{\partial p_m}\frac{\partial L_i}{\partial r_m} = -\delta_{lm}\varepsilon_{ijk}\,\delta_{jm}p_k = \varepsilon_{lik}p_k$$

It follows from the properties of the Levi–Civita tensor that the brackets are zero when $l = i$. The nonzero brackets are

$$[p_1, L_2] = p_3, \qquad [p_2, L_3] = p_1, \qquad [p_3, L_1] = p_2$$

Example 4.8. A particle of mass m moves under the influence of a cylindrically symmetric potential. (a) Determine the Hamilton equations of motion relative to axes which are rotating uniformly about the symmetry axis with an angular velocity ω. (b) Determine the generating function of the canonical transformation to the rotating axes.

Solution. (a) We shall obtain the equations of motion in both Cartesian and cylindrical coordinates. The kinetic energy and Hamiltonian relative to the inertial unprimed system are simply

$$T = \tfrac{1}{2}m(\dot{x}^2 + \dot{y}^2 + \dot{z}^2), \qquad H = (1/2m)(p_x{}^2 + p_y{}^2 + p_z{}^2) + V(\varrho, z)$$

The equations relating the primed and unprimed coordinates are easily obtained employing the inverse of the rotation matrix

$$\begin{bmatrix} x \\ y \\ z \end{bmatrix} = \begin{bmatrix} \cos\omega t & -\sin\omega t & 0 \\ \sin\omega t & \cos\omega t & 0 \\ 0 & 0 & 1 \end{bmatrix} \begin{bmatrix} x' \\ y' \\ z' \end{bmatrix} \tag{4.171}$$

The kinetic energy in primed cartesian coordinates is

$$T = \tfrac{1}{2}m[\dot{x}'^2 + \dot{y}'^2 + \dot{z}'^2 + \omega^2(x'^2 + y'^2) + 2\omega(\dot{y}'x' - \dot{x}'y')]$$

The Hamiltonian in primed Cartesian coordinates is

$$H' = (1/2m)(p_x'^2 + p_y'^2 + p_z'^2) - \omega(x'p_y' - y'p_x') + V(x', y', z')$$

The Hamiltonian H' is a constant of the motion since it does not depend explicitly on the time. It is not, however, the total energy of the system;

this could have been anticipated by the fact that the transformation equations connecting the primed and unprimed coordinates depends explicitly on the time. The second term in H' is just the energy associated with the apparent rotation of the particle when observed from the noninertial rotating system. The dynamical equations are

$$\dot{p}_x' = -\frac{\partial H'}{\partial x'} = \omega p_y' - \frac{\partial V}{\partial x'}$$

$$\dot{p}_y' = -\frac{\partial H'}{\partial y'} = -\omega p_x' - \frac{\partial V}{\partial y'}$$

$$\dot{p}_z = -\frac{\partial V}{\partial z'}$$

To express the Hamiltonian in cylindrical coordinates we use the transformation equations

$$x' = \varrho' \cos \varphi', \qquad y' = \varrho' \sin \varphi'$$

$$\dot{x}' = \dot{\varrho}' \cos \varphi' - \varrho'\dot{\varphi}' \sin \varphi', \qquad \dot{y}' = \dot{\varrho}' \sin \varphi' + \varrho'\dot{\varphi}' \cos \varphi'$$

The momenta conjugate to the cylindrical coordinates are (since the potential is independent of the velocities)

$$p_\varrho' = \frac{\partial T}{\partial \dot{\varrho}'} = \frac{\partial T}{\partial \dot{x}'}\frac{\partial \dot{x}'}{\partial \dot{\varrho}'} + \frac{\partial T}{\partial \dot{y}'}\frac{\partial \dot{y}'}{\partial \dot{\varrho}'} = p_x' \cos \varphi' + p_y' \sin \varphi'$$

$$p_\varphi' = \frac{\partial T}{\partial \dot{\varphi}'} = \frac{\partial T}{\partial \dot{x}'}\frac{\partial \dot{x}'}{\partial \dot{\varphi}'} + \frac{\partial T}{\partial \dot{y}'}\frac{\partial \dot{y}'}{\partial \dot{\varphi}'} = -p_x'\varrho' \sin \varphi' + p_y'\varrho' \cos \varphi'$$

or

$$p_y' = p_\varrho' \sin \varphi' + (1/\varrho')p_\varphi' \cos \varphi', \qquad p_x' = p_\varrho' \cos \varphi' - (1/\varrho')p_\varphi' \sin \varphi'$$

Thus H' becomes

$$H' = (1/2m)(p_\varrho'^2 + (1/\varrho'^2)p_\varphi'^2 + p_z'^2) - \omega p_\varphi' + V(\varrho', z')$$

and the dynamical equations are

$$\dot{p}_\varrho' = -\frac{\partial H'}{\partial \varrho'} = \frac{1}{m\varrho'^3}\, p_\varphi'^2 - \frac{\partial V}{\partial \varrho'}, \qquad \dot{p}_z' = -\frac{\partial V}{\partial z'}, \qquad \dot{p}_\varphi' = 0$$

$$\dot{\varrho} = \frac{\partial H}{\partial p_\varrho'} = \frac{p_\varrho'}{m}, \qquad \dot{\varphi} = \frac{\partial H}{\partial p_\varphi'} = \frac{p_\varphi'}{m\varrho'^2} - \omega, \qquad \dot{z}' = \frac{p_z'}{m}$$

That the angular momentum p_φ' is a constant of the motion follows from the cylindrical symmetry of the potential, i.e., H' is cyclic in φ.

(b) The new Hamiltonian H' may also be obtained from the generator of the canonical transformation connecting the primed and unprimed coordinates. Consider the generator $W_2 = W_2(x, y, z, p_x', p_y', p_z')$. Then from (4.27b)

$$x' = \frac{\partial W_2}{\partial p_x'} = x \cos \omega t + y \sin \omega t \tag{4.172a}$$

$$y' = \frac{\partial W_2}{\partial p_y'} = -x \sin \omega t + y \cos \omega t \tag{4.172b}$$

where the right side of the equation follows from the inverse of (4.171). It follows from (4.172) and $z' = z$ that the generator is

$$W_2 = p_x'(x \cos \omega t + y \sin \omega t) + p_y'(y \cos \omega t - x \sin \omega t) + p_z'z \tag{4.173}$$

Now using (4.173) in (4.27c) we find

$$H' = H + \frac{\partial W_2}{\partial t} = H - \omega(x'p_y' - y'p_x')$$

But from (4.27a)

$$p_x = \frac{\partial W_2}{\partial x} = p_x' \cos \omega t - p_y' \sin \omega t, \qquad p_z = p_z'$$

$$p_y = \frac{\partial W_2}{\partial y} = p_x' \sin \omega t + p_y' \cos \omega t$$

and therefore

$$(1/2m)(p_x^2 + p_y^2 + p_z^2) = (1/2m)(p_x'^2 + p_y'^2 + p_z'^2)$$

Thus the H' obtained by this technique agrees with that obtained in (a).

Example 4.9. Determine whether the angular momentum of the particle in Example 4.2 is a constant of the motion when it is defined relative to inertial coordinates at the center of the earth, i.e., $\mathbf{L} = \boldsymbol{\rho} \times \mathbf{p}$, and when it is defined relative to an observer on the surface of the rotating earth, i.e., $\mathbf{L}' = \mathbf{r}' \times \mathbf{p}'$. (a) Assume a central force potential with the center of the force located at the center of the earth. (b) Assume a potential of the form

$V(\varrho) + V_1(\varrho_1)$, where ϱ is the distance of the particle from the center of the earth and ϱ_1 is the distance of the particle from a second source of interaction which is located at $\mathbf{R}'$ relative to the center of the earth. Consider only $\mathbf{L}$.

Solution. It is advantageous to write the angular momenta in index form

$$L_i = \varepsilon_{ijk} x_j p_k, \qquad L_i' = \varepsilon_{ijk} x_j' p_k'$$

L_i and L_i' are constants of the motion if they satisfy

$$[L_i, H] = 0, \qquad [L_i', H'] = 0$$

(a) The Hamiltonian H expressed in terms of the inertial system at the center of the earth is

$$H = p_l^2/2m + V(\varrho)$$

$$[L_i, H] = [\varepsilon_{ijk} x_j p_k, H] = \varepsilon_{ijk}\left(\frac{\partial(x_j p_k)}{\partial x_m}\frac{\partial H}{\partial p_m} - \frac{\partial(x_j p_k)}{\partial p_m}\frac{\partial H}{\partial x_m}\right)$$

$$= \frac{1}{m}\varepsilon_{ijk}\left(p_k\,\delta_{jm} p_l\,\delta_{lm} - m x_j\,\delta_{km}\frac{\partial V}{\partial \varrho}\frac{x_m}{\varrho}\right)$$

$$= \frac{1}{m}\varepsilon_{ijk}\left(p_k p_j - \frac{m}{\varrho}\frac{\partial V}{\partial \varrho}x_j x_k\right) = 0$$

That the bracket is zero follows from the fact that the term in the parenthesis is symmetric in (jk), whereas ε_{ijk} is antisymmetric in (jk). Thus $\mathbf{L}$ is a constant of the motion.

Now when the system is described relative to an observer on the surface of the rotating earth the Hamiltonian of (4.172) applies. If the "angular momentum" is defined by this observer to be

$$L_i' = \varepsilon_{ijk} x_j' p_k'$$

the Poisson bracket of L_i' with H' is

$$[L_i', H'] = \varepsilon_{ijk}\left(\frac{\partial(x_j' p_k')}{\partial x_m'}\frac{\partial H'}{\partial p_m'} - \frac{\partial(x_j' p_k')}{\partial p_m'}\frac{\partial H'}{\partial x_m'}\right) \neq 0$$

(b) The Hamiltonian is now

$$H = p_l^2/2m + V(\varrho) + V_1(\varrho_1)$$

But

$$\boldsymbol{\rho}_1 = \boldsymbol{\rho} - \mathbf{R}', \qquad \varrho_1[(x_l - R_l')^2]^{1/2}$$

Therefore,

$$\frac{\partial H}{\partial x_m} = \frac{\partial V}{\partial x_m} + \frac{\partial V_1}{\partial x_m} = \frac{\partial V}{\partial \varrho}\frac{\partial \varrho}{\partial x_m} + \frac{\partial V_1}{\partial \varrho_1}\frac{\partial \varrho_1}{\partial x_m}$$

$$= \frac{\partial V}{\partial \varrho}\frac{x_m}{\varrho} + \frac{\partial V_1}{\partial \varrho_1}\frac{(x_m - R_m')}{\varrho_1}$$

To determine whether $\mathbf{L}$ is a constant of the motion we evaluate the Poisson bracket

$$[L_i, H] = \varepsilon_{ijk}\left(\frac{\partial(x_j p_k)}{\partial x_m}\frac{\partial H}{\partial p_m} - \frac{\partial(x_j p_k)}{\partial p_m}\frac{\partial H}{\partial x_m}\right)$$

$$= \varepsilon_{ijk}\left(\frac{p_j p_k}{m} - x_j\frac{\partial V}{\partial \varrho}\frac{x_k}{\varrho} - x_j\frac{\partial V_1}{\partial \varrho_1}\frac{x_k - R_k'}{\varrho_1}\right)$$

$$= \varepsilon_{ijk}\frac{\partial V_1}{\partial \varrho_1}\frac{x_j R_k'}{\varrho_1}$$

or

$$[\mathbf{L}, H] = \frac{1}{\varrho_1}\frac{\partial V_1}{\partial \varrho_1}\boldsymbol{\rho}\times\mathbf{R}'$$

The angular momentum $\mathbf{L}$ is conserved only for those source-particle orientations for which $\boldsymbol{\rho}\times\mathbf{R}' = 0$.

$$\left(\frac{dS_\varphi}{d\varphi}\right)^2 + 2mf_3(\varphi) = Q_1^2$$

$$r^2\left(\frac{dS_r}{dr}\right)^2 + 2mr^2 f_1 - 2m\alpha_1 r^2 + \left(\frac{dS_\theta}{d\theta}\right)^2 + 2mf_2 = -\frac{a_1^2}{\sin^2\theta}$$

where a_1^2 is the *separation constant*. Similarly, the second equation can be separated into

$$\left(\frac{dS_\theta}{d\theta}\right)^2 + 2mf_2(\theta) + \frac{a_1^2}{\sin^2\theta} = a_2^2$$

$$\left(\frac{dS_r}{dr}\right)^2 + 2mf_1(r) - 2m\alpha_1 + \frac{a_2^2}{r^2} = 0$$

where a_2^2 is the separation constant. Thus we have shown that the Hamilton–Jacobi equation is completely separable under the conditions of this problem.

Notice that each of the separated equations corresponds to a conservation law of the motion; e.g., if $f_3 = 0$, it follows from (4.67a) that the φ equation simply expresses the conservation of the angular momentum p_φ.

Example 4.10. Show the Hamilton–Jacobi equation of a particle moving under the influence of a potential V is separable when it is expressed in terms of spherical coordinates and the potential is of the form

$$V = f_1(r) + f_2(\theta)/r^2 + f_3(\varphi)/r^2 \sin^2\theta$$

where f_1, f_2, and f_3 are arbitrary functions of r, θ, and φ, respectively.

Solution. The kinetic energy of a particle in spherical polar coordinates is presented in (2.58). Using this expression we find for the Hamiltonian

$$H = \frac{1}{2m}\left(p_r{}^2 + \frac{1}{r^2} p_\theta{}^2 + \frac{1}{r^2 \sin^2\theta} p_\varphi{}^2\right)$$
$$+ f_1(r) + \frac{1}{r} f_2(\theta) + \frac{1}{r^2 \sin^2\theta} f_3(\theta)$$

Since the Hamiltonian is not an explicit function of the time, we may use the time separated form of the Hamilton–Jacobi equation, viz., (4.73):

$$\frac{1}{2m}\left[\left(\frac{\partial S_1}{\partial r}\right)^2 + \frac{1}{r^2}\left(\frac{\partial S_1}{\partial \theta}\right)^2 + \frac{1}{r^2 \sin^2\theta}\left(\frac{\partial S_1}{\partial \varphi}\right)^2\right]$$
$$+ f_1(r) + \frac{1}{r^2} f_2(\theta) + \frac{1}{r^2 \sin^2\theta} f_3(\theta) = \alpha_1$$

Let us assume a separation of the form

$$S_1 = S_r(r) + S_\theta(\theta) + S_\varphi(\varphi)$$

and substitute this into the Hamilton–Jacobi equation. We obtain

$$\sin^2\theta\left[r^2\left(\frac{\partial S_r}{\partial r}\right)^2 + \left(\frac{\partial S_\theta}{\partial \theta}\right)^2 + 2mr^2 f_1(r) + 2mf_2(\theta) - 2m\alpha_1 r^2\right]$$
$$+ \left[\left(\frac{\partial S_\varphi}{\partial \varphi}\right)^2 + 2mf_3(\varphi)\right] = 0$$

The first term is independent of φ, whereas the second term depends only on φ. We conclude that the equation can be satisfied only if

$$\left(\frac{dS_\varphi}{d\varphi}\right)^2 + 2mf_3 = a_1^2,$$

$$\left(\frac{dS_\theta}{d\theta}\right)^2 + 2mf_2 - \frac{a_1^2}{\sin^2\theta} = a_2^2,$$

$$\left(\frac{dS_r}{dr}\right)^2 + 2mf_1 - 2m\alpha_1 = -\frac{a_2^2}{r^2}$$

where a_1^2 and a_2^2 are separation constants.

Example 4.11. Apply the results of the previous problem to the bound orbit of a particle moving under the influence of an inverse square law force to determine the fundamental frequencies of the system.

Solution. A discussion of the conditions necessary for a bound orbit is presented by Pars [11]. Here we proceed on the assumption that the conditions necessary for a Keplerian bound orbit hold. The motion is then periodic and it is advantageous to introduce action-angle variables. From the previous problem we obtain the separated equations

$$\left(\frac{dS_\varphi}{d\varphi}\right)^2 = a_1^2 \tag{4.174a}$$

$$\left(\frac{dS_\theta}{d\theta}\right)^2 + \frac{a_1^2}{\sin^2\theta} - a_2^2 = 0 \tag{4.174b}$$

$$\left(\frac{dS_r}{dr}\right)^2 - \frac{2mk}{r} - 2m\alpha_1 + \frac{a_2^2}{r^2} = 0 \tag{4.174c}$$

where we have used

$$f_1 = -k/r, \qquad f_2 = f_3 = 0$$

Equation (4.174a) expresses the conservation of the angular momentum p_φ about the polar axis (the z axis). To interpret the conservation law contained in (4.174b) and to prepare for the introduction of action-angle variables we note that since the orbit of a particle moving under the influence of a central force is confined to a plane it is possible to describe its motion

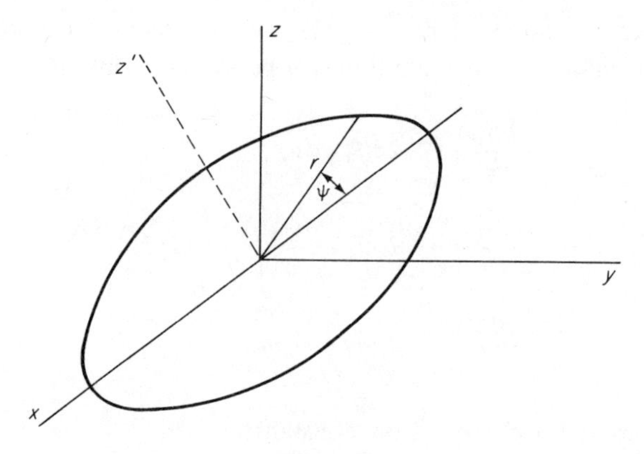

Fig. 4.2

in terms of the plane polar coordinates (r, ψ) as indicated in Fig. 4.2. In this case, it is easy to show that the Hamiltonian is

$$H = (1/2m)[p_r^2 + (1/r^2)p_\psi^2] - k/r$$

where p_ψ is the momentum conjugate to the polar angle ψ, i.e., the total angular momentum of the particle. This Hamiltonian represents the same system as the one expressed in spherical polar coordinates (see previous example) and by comparing the two we conclude

$$p_\psi^2 = p_\theta^2 + (1/\sin^2\theta)p_\varphi^2 \tag{4.175}$$

But (4.174b) may be written as

$$p_\theta^2 + p_\varphi^2/\sin^2\theta = a_2^2 \tag{4.176}$$

Comparing (4.170) and (4.175), we see that (4.170) or equivalently (4.174b), is a statement of the conservation of the total angular momentum.

The action variables corresponding to the spherical polar coordinates are

$$J_\varphi = \oint p_\varphi \, d\varphi = \oint \frac{\partial S_\varphi}{d\varphi} \, d\varphi \tag{4.177a}$$

$$J_\theta = \oint p_\theta \, d\theta = \oint \frac{\partial S_\theta}{\partial\theta} \, d\theta \tag{4.177b}$$

$$J_r = \oint p_r \, dr = \oint \frac{\partial S_r}{\partial r} \, dr \tag{4.177c}$$

To evaluate the integrals in (4.177) it is necessary to know the period of the libration. From the diagram of the orbit it is clear that as the particle performs one complete orbit, φ changes by 2π, as does ψ, but the change in θ is more complicated and not, in general, 2π. It is easy to evaluate (4.177b) if we realize that it may be expressed as integrations over φ and ψ whose periods we know. In (3.22) it was shown that if the coordinate transformation equations are not explicit functions of the time, the kinetic energy may be written

$$2T = p_a \dot{q}_a$$

In spherical coordinates

$$2T = p_r \dot{r} + p_\theta \dot{\theta} + p_\varphi \dot{\varphi} \tag{4.178}$$

In polar coordinates

$$2T = p_r \dot{r} + p_\psi \dot{\psi} \tag{4.179}$$

Comparing (4.179) and (4.178) it follows that

$$p_\theta \, d\theta = p_\psi \, d\psi - p_\varphi \, d\varphi$$

and therefore (4.177b) may be written

$$J_\theta = \oint p_\psi \, d\psi - \oint p_\varphi \, d\varphi = 2\pi(p_\psi - p_\varphi) = 2\pi(a_2 - a_1) \tag{4.180}$$

since it was previously shown that the total angular momentum p_ψ and the angular momentum about the z axis, p_φ, are constants of the motion. Equation (4.177a) is simply

$$J_\varphi = 2\pi p_\varphi = 2\pi a_1 \tag{4.181}$$

It remains to evaluate (4.177c) which with (4.181) and (4.180) may be written

$$J_r = \oint [2mEr^2 + 2mkr - (1/4\pi^2)(J_\theta + J_\varphi)^2]^{1/2} \, dr/r \tag{4.182}$$

where we have used (4.115) which states that $\alpha_1 = H = E$. The roots of the integrand correspond to the values of r for which the radial momentum p_r is zero. For a closed orbit there are two roots r_1 and r_2, the perigee and apogee of the orbit. Thus the radial cycle is characterized by r going from r_1 to r_2 and back to r_1. Thus we may write

$$J_r = 2(2mE)^{1/2} \int_{r_1}^{r_2} (1/r)[(r_2 - r)(r - r_1)]^{1/2} \, dr \tag{4.183}$$

An elegant technique for evaluating the integral was introduced by Born and consists in extending r into the complex plane. The integrand then has an interesting mathematical character; it is a multiple-valued function having the branch points p_1 and r_2 and a simple pole at the origin. In the usual fashion we insert a branch cut between r_1 and r_2 and represent the integral of (4.183) by an integration around the contour C as shown in Fig. 4.3. The branches of the integrand are chosen such that the portion of the contour above the real axis corresponds to the positive square root and that below the real axis the negative square root. Introducing the small contour C_1 enclosing the origin and the large circular contour C_2 whose radius ap-

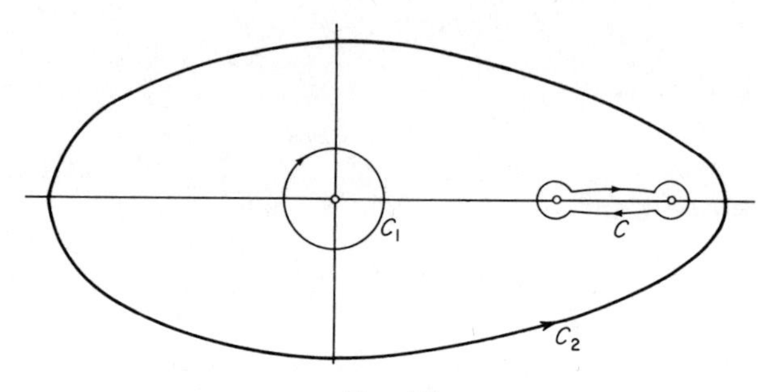

Fig. 4.3

proaches infinity, we establish the multiply connected domain bounded by the contours C, C_1, and C_2. The integrand of (4.183) is analytic within this multiply connected domain and on its boundaries and we may, therefore, apply Cauchy's theorem.

$$\oint_{C+C_1+C_2} (1/r)[(r_2 - r)(r - r_1)]^{1/2}\, dr = 0$$

or

$$\int_C (1/r)[(r_2 - r)(r - r_1)]^{1/2}\, dr = -\int_{C_1} (1/r)[(r_2 - r)(r - r_1)]^{1/2}\, dr - \int_{C_2} (1/r)[(r_2 - r)(r - r_1)]^{1/2}\, dr \quad (4.184)$$

Consider first the C_1 integration. It encloses a simple pole of order one and from the residue theorem

$$\int_{C_1} (1/r)[(r_2 - r)(r - r_1)]^{1/2}\, dr = -2\pi i \underset{r=0}{(\text{Res})} = -2\pi i[i(r_2 r_1)^{1/2}] = 2\pi(r_1 r_2)^{1/2} \quad (4.185)$$

The minus appears because of the direction of the contour.

The C_2 integration is obtained by introducing the transformation $r = 1/t$. Then we must evaluate

$$-\int_{C_2} (1/t^2)[(r_2 t - 1)(1 - r_1 t)]^{1/2}\, dt$$

This integral has a pole of order two at $t = 0$, i.e., at $r = \infty$. Using the formula for a residue of order two,

$$\text{Res} = \lim_{t \to 0}\left(\frac{d}{dt}\,[(r_2 t - 1)(1 - r_1 t)]^{1/2}\right)$$

we obtain

$$\int_{C_2} (1/r)[(r_2 - r)(r - r_1)]^{1/2}\, dr = -\pi(r_1 + r_2) \tag{4.186}$$

Putting (4.186), (4.185), and (4.184) into (4.183), we find

$$J_r = \pi(r_1 + r_2) - 2\pi(r_1 r_2)^{1/2} = -(J_\theta + J_\varphi) - k\pi(2m/-E)^{1/2}$$

$$E = -\frac{2mk^2\pi^2}{(J_r + J_\theta + J_\varphi)^2} = H \tag{4.187}$$

The frequencies are just

$$\nu_r = \frac{\partial H}{\partial J_r} = \frac{4\pi^2 mk^2}{(J_r + J_\varphi + J_\theta)^2} = \frac{1}{\pi k}\left(\frac{-2E^3}{m}\right)^{1/2} = \nu_\theta = \nu_\varphi \tag{4.188}$$

If two or more of the fundamental frequencies are equal the system is *degenerate*. When, as in the Kepler problem, all the frequencies are equal, the system is *completely degenerate*.

Example 4.12. Determine the Hamilton equations of motion of a particle of mass m and charge e moving in an electromagnetic field whose potential is $A^\mu(x)$. The equations should be compatible with ESR.

Solution. The Lagrangian for this system is given in (2.54). Thus the modified Lagrangian is

$$\bar{L} = mc + (e/c)A^\mu \dot{x}_\mu + \tfrac{1}{2}\lambda(\dot{x}^\mu \dot{x}_\mu - 1) \tag{4.189}$$

The conjugate momenta are

$$\bar{p}^\mu = \partial\bar{L}/\partial\dot{x}^\mu = (e/c)A^\mu + \lambda\dot{x}^\mu \tag{4.190}$$

Inserting (4.189) into (4.89), the defining equation of the Hamiltonian, and using (4.190) to eliminate the velocities, we obtain

$$\bar{H} = \frac{1}{2\lambda}\left(\bar{p}^\mu - \frac{e}{c} A^\mu\right)^2 + \tfrac{1}{2}\lambda - mc \tag{4.191}$$

The dynamical equations are

$$\dot{x}^\mu = \frac{\partial \bar{H}}{\partial \bar{p}_\mu} = \frac{1}{\lambda}\left(\bar{p}^\mu - \frac{e}{c} A^\mu\right) \tag{4.192a}$$

$$\dot{\bar{p}}^\mu = -\frac{\partial \bar{H}}{\partial x_\mu} = \frac{e}{c}\frac{1}{\lambda}\left(\bar{p}^\nu - \frac{e}{c} A^\nu\right)A_{\nu,}{}^{\mu} \tag{4.192b}$$

Inserting the constraint on the velocities into the scalar product of (4.192a) with itself yields

$$\lambda^2 = [\bar{p}^\mu - (e/c)A^\mu]^2 \tag{4.193}$$

If these equations are to reduce to the free particle equations as $A^\mu \to 0$, it is necessary that $\lambda = mc$ and (4.192b) is

$$\dot{\bar{p}}^\mu = (e/mc^2)[\bar{p}^\nu - (e/c)A^\nu]A_{\nu,}{}^{\mu} \tag{4.194}$$

It is a simple matter to show that (4.199) is equivalent to (2.55), the Lagrangian solution.

Example 4.13. A particle is moving freely along the x axis except for perfectly elastic collisions occurring at walls located at $x = 0$ and $x = L$. Determine how an adiabatic change in the location of the wall at $x = L$ will modify the momentum and energy of the particle.

Solution. Since the particle is free, the momentum changes only at the walls where it goes from $+p$ to $-p$. Thus the action variable is

$$J = \int p\, dx = p\int_0^L dx + (-p)\int_L^0 dx = 2pL$$

For an adiabatic change in L

$$J_i = J_f$$

or

$$2p_iL_i = 2p_fL_f, \qquad p_f = (L_i/L_f)p_i$$

The energy is

$$E_f = \frac{p_f^2}{2m} = \left(\frac{L_i}{L_f}\right)^2\frac{p_i^2}{2m} = \left(\frac{L_i}{L_f}\right)^2 E_i$$

Example 4.14. Apply canonical perturbation theory to determine the motion of the one dimensional anharmonic oscillator whose Hamiltonian is

$$H = p^2/2m + \tfrac{1}{2}m\omega^2 q^2 + \lambda q^3$$

where λ is sufficiently small.

Solution. We first identify the unperturbed Hamiltonian. In this case $H_0 = p^2/2m + \frac{1}{2}m\omega^2 q^2$, i.e., the unperturbed system is just that of the harmonic oscillator whose motion and action-angle variables have already been obtained and are reproduced here for convenience

$$H_0'(J_0) = J_0\omega/2\pi = \nu_0 J_0 \tag{4.195}$$

$$q = A\cos\omega t = (J_0/\pi m\omega)^{1/2}\cos 2\pi w_0 \tag{4.196}$$

Since in this problem the Hamiltonian is initially expressed in powers of λ, the coefficients of the expansion of (4.156) are obtained by merely equating coefficients of like powers of λ in the original Hamiltonian and (4.156). We find with the help of (4.195) and (4.196)

$$H_0'(J_0) = \nu_0 J_0 \tag{4.197}$$

$$H_1'(J_0) = (J_0/\pi m\omega)^{3/2}\cos^3 2\pi w_0 \tag{4.198}$$

$$H_j'(J_0) = 0, \qquad j = 2, 3, 4, \ldots \tag{4.199}$$

We will first obtain the corrections to the energy from (4.167). To this end consider

$$\langle H_1'\rangle = \int_0^1 H_1'\, dw_0 = \left(\frac{J_0}{\pi m\omega}\right)^{3/2}\left(\frac{\sin 2\pi w_0}{2\pi} - \frac{\sin^3 2\pi w_0}{6\pi}\right)_0^1 = 0 \tag{4.200}$$

We see immediately that the first-order correction to the energy is zero. The second-order term in (4.167) has a nonzero contribution only from the term

$$\left\langle \frac{\partial H_1'}{\partial J_0} H_1'\right\rangle_{J_0=J} = \frac{3}{2}\frac{J^2}{(\pi m\omega)^3}\int_0^1 \cos^6 2\pi w_0\, dw_0 = \frac{15}{32}\frac{J^2}{(\pi m\omega)^3}$$

Therefore, the energy of the anharmonic oscillator is to second order in λ

$$H''(J, \lambda) = \frac{\omega J}{2\pi} - \lambda^2\frac{15J^2}{16\pi^2 m^3\omega^4} + O(\lambda^3)$$

where $O(\lambda^3)$ signifies terms in λ to the third, or higher, power. From (4.168) it follows that the frequency of the anharmonic oscillator is

$$\nu = \frac{\partial H''}{\partial J} = \nu_0 - \lambda^2 \frac{15J}{8\pi^2 m^3 \omega^4}$$

and

$$w = \nu t = \left(\nu_0 - \lambda^2 \frac{15J}{8\pi^2 m^3 \omega^4}\right)t \tag{4.201}$$

It has been fairly simple to obtain the energy and frequency to the above order of accuracy; however, it is considerably more tedious to obtain the generator W_2 and the position to the same accuracy.

If we can obtain the generator W_2 which defines the transformation from (w_0, J_0) variables to (w, J) variables, we can obtain $q = q(w, J)$ from (4.196). To use (4.156) and (4.157) we evaluate from (4.165)

$$\frac{\partial W_{21}}{\partial w_0} = -\frac{1}{\nu_0}\left[\frac{J}{\pi m\omega}\right]^{3/2} \cos^3 2\pi w_0$$

Therefore,

$$W_{21} = -\frac{1}{2\pi\nu_0}\left[\frac{J}{\pi m\omega}\right]^{3/2} (\sin 2\pi w_0 - \tfrac{1}{3}\sin^3 2\pi w_0)$$

and

$$\frac{\partial W_{21}}{\partial J} = -\frac{3}{8\pi m\omega^2}\left[\frac{J}{\pi m\omega}\right]^{1/2} (3\sin 2\pi w_0 + \tfrac{1}{3}\sin 6\pi w_0)$$

Writing (4.154) and (4.155) to first order in λ, we obtain

$$J_0 = J - \lambda\frac{2\pi}{\omega}\left[\frac{J}{\pi m\omega}\right]^{3/2} \cos^3 2\pi w_0$$

$$w = w_0 - \lambda\frac{3}{8\pi m\omega^2}\left[\frac{J}{\pi m\omega}\right]^{1/2} (3\sin 2\pi w_0 + \tfrac{1}{3}\sin 6\pi w_0)$$

Having the equations connecting (w_0, J_0) with (w, J) we can find the equations connecting q with (w, J) by using (4.191). Thus

$$[J_0]^{1/2} = [J]^{1/2}\left(1 - \frac{\lambda}{m\omega^2}\left[\frac{J}{\pi m\omega}\right]^{1/2} \cos^3 2\pi w + O(\lambda^2)\right) \tag{4.202}$$

$$\cos 2\pi w_0 = \cos 2\pi w - \lambda\frac{3}{4m\omega^2}\left[\frac{J}{\pi m\omega}\right]^{1/2} \times \sin 2\pi w(3\sin 2\pi w + \tfrac{1}{3}\sin 6\pi w) + O(\lambda^2) \tag{4.203}$$

Inserting (4.202) and (4.203) into (4.196), we obtain

$$q = \left[\frac{J}{\pi m\omega}\right]^{1/2} \cos 2\pi w - \frac{\lambda}{2m\omega^2}\frac{J}{\pi m\omega}(3 - \cos 4\pi w) + O(\lambda^2) \quad (4.204)$$

Of course (4.204) is the desired equation of motion correct to first order in λ since w may be expressed as a function of time through Eq. (4.201).

PROBLEMS

4.1 Find the Hamilton equations of motion for Problems 2.1–2.4.

4.2 Find the Hamilton equations of motion for the mass–pulley system discussed in Example 2.6, first by using the constraint equations to eliminate the dependent variables, and second by recourse to Dirac's method.

4.3 A particle of mass m moves under the influence of gravity along the spiral $z = k\theta$, $r =$ const, where k is a constant and z is the vertical coordinate. Determine the Hamilton equations of motion.

4.4 The Hamiltonian of a system having two degrees of freedom is given by

$$H = \tfrac{1}{2}(p_1^2 x_1^4 + p_2^2 x_1^2 - 2ax_1)$$

where "a" is a constant. Prove that

$$x_1 = A \cos x_2 + B \sin x_2 + C$$

where A, B, and C are constants.

4.5 Show that the transformation

$$Q_1 = q_1^2 + p_1^2, \qquad Q_2 = \tfrac{1}{2}(q_1^2 + q_2^2 + p_1^2 + p_2^2)$$

$$P_1 = \tfrac{1}{2}\tan^{-1} q_2/p_2 - \tfrac{1}{2}\tan^{-1} q_1/p_1, \qquad P_2 = -\tan^{-1} q_2/p_2$$

is a canonical transformation. Use this transformation to solve the equations of motion for a system whose Hamiltonian is

$$H = \tfrac{1}{2}(p_1^2 + p_2^2 + q_1^2 + q_2^2),$$

and compare this solution with that obtained in terms of the original variables.

4.6 Prove that the Poisson brackets are invariant under a canonical transformation.

4.7 Prove that the Poisson bracket of two constants of the motion is itself a constant of the motion even when the constants depend on time explicitly.

4.8 A system of n degrees of freedom has a Hamiltonian given by

$$H = \tfrac{1}{2} \sum_{a=1}^{n} p_a{}^2 + \sum_{a=1}^{n+1} (q_a - q_{a-1})^2 \qquad (q_0 = q_{n+1} = 0)$$

Use the canonical transformation obtained from the generator

$$W_1 = \sum_{a=1}^{n} \tfrac{1}{4}\, \alpha_a{}^2 \left[\sum_{b=1}^{n} A_{ba} q_b \right]^2 \cot Q_a$$

with

$$A_{ba} = \frac{2}{[(n+1)\alpha_a]^{1/2}} \sin \frac{ba\pi}{n+1}, \qquad \alpha_a = 2 \sin \frac{a\pi}{2(n+1)}$$

to solve the equations of motion of the q_a and to find the normal coordinates of the problem.

4.9 Determine the Poisson brackets of the components of angular momentum with each other.

4.10 Show that $[V, L_3] = 0$, where V is any spherically symmetric function of the coordinates and momentum of a particle and L_3 is the z component of angular momentum.

4.11 Show that $[\mathbf{A}, L_3] = \mathbf{n} \times \mathbf{A}$, where $\mathbf{n}$ is a unit vector parallel to the z axis and $\mathbf{A}$ is a vector function of the coordinates and momenta of a particle.

4.12 (a) Show that the Hamilton–Jacobi equation of a particle is separable when it is expressed in parabolic coordinates and the potential energy has the form

$$V = \frac{a(\xi) + b(\eta)}{\xi + \eta}$$

where the parabolic coordinates ξ, η, φ are related to the cylindrical coordinates ϱ, φ, z by the transformation equations

$$z = \tfrac{1}{2}(\xi - \eta), \qquad \varrho = (\xi\eta)^{1/2}, \qquad \varphi = \varphi$$

and $a(\xi)$ and $b(\eta)$ are arbitrary functions.
(b) Use (a) to consider the motion of a particle in a field $V = \alpha/r - Fz$ (Coulomb field plus a uniform field in z direction).

4.13 (a) Show that the Hamilton–Jacobi equation of a particle is separable when it is expressed in elliptic coordinates and the potential energy has the form

$$V = \frac{a(\xi) + b(\eta)}{\xi^2 - \eta^2}$$

where $a(\xi)$ and $b(\eta)$ are arbitrary functions and the elliptic coordinates (ξ, η, φ) are related to the cylindrical coordinates (ϱ, φ, z) by

$$\varrho = \sigma[(\xi^2 - 1)(1 - \eta^2)]^{1/2}, \qquad z = \sigma\xi\eta, \qquad \varphi = \varphi$$

The constant σ is a parameter of the transformation.
(b) Use (a) to consider the motion of a particle in a field $V = k_1/r_1 + k_2/r_2$ (Coulomb field of two fixed sources separated by a distance 2σ).

4.14 Use the Hamilton–Jacobi equation to obtain an equation for the orbit of a particle moving in a two-dimensional $V = \frac{1}{2}kr^2$. Express the motion in terms of the coordinates u and v defined by $x = \cosh u \cos v$, $y = \sinh u \sin v$.

4.15 Use the Hamilton–Jacobi method to obtain the formal solution of the motion of the symmetrical top discussed in Example 4.4.

4.16 Use the action-angle variable method to determine the frequencies of a three-dimensional harmonic oscillator with unequal force constants.

4.17 Obtain within the framework of ESR the Hamiltonian and the Hamilton–Jacobi equation for a particle moving in a scalar field φ. The Lagrangian is

$$L = mc - \varphi(x^\mu)$$

Discuss the motion.

4.18 (a) The length of a simple pendulum is shortened by moving the point of suspension, adiabatically, toward the suspended mass. Assume that initially the amplitude of oscillation is small. Determine how the frequency, energy, and amplitude change as the length of the pendulum decreases.

(b) Repeat (a), but now change the length of the pendulum by keeping the point of suspension fixed and allowing the cord to be drawn up through a hole at the point of suspension. How do these adiabatic results compare to the frequency of a simple pendulum whose length is equal to the final length of the adiabatically changing pendulum?

4.19 A particle slides down a frictionless inclined plane and is reflected elastically at a wall which is perpendicular to the inclined plane and located at the bottom of the plane. Determine how the maximum rebound height varies if the angle of inclination of the plane is changed adiabatically.

4.20 Apply canonical perturbation theory of determine the motion of a system having a Hamiltonian $H = p^2/2m + \frac{1}{2}m\omega^2 q^2 + \lambda q^4$ and compare the results with those of Example 4.14.

4.21 Apply canonical perturbation theory to determine the motion of a system having a Hamiltonian $H = p^2/2m + \lambda q + \frac{1}{2}m\omega^2 q^2$ and compare the results with the exact solution.

REFERENCES

1. D. Ter Haar, "Elements of Hamiltonian Dynamics." North-Holland Publ., Amsterdam, 1964.
2. A. O. Barut, "Electrodynamics and Classical Theory of Fields and Particles." Macmillan, New York, 1964.
3. P. A. M. Dirac, "Lectures on Quantum Mechanics." Yeshiva Univ. Press, New York, 1964; "Lectures on Quantum Field Theory." Yeshiva Univ. Press, New York, 1966.
4. Y. Hagihara, "Celestial Mechanics." Vol. 2, "Perturbation Theory." MIT Press, Cambridge, Massachusetts, 1972.
5. F. T. Geyling and H. R. Westerman, "Introduction to Orbital Mechanics." Addison-Wesley, Reading, Massachusetts, 1971.
6. R. Bellman, "Perturbation Technique in Mathematics, Physics and Engineering." Holt, New York, 1963.
7. T. E. Sterne, "An Introduction to Celestial Mechanics." Wiley, New York, 1960.

8. L. A. Pipes and L. R. Harvill, "Applied Mathematics for Engineers and Physicists," 3rd ed. McGraw-Hill, New York, 1970.
9. S. S. Kuo, "Numerical Methods and Computers." Addison-Wesley, Reading, Massachusetts, 1965.
10. N. W. McLachlan, "Ordinary Nonlinear Differential Equations in Engineering and Physical Science." Oxford Univ. Press, London and New York, 1956.
11. L. A. Pars, "A Treatise on Analytical Dynamics." Heinemann, London, 1965.

BIBLIOGRAPHY

H. Goldstein, "Classical Mechanics." Addison-Wesley, Reading, Massachusetts, 1950.

S. W. Groesberg, "Advanced Mechanics." Wiley, New York, 1968.

C. Lanczos, "The Variational Principles of Mechanics." Univ. of Toronto, Toronto, 1966.

L. D. Landau and E. M. Lifshitz, "Mechanics." Pergamon, Oxford, 1960.

L. Meirovitch, "Methods of Analytical Dynamics." McGraw-Hill, New York, 1970.

T. G. Northrop, "The Adiabatic Motion of Charged Particles." Wiley (Interscience), New York, 1963.

H. Rund, "The Hamilton-Jacobi Theory in the Calculus of Variations." Van Nostrand-Reinhold, Princeton, New Jersey, 1966.

E. J. Saletan and A. H. Cromer, "Theoretical Mechanics." Wiley, New York, 1971.

5

MANY-PARTICLE DYNAMICS

5.1. CONSTRUCTION OF A MANY-PARTICLE DYNAMICS COMPATIBLE WITH EINSTEIN'S SPECIAL THEORY OF RELATIVITY

In Chapter 2 and 4 we encountered rather severe difficulties in attempting to construct a dynamics of interacting point particles consistent with Einstein's relativity. In fact, we were able to proceed only by introducing the "field" into our formalism, thereby removing ourselves from the realm of a pure particle mechanics. In this chapter we consider the question of whether it is possible to develop a pure particle dynamics that is compatible with Einstein's relativity.

A. The No-Interaction Theorem

Characteristic of the problems which must be surmounted is the "No-Interaction Theorem." This theorem has been presented in many forms all of which may be summarized as follows: Given an apparently benign set of postulates, a dynamical theory describing interactions between two or more particles is incompatible with Einstein's relativity. We now present a proof of the theorem presented by Van Dam and Wigner [1]. A very elegant group theoretic proof is presented in Appendix C.

We consider a system of N particles whose space–time locations relative to an inertial observer are given by

$$ {}_a x^\mu, \qquad a = 1, 2, \ldots, N \tag{5.1} $$

where a is the particle index and μ the contravariant tensor index. The history of a particle may be summarized by its "world line" (orbit) in

Lorentz space which may be described parametrically in terms of its proper time s_a which may be taken to be the same as the arc length of the world line. Thus

$$(ds_a)^2 = d\, {}_a x^\mu \, d\, {}_a x_\mu \tag{5.2}$$

and

$$({}_a\dot{x}^\mu)^2 = (d\, {}_a x^\mu / ds_a)(d\, {}_a x_\mu / ds_a) = 1 \tag{5.3}$$

where the summation convention applies only to the tensor indices. Note that each particle has its *own* proper time.

The momentum of the ath particle at the observer's time t_1 is defined as

$${}_a P^\mu(t_1) = m_a \, {}_a\dot{x}^\mu(s_{a_1}) \tag{5.4}$$

where m_a is the mass of the ath particle and s_{a_1} is the ath particle's proper time at the observer's time t_1. s_{a_1} is the solution of the equation

$${}_a x^0(s_{a_1}) = t_1 \tag{5.5}$$

We now introduce the following three postulates.

Postulate A. The components of the total linear momentum of the system of particles at observer time t_1 is given by

$$P^\mu(t_1) = \sum_{a=1}^{N} {}_a P^\mu(t_1) \tag{5.6}$$

This postulate also assumes that the components of total linear momentum are conserved, i.e.,

$$P^\mu(t_1) = P^\mu(t_2) = \text{const} \tag{5.7}$$

Postulate B. The particles do not coincide, i.e., there are no pairs of particles a and b and proper times s_a and s_b such that

$${}_a x^\mu(s_a) = {}_b x^\mu(s_b) \tag{5.8}$$

for all μ.

Postulate C. Asymptotically, i.e., either for $s_a \to \infty$ or $s_a \to -\infty$, the world lines are straight. In other words

$${}_a\dot{x}^\mu \to \text{const} \qquad \text{as} \quad s_a \to \pm\infty \tag{5.9}$$

With the preceding three postulates it is not difficult to prove that for $N = 2$, 3, or 4 the world lines are straight throughout, and this is the essence of the No-Interaction Theorem. Before proceeding to the proof of this theorem, we demonstrate that Postulate C implies that the total linear momentum ${}_aP^\mu$ transforms as a vector under a Lorentz transformation. In order to verify this we introduce a second Lorentz observer who designates the space–time locations of the particles by ${}_ix^{\mu\prime}$ where

$$x^{\mu\prime} = L^\mu{}_\nu \, x^\nu \tag{5.10}$$

and the $L^\mu{}_\nu$ are the Lorentz transformations as defined in Chapter 1. From (5.9) it follows that the unprimed observer may select a time t_1 for which the ${}_a\dot{x}^\mu$ become independent of the proper time s_a, i.e., for $s_a \geq s_{a_1}$, ${}_a\dot{x}^\mu$ is not a function of s_a. The primed observer may also select a time $t_1{}'$ for which ${}_a\dot{x}^{\mu\prime}$ is independent of s_a and furthermore he may choose his $t_1{}'$ such that $s'_{a_1} > s_{a_1}$ for all a. Asymptotically the linear momentum in the primed frame is

$$P^{\mu\prime} = \sum_{a=1}^{N} m_a \, {}_a\dot{x}^{\mu\prime}(s'_{a_1}) = \sum_{a=1}^{N} m_a L^\mu{}_\nu \, {}_a\dot{x}^\nu(s'_{a_1}) \tag{5.11}$$

But since ${}_a\dot{x}^\nu$ is independent of the proper time in this asymptotic region, we may replace s'_{a_1} with s_{a_1} and write

$$P^{\mu\prime} = \sum_{a=1}^{N} m_a L^\mu{}_\nu \, {}_a\dot{x}^\nu(s_{a_1}) = L^\mu{}_\nu \, P^\nu \tag{5.12}$$

Therefore the total linear momentum transforms as a vector. It is in order to assure this that Postulate C is introduced.

Figure 5.1 presents the world lines of two particles. The primed and unprimed observers are designated by the $t' - x'$ and $t - x$ axes, respectively.

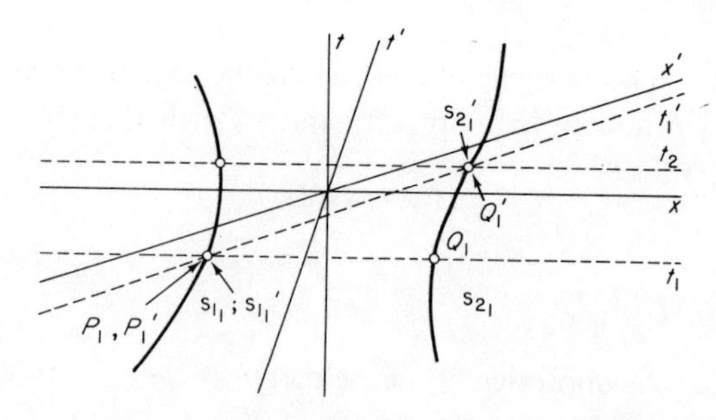

Fig. 5.1. The world lines of two particles and their coordinates relative to primed and unprimed observers.

The lines t_1 and t_2 connect those events which are simultaneous relative to the unprimed observer at times t_1 and t_2, respectively. t_1' performs the same function for the primed observer. Thus, the unprimed observer observes the events P_1 and Q_1 to be simultaneous at t_1 and the primed observer finds P_1' and Q_1' to be simultaneous at t_1'.

The unprimed observer would record the total momentum at time t_1 as the sum of the momentum of the particles when they are at P_1 and Q_1, i.e.,

$$P^\mu = \sum_{a=1}^{2} {}_aP^\mu(t_1) = \sum_{a=1}^{2} m_a\, {}_a\dot{x}^\mu(s_{a_1}) \tag{5.13}$$

The primed observer would record the total momentum at time t_1' as

$$P^{\mu\prime} = \sum_{a=1}^{2} m_a\, {}_a\dot{x}^{\mu\prime}(s'_{a_1}) \tag{5.14}$$

According to Postulate A the total momentum is conserved in each frame and if (5.12) holds in the asymptotic region it must hold everywhere. Therefore,

$$\sum_{a=1}^{2} m_a\, {}_a\dot{x}^{\mu\prime}(s'_{a_1}) = \sum_{a=1}^{2} L^\mu{}_\nu\, m_a\, {}_a\dot{x}^\nu(s_{a_1}) \tag{5.15}$$

But

$${}_a\dot{x}^{\mu\prime}(s_{a_1}) = L^\mu{}_\nu\, {}_a\dot{x}^\nu(s_{a_1}) \tag{5.16}$$

i.e., the Lorentz transformation connects the velocity of a particular event (proper time) as measured by the unprimed observer with the velocity of that same event as measured by the primed observer. Therefore, (5.15) becomes

$$\sum_{a=1}^{2} m_a\, {}_a\dot{x}^{\mu\prime}(s'_{a_1}) = \sum_{a=1}^{2} m_a\, {}_a\dot{x}^{\mu\prime}(s_{a_1}) \tag{5.17}$$

or equivalently

$$\sum_{a=1}^{2} m_a\, {}_a\dot{x}^{\mu}(s'_{a_1}) = \sum_{a=1}^{2} m_a\, {}_a\dot{x}^{\mu}(s_{a_1}) \tag{5.18}$$

Let us now apply (5.18) to the situation represented in Fig. 5.1 by the events P_1', Q_1', and Q_1. At time t_1, P_1', and Q_1 are simultaneous for the unprimed observer and for him the right side of (5.18) becomes

$$\underset{\substack{\uparrow \\ P_1'}}{m_1\, {}_1\dot{x}^\mu(s_{1_1})} + \underset{\substack{\uparrow \\ Q_1}}{m_2\, {}_2\dot{x}^\mu(s_{2_1})} \tag{5.19}$$

Now the primed observer at time t_1 observes that P_1' and Q_1' are simultaneous, and therefore the left side of (5.18) is

$$\underset{\substack{\uparrow\\ P_1'}}{m_1\,{}_1\dot{x}^\mu(s_{1_1}')} + \underset{\substack{\uparrow\\ Q_1'}}{m_2\,{}_2\dot{x}^\mu(s_{2_1}')} \tag{5.20}$$

Equating (5.19) and (5.20) and noting that $m_1\,{}_1\dot{x}^\mu(s_{1_1}) = m_1\,{}_1\dot{x}^\mu(s_{1_1}')$, we obtain

$$m_2\,{}_2\dot{x}^\mu(s_{2_1}') = m_2\,{}_2\dot{x}^\mu(s_{2_1}) \tag{5.21}$$

Equation (5.21) states that the momentum of particle 2 at Q_1' must be the same as the momentum at Q_1. Since the choice of the event P_1' and Lorentz observers was arbitrary, we conclude that (5.21) holds for arbitrary Q_1' and Q_1, i.e., the momentum of the particle does not change, and therefore its world line must be straight. We have, therefore, derived the No-Interaction Theorem which states that from Postulates A, B, and C it follows that there can be no interaction between two particles that is compatible with Einstein's relativity. It is a simple exercise to extend the above argument to three or four particles.

Appendix C presents a very elegant and general group theoretic proof of a No-Interaction Theorem within the framework of classical Hamiltonian dynamics. To accomplish our objective of constructing a pure particle dynamics that is compatible with Einstein's relativity we must circumvent these No-Interaction Theorems. Several dynamical particle theories have been developed which do successfully circumvent the No-Interaction Theorems and their difference is primarily determined by whether they seek to circumvent the Wigner–Van Dam type of theorem presented in this chapter or the group theoretic type theorem of Appendix C. The theories which are constructed to avoid the consequences of the Wigner–Van Dam theorem will be called Newtonian type theories and the others will be called Hamiltonian-type theories.

B. Newtonian-Type Theory of Wigner and Van Dam

Theories of the Newtonian type are constructed so that the equations of motion determining the orbits of the particles are given in terms of the accelerations which are defined as

$${}_aA^\mu = d^2\,{}_ax^\mu/ds_a^{\,2} \tag{5.22}$$

Now following Van Dam and Wigner [2], we proceed as follows. We wish to construct the equations of motion so that they may easily be transformed from one Lorentz frame to another thereby making their Lorentz invariance readily apparent. In Galilean relativity the procedure was to express the accelerations in terms of the position and velocity of the N particles that were simultaneous, i.e., the position and velocity of particle 1 at time t_1, the position and velocity of particle 2 at time t_2, and so on. Now the important point is that these same positions and velocities were simultaneous for all Galilean observers. But because of the relativity of simultaneity in Einstein relativity, various observers would disagree on which positions of the particles were simultaneous. It is therefore a serious question as to which ones should be used in writing the equations of motion.

Wigner and Van Dam suggest that the acceleration of a particle should be expressed in terms of integrals over the paths of the other particles. Assuming that the force due to several particles is the sum of the forces due to these particles individually, we may write

$$m_a\, {}_aA^\mu = \sum_{b \neq a} {}_{ab}F^\mu(s_a) \tag{5.23}$$

where the force ${}_{ab}F^\mu$ will be expressed in terms of the positions and velocities of the other particles. And now we further assume that this force on particle "a" when it is located at ${}_ax^\mu(s_a)$ depends only on those positions of the "b" particle, ${}_bx^\mu(s_b)$, which have a spacelike relation to ${}_ax^\mu(s_a)$. This is illustrated in Fig. 5.2.

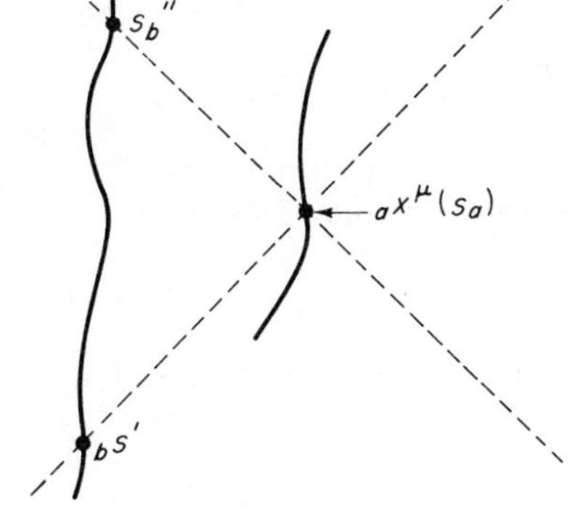

Fig. 5.2. All points between s_b' and s'' and spacelike relative to ${}_ax^\mu(s_a)$.

Thus far we have replaced the Galilean requirement that the forces depend on the simultaneous positions of the particles (which is an absolute concept —the same for all Galilean observers) with the requirement that the forces depend on positions that have a spacelike relationship to one another and this is an "absolute" relationship within the framework of Einstein relativity (same for all Lorentz observers). Next we postulate that this dependence is of the following form:

$$ {}_{ab}F^{\mu}(s_a) = \int_{-\infty}^{\infty} ds_b \, {}_{ab}\hat{F}^{\mu}({}_a x^{\nu}(s_a), {}_a\dot{x}^{\nu}(s_a), {}_b x^{\nu}(s_b), {}_b\dot{x}^{\nu}(s_b)) \tag{5.24} $$

The form of the force is further restricted if we require that the integrand of (5.24) transforms as a vector under Lorentz transformations. Lorentz invariants which can be constructed from ${}_a x^{\nu}(s_a)$, ${}_a\dot{x}^{\nu}(s_a)$, ${}_b x^{\nu}(s_b)$, ${}_b\dot{x}^{\nu}(s_b)$ are

$$ \varrho_{ab} = \{[{}_a x^{\mu}(s_a) - {}_b x^{\mu}(s_b)]^2\}^{1/2} \tag{5.25a} $$

$$ \xi_{ab} = {}_a\dot{x}^{\mu}(s_a)\, {}_b\dot{x}_{\mu}(s_b) \tag{5.25b} $$

$$ \sigma_{ab} = {}_a\dot{x}^{\mu}(s_a)[{}_a x_{\mu}(s_a) - {}_b x_{\mu}(s_b)] \tag{5.25c} $$

$$ \sigma_{ba} = {}_b\dot{x}^{\mu}(s_b)[{}_b x_{\mu}(s_b) - {}_a x_{\mu}(s_a)] \tag{5.25d} $$

Using the invariants in (5.25), a general expression for ${}_{ab}\hat{F}^{\mu}$ is

$$ {}_{ab}\hat{F}^{\mu} = ({}_a x^{\mu} - {}_b x^{\mu}) f_{ab} + {}_a\dot{x}^{\mu} g_{ab} + {}_b\dot{x}^{\mu} h_{ab} + \varepsilon^{\mu\nu\varrho\sigma}({}_a x_{\nu} - {}_b x_{\nu})\, {}_a\dot{x}_{\varrho}\, {}_b\dot{x}_{\sigma} j_{ab} \tag{5.26} $$

where $\varepsilon^{\mu\nu\varrho\sigma}$ is the Levi–Civita tensor and f_{ab}, g_{ab}, h_{ab}, and j_{ab} are functions of the invariants appearing in (5.25).

We now impose several additional restrictions on (5.26) the motivation for which will be discussed in Section 5.2.

We require

$$ g_{ab} = h_{ab} = j_{ab} = 0 \tag{5.27} $$

and

$$ {}_{ab}\hat{F}^{\mu}(s_a, s_b) = -\, {}_{ba}\hat{F}^{\mu}(s_b, s_a) \tag{5.28} $$

It follows from (5.28) that

$$ f_{ab} = f_{ba} \tag{5.29} $$

Differentiating (5.3), we obtain

$$ {}_a\dot{x}_{\mu}\, {}_a\ddot{x}^{\mu} = 0 \tag{5.30} $$

Multiplying (5.23) by ${}_a\dot{x}_{\mu}$ and using (5.24), (5.25), (5.26), (5.27), and (5.30), we obtain

$$ 0 = \sum_{b \neq a} \int_{-\infty}^{\infty} ds_b \, {}_a\dot{x}_{\mu}({}_a x^{\mu} - {}_b x^{\mu}) f_{ab} = \sum_{b \neq a} \int_{-\infty}^{\infty} ds_b \, \sigma_{ab} f_{ab} \tag{5.31} $$

It follows from (5.25) that

$$d\sigma_{ab}/ds_b = -\xi_{ab} \tag{5.32}$$

If now we choose

$$f_{ab} = 2\varphi_{ab}(\varrho_{ab})\xi_{ab} - \sigma_{ab}\frac{d}{ds_b}\varphi_{ab}(\varrho_{ab}) \tag{5.33}$$

the integrand in (5.31) becomes

$$-d(\sigma_{ab}^2\varphi_{ab})/ds_b \tag{5.34}$$

where φ_{ab} is a function of ϱ_{ab} which is restricted by (5.29) to be symmetric in "a" and "b." We further require that

$$\varphi_{ab} \to 0 \quad \text{as} \quad \varrho_{ab}^2 \geq 0 \tag{5.35}$$

This assures that the force is nonzero only when ${}_ax^\mu(s_a)$ and ${}_bx^\mu(s_b)$ have a spacelike relationship. Using (5.34) and (5.35), (5.31) becomes

$$0 = -\sum_{b\neq a} \sigma_{ab}^2\varphi_{ab}\Big|_{s_b'}^{s_b''} \tag{5.36}$$

Now (5.35) is sufficient to guarantee that each term in the sum on the right side of (5.36) goes to zero because at s_b' and s_b'', $\varrho_{ab} = 0$.

An alternative form for f_{ab} is

$$f_{ab} = 2\varphi_{ab}(\varrho_{ab})\xi_{ab} + \left(\frac{\sigma_{ab}\sigma_{ba}}{\varrho_{ab}}\right)\frac{d}{d\varrho_{ab}}\varphi_{ab}(\varrho_{ab}) \tag{5.37}$$

which follows from (5.33) if we note that

$$d\varrho_{ab}/ds_b = -\sigma_{ba}/\varrho_{ab} \tag{5.38}$$

Thus the equations of the orbits of the particles which incorporate the restrictions and assumptions discussed between Eqs. (5.23) and (5.38) are

$$m_a\,{}_a\ddot{x}^\mu(s_a) = \sum_{b\neq a}\int_{-\infty}^{\infty} ds_b\,[{}_ax^\mu(s_a) - {}_bx^\mu(s_b)] \times \left[2\varphi_{ab}(\varrho_{ab})\xi_{ab} + \left(\frac{\sigma_{ab}\sigma_{ba}}{\varrho_{ab}}\right)\frac{d}{d\varrho_{ab}}\varphi_{ab}(\varrho_{ab})\right] \tag{5.39a}$$

A useful alternative form for the equations of motion is obtained by using (5.33) instead of (5.37) in (5.31). The result may be written

$$m_a\,{}_a\ddot{x}^\mu(s_a) = \sum_{b\neq a}\int_{-\infty}^{\infty} ds_b\,\varphi_{ab}(\varrho_{ab})\{\xi_{ab}[{}_ax^\mu(s_a) - {}_bx^\mu(s_b)] - \sigma_{ab}\,{}_b\dot{x}^\mu\} \tag{5.39b}$$

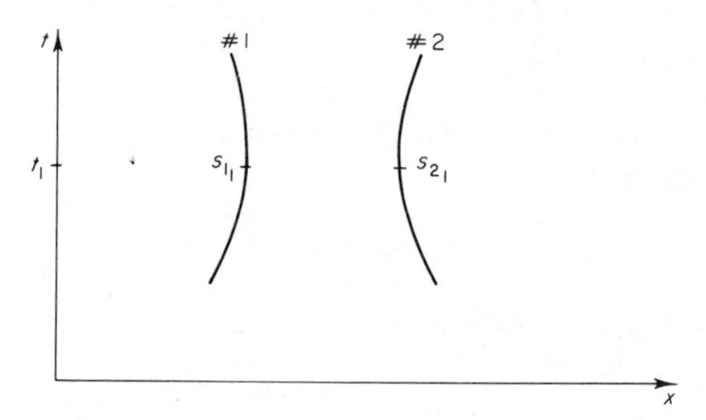

Fig. 5.3. The world lines of two interacting particles.

Consider now two particles whose world lines are shown in Fig. 5.3. The momentum of particle 1 at time t_1 is determined by multiplying the equation of motion by ds_1 and integrating.

$$ {}_1P^\mu(s_{1_1}) - {}_1P^\mu(-\infty) = \int_{-\infty}^{s_{1_1}} ds_1 \int_{-\infty}^{\infty} ds_2\ {}_{12}\hat{F}(s_1, s_2) \tag{5.40} $$

A similar expression holds for particle 2. At time t_1 the total particle momentum is

$$ \begin{aligned} &{}_1P^\mu(s_{1_1}) + {}_2P^\mu(s_{2_1}) - {}_1P^\mu(-\infty) - {}_2P^\mu(-\infty) \\ &\quad = \int_{-\infty}^{s_{1_1}} ds_1 \int_{-\infty}^{\infty} ds_2\ {}_{12}\hat{F}^\mu(s_1, s_2) + \int_{-\infty}^{s_{2_1}} ds_2 \int_{-\infty}^{\infty} ds_1\ {}_2\hat{F}^\mu(s_2, s_1) \end{aligned} \tag{5.41} $$

If now we add and subtract

$$ \int_{s_{1_1}}^{\infty} ds_1 \int_{-\infty}^{\infty} ds_2\ {}_{12}\hat{F}^\mu(s_1, s_2) + \int_{s_{2_1}}^{\infty} ds_2 \int_{-\infty}^{\infty} ds_1\ {}_{21}\hat{F}^\mu(s_2, s_1) $$

to the right side of (5.41) we obtain

$$ \begin{aligned} &{}_1P^\mu(s_{1_1}) + {}_2P^\mu(s_{2_1}) - {}_1P^\mu(-\infty) - {}_2P^\mu(-\infty) \\ &\quad = \int_{-\infty}^{\infty} ds_1 \int_{-\infty}^{\infty} ds_2\ {}_{12}\hat{F}^\mu(s_1, s_2) + \int_{-\infty}^{\infty} ds_2 \int_{-\infty}^{\infty} ds_1\ {}_{21}\hat{F}^\mu(s_1, s_2) \\ &\qquad - \int_{s_{1_1}}^{\infty} ds_1 \int_{-\infty}^{\infty} ds_2\ {}_{12}\hat{F}^\mu(s_1, s_2) - \int_{s_{2_1}}^{\infty} ds_2 \int_{-\infty}^{\infty} ds_1\ {}_{21}\hat{F}^\mu(s_2, s_1) \end{aligned} \tag{5.42} $$

It follows from (5.28) that the first two integrals on the right side of (5.42) cancel each other. Thus (5.42) may be written

$$
\begin{aligned}
{}_1P^\mu(s_{1_1}) &+ {}_2P^\mu(s_{2_1}) - {}_1P^\mu(-\infty) - {}_2P^\mu(-\infty) \\
&= -\int_{s_{1_1}}^{\infty} ds_1 \int_{-\infty}^{\infty} ds_2\ {}_{12}\hat{F}^\mu(s_1, s_2) - \int_{s_{1_1}}^{\infty} ds_1 \int_{s_{2_1}}^{\infty} ds_2\ {}_{12}\hat{F}^\mu(s_1, s_2) \\
&\quad - \int_{s_{2_1}}^{\infty} ds_2 \int_{-\infty}^{s_{1_1}} ds_1\ {}_{21}\hat{F}^\mu(s_2, s_1) - \int_{s_{2_1}}^{\infty} ds_2 \int_{s_{1_1}}^{\infty} ds_1\ {}_{21}\hat{F}^\mu(s_2, s_1) \qquad (5.43)
\end{aligned}
$$

Once again it follows from (5.28) that the second and fourth integrals cancel, leaving us with

$$
\begin{aligned}
{}_1P^\mu(s_{1_1}) &+ {}_2P^\mu(s_{2_1}) - {}_1P^\mu(-\infty) - {}_2P^\mu(-\infty) \\
&= -\int_{s_{1_1}}^{\infty} ds_1 \int_{-\infty}^{s_{2_1}} ds_2\ {}_{12}\hat{F}^\mu(s_1, s_2) - \int_{s_{2_1}}^{\infty} ds_2 \int_{-\infty}^{s_{1_1}} ds_1\ {}_{21}\hat{F}^\mu(s_2, s_1) \qquad (5.44)
\end{aligned}
$$

The left side of (5.44) is the change in the total particle momentum in going from time $t = -\infty$ to time $t = t_1$. The right side of (5.44) will not in general be zero, and therefore the total particle momentum is not conserved. Suppose, however, that we define the *total momentum of the system* at time t_1, $P^\mu(t_1)$, as

$$P^\mu(t_1) = {}_1P^\mu(s_1) + {}_2P^\mu(s_2) + P_I{}^\mu(t_1) \qquad (5.45)$$

where $P_I{}^\mu(t_1)$ is the *interaction momentum* at time t_1 defined as

$$
\begin{aligned}
P_I{}^\mu(t_1) &\equiv {}_{12}P^\mu(t_1) + {}_{21}P^\mu(t_1) \\
&= \int_{s_{1_1}}^{\infty} ds_1 \int_{-\infty}^{s_{2_1}} ds_2\ {}_{12}\hat{F}^\mu(s_1, s_2) + \int_{s_{2_1}}^{\infty} ds_2 \int_{-\infty}^{s_{1_1}} ds_1\ {}_{21}\hat{F}^\mu(s_2, s_1) \qquad (5.46)
\end{aligned}
$$

Thus in this theory conservation of the system's momentum is preserved only by introducing a "system" or interaction momentum. The definition of total momentum presented in (5.45) and (5.46) is contrary to Postulate A, i.e., Eq. (5.6), of the No-Interaction Theorem; this is why it has been possible to circumvent the theorem, i.e., to construct a nontrivial dynamical theory of directly interacting particles.

The limits on the integrals appearing in (5.46) suggest the following interpretation. For ${}_{12}P^\mu(t_1)$ the integration over s_2 is the contribution of particle 2 to the momentum as it traverses its world line from $-\infty$ to s_{21}. The integration over s_1 indicates that portion of the world line of particle 1 which receives the momentum contributed by particle 2. In this case it follows that there is momentum contributed by particle 2 prior to time t_1 which arrives at (i.e., is "picked up" by) particle 1 after time t_2. This provides us with the concept of energy momentum in transit analogous to the energy and momentum of the field in electromagnetic field theory, but here of course there is no field. The interaction momentum is then the momentum that has left one particle but has not yet arrived at the other particle, and we were led to introduce this concept to preserve conservation of momentum.

The results obtained for two particles are readily extended to N particles. Thus the total momentum of an N particle system at time t_1 is

$$P^\mu(t_1) = \sum_{a=1}^{N} {}_aP^\mu(s_a) + \sum_{a \neq b} {}_{ab}P^\mu(t_1) \tag{5.47}$$

where

$$ {}_{ab}P^\mu(t_1) = \int_{s_{a_1}}^{\infty} ds_a \int_{-\infty}^{s_{b_1}} ds_b \, {}_{ab}\hat{F}^\mu(s_a, s_b) \tag{5.48}$$

The chief merit of the Wigner–Van Dam theory is that it demonstrates the possibility of constructing a many-particle direct interaction dynamics that is compatible with Einstein relativity. As with other classical theories the force is not determined by the theory, but its possible form is restricted by the relativity to be satisfied by the dynamics. In Section 5.2 it will be shown that the Wigner–Van Dam theory is capable of yielding inverse-square-law-type motion in the nonrelativistic limit. A major liability has been the inability to construct a corresponding Hamiltonian dynamics which would permit the theory to be quantized and applied to atomic or nuclear systems. Currie [3], Kerner [4], and Hill [5] have also been able to construct Newtonian-type dynamical theories compatible with Einstein relativity. While these two theories are similar to each other, they are apparently quite different from that of Wigner and Van Dam. We will now present Currie's theory; the reader may consult the references at the end of this chapter for the contributions of Kerner and Hill.

C. Newtonian-Type Theory of Currie

As in the Wigner–Van Dam theory, Currie presents dynamical equations in the Newtonian form $\mathbf{a} = \mathbf{F}$, but his dynamical variables are defined at a given instant of time t in an arbitrary Lorentz frame rather than the covariant variables of Wigner and Van Dam which are defined in terms of the proper time S, a Lorentz scalar. It is then necessary to inquire whether there exist nontrivial force functions **F** such that the dynamical equations are Lorentz invariant. The discussion will be restricted to two-particle systems, but the generalization to an arbitrary number of particles is straightforward.

The basic kinematical quantities, defined in a given frame, are the three vectors locating the position, velocity, and acceleration of the particles at a time t.

$$ {}_1x^i = {}_1x^i(t), \qquad {}_2x^i = {}_2x^i(t) \tag{5.49a} $$

$$ {}_1v^i \equiv d\,{}_1x^i/dt = {}_1\dot{x}^i, \qquad {}_2v^i \equiv d\,{}_2x^i/dt = {}_2\dot{x}^i \tag{5.49b} $$

$$ {}_1a^i \equiv d\,{}_1v^i/dt = {}_1\dot{v}^i, \qquad {}_2a^i \equiv d\,{}_2v^i/dt = {}_2\dot{v}^i \tag{5.49c} $$

It is advantageous to express the subsequent equations in terms of the average coordinates and relative coordinates defined, respectively, by

$$ \begin{aligned} X^i &= \tfrac{1}{2}({}_1x^i + {}_2x^i), \qquad & V^i &= \tfrac{1}{2}({}_1v^i + {}_2v^i), \\ A^i &= \tfrac{1}{2}({}_1a^i + {}_2a^i), \qquad & F^i &= \tfrac{1}{2}({}_1f^i + {}_2f^i) \end{aligned} \tag{5.50a} $$

$$ \begin{aligned} x^i &= \tfrac{1}{2}({}_1x^i - {}_2x^i), \qquad & v^i &= \tfrac{1}{2}({}_1v^i - {}_2v^i), \\ a^i &= \tfrac{1}{2}({}_1a^i - {}_2a^i), \qquad & f^i &= \tfrac{1}{2}({}_1f^i - {}_2f^i) \end{aligned} \tag{5.50b} $$

The equations of motion will in general have the form

$$ a^i = \ddot{x}^i = f^i(\mathbf{x}, \mathbf{X}, \mathbf{v}, \mathbf{V}) \tag{5.51a} $$

$$ A^i = \ddot{X}^i = F^i(\mathbf{x}, \mathbf{X}, \mathbf{v}, \mathbf{V}) \tag{5.51b} $$

At this point we make the crucial assumption that the positions, velocities, and accelerations transform in the usual way under Lorentz transformation. Then applying the transformations to (5.51) we obtain certain restrictions on the force functions. It is instructive to apply the various subgroups of the Poincaré group to (5.51) rather than the most general transformation.

Consider the invariance of the system under a spatial displacement, i.e., $\delta\,{}_\alpha x^i = b^i$ and $\delta t = 0$. Then it follows from (5.50) and (5.49) that

$$ \delta x^i = 0, \qquad \delta v^i = 0, \qquad \delta a^i = 0 \tag{5.52a} $$

$$ \delta X^i = b^i, \qquad \delta V^i = 0, \qquad \delta A^i = 0 \tag{5.52b} $$

The equations of motion (5.51a) become

$$ a^{i\prime} = a^i + \delta a^i = f^i(\mathbf{x} + \delta\mathbf{x}, \mathbf{X} + \delta\mathbf{X}, \mathbf{v} + \delta\mathbf{v}, \mathbf{V} + \delta\mathbf{V}) \tag{5.53a} $$

$$ A^{i\prime} = A^i + \delta A^i = F^i(\mathbf{x} + \delta\mathbf{x}, \mathbf{X} + \delta\mathbf{X}, \mathbf{v} + \delta\mathbf{v}, \mathbf{V} + \delta\mathbf{V}) \tag{5.53b} $$

Taylor expanding (5.53a) and (5.53b) we obtain

$$a^i + \delta a^i = f^i + \frac{\partial f^i}{\partial x_k}\,\delta x_k + \frac{\partial f^i}{\partial X_k}\,\delta X_k + \frac{\partial f^i}{\partial v_k}\,\delta v_k + \frac{\partial f^i}{\partial V_k}\,\delta V_k + \cdots \tag{5.54a}$$

$$A^i + \delta A^i = F^i + \frac{\partial F^i}{\partial x_k}\,\delta x_k + \frac{\partial F^i}{\partial X_k}\,\delta X_k + \frac{\partial F^i}{\partial v_k}\,\delta v_k + \frac{\partial F^i}{\partial V_k}\,\delta V_k + \cdots \tag{5.54b}$$

Applying (5.52) to (5.54) we conclude that

$$\partial f^i/\partial X_k = 0, \qquad \partial F^i/\partial X_k = 0 \tag{5.55}$$

or, in other words, that the force functions are independent of the average positions.

Next we consider the invariance of the system under the homogeneous part of the Poincaré group. It is essential in this three-vector formulation to define the positions of the particles entering into the equation of motion in the different Lorentz frames. This is done in Fig. 5.4. We wish to express our equations of motion in a given Lorentz frame in the usual Newtonian manner, i.e., we define the positions of the two particles at a given instant of time in that frame. Thus the unprimed observer records the positions A and B of the particles as being simultaneous, likewise A' and C. We proceed under the assumption that the usual Newtonian data is available, viz.,

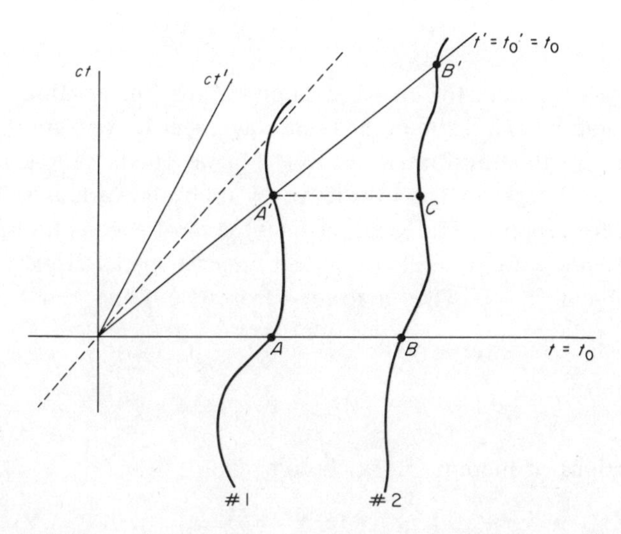

Fig. 5.4. Definitions of the position functions as measured in the primed and unprimed frames.

the position of the particles at a given time t and all of its time derivatives at some initial time $t = t_0$. From this information we seek the position of the particles in the primed frame which are simultaneous in the primed frame at the time t' which is numerically equal to the time t_0 in the unprimed frame when the initial data is available, i.e., the positions A' and B'.

The transformation equations are

$$ {}_\alpha x^{i'}(t_\alpha') = L^i{}_0\, ct_\alpha + L^i{}_j\, {}_\alpha x^j(t_\alpha) \tag{5.56a}$$

$$ t_\alpha = (1/c) L^\varrho{}_0\, {}_\alpha x_\varrho'(t_\alpha') \tag{5.56b}$$

The t_α which appear in (5.56a) are obtained from (5.56b). The velocity in the primed frame is

$$ \frac{d\, {}_\alpha x^{i'}}{dt_\alpha'} = \frac{dt_\alpha}{dt_\alpha'} \left(L_0{}^i\, c + L_j{}^i \frac{d\, {}_\alpha x^i(t_\alpha)}{dt_\alpha} \right) \tag{5.57}$$

where from (5.56b) it follows that

$$ \frac{dt_\alpha}{dt_\alpha'} = \frac{1}{c} L^\varrho{}_0 \frac{d\, {}_\alpha x_\varrho'(t_\alpha')}{dt_\alpha'} \tag{5.58}$$

We restrict ourselves to infinitesmal transformations of the form

$$ L^\mu{}_\nu = g^\mu{}_\nu + \omega^\mu{}_\nu \tag{5.59}$$

where $\omega^\mu{}_\nu$, the infinitesmal part of the transformation is antisymmetric

$$ \omega^\mu{}_\nu = -\omega^\nu{}_\mu \tag{5.60}$$

Equation (5.56b) becomes

$$ t_\alpha = t_\alpha' + (1/c)\omega_0{}^i\, {}_\alpha x_i'(t_\alpha') \tag{5.61}$$

Since t_α differs from t_α' only by an infinitesmal, it is advantageous to perform a Taylor expansion of the ${}_\alpha x^j(t_\alpha)$ in (5.56a)

$$ {}_\alpha x^{j'}(t_\alpha') = L_0{}^i\, c[t_\alpha' + (1/c)\omega^{j0}\, {}_\alpha x_j'(t_\alpha')] $$
$$ + L^i{}_j[{}_\alpha x^j(t_\alpha') t_\alpha \dot{x}^j\, |_{t_\alpha = t_\alpha'} (t_\alpha - t_\alpha') + \tfrac{1}{2}\, {}_\alpha \ddot{x}^j\, |_{t_\alpha = t_\alpha'} (t_\alpha - t_\alpha')^2 + \cdots] $$

Using (5.61) to replace $(t_\alpha - t_\alpha')$ and employing (5.59), we obtain to first order in the infinitesimals

$$ {}_\alpha x^{i'}(t_\alpha') = {}_\alpha x^i(t_\alpha') + \omega_0{}^i c t_\alpha' + \omega_j{}^i\, {}_\alpha x^j(t_\alpha') + {}_\alpha \dot{x}^i(t_\alpha')(1/c)\omega_0{}^k\, {}_\alpha x_k'(t_\alpha') $$

This equation can be iterated to eliminate the ${}_\alpha x_k'(t_\alpha')$ in the last term on the right. We obtain to first order

$$ {}_\alpha x^{i'}(t_\alpha') = {}_\alpha x^i(t_\alpha') + \omega_0{}^i\, ct_\alpha' + \omega_j{}^i\, {}_\alpha x^j(t_\alpha') + (1/c)\omega_0{}^k\, {}_\alpha x_k(t_\alpha')\, {}_\alpha \dot{x}^i(t_\alpha') \tag{5.62a} $$

The significant feature of this equation is that the unprimed position functions are defined in terms of t_α'. It is now a simple matter to write expressions for the velocity and acceleration.

$$ \frac{d\, {}_\alpha x^{i'}(t_\alpha'')}{dt_\alpha'} = {}_\alpha v^{i'}(t_\alpha') = \omega_0{}^i\, c + \frac{d\, {}_\alpha x^i(t_\alpha')}{dt_\alpha'} + \omega_j{}^i \frac{d\, {}_\alpha x^j(t_\alpha')}{dt_\alpha'} $$
$$ + (1/c)\omega_0{}^k[{}_\alpha \ddot{x}^i(t_\alpha')\, {}_\alpha x_k(t_\alpha') + {}_\alpha \dot{x}^i(t_\alpha')\, {}_\alpha \dot{x}_k(t_\alpha')] \tag{5.62b} $$

$$ \frac{d^2\, {}_\alpha x^{i'}(t_\alpha')}{dt_\alpha'} = {}_\alpha a^{i'}(t_\alpha') $$
$$ = \frac{d^2\, {}_\alpha x^i(t_\alpha')}{dt_\alpha'^2} + \omega_j{}^i \frac{d^2\, {}_\alpha x^j(t_\alpha')}{dt_\alpha'^2} $$
$$ + (1/c)\omega^{k0}[{}_\alpha \dddot{x}^i(t_\alpha')\, {}_\alpha x_k(t_\alpha') + 2\, {}_\alpha \ddot{x}^i(t_\alpha')\, {}_\alpha \dot{x}_k(t_\alpha')\, {}_\alpha \dot{x}^i(t_\alpha')\, {}_\alpha \ddot{x}_k(t_\alpha')] \tag{5.62c} $$

We now specialize these results as discussed previously, i.e., we require the primed positions to be defined at the same time in the primed system and also that this time be numerically equal to the unprimed time t_0, which we may conveniently set equal to zero.

$$ t_1' = t_2' = t_0 = 0 \tag{5.63} $$

Substituting (5.63) and (5.62) into (5.50) we find

$$ X^{i'} = X^i + \omega^i{}_j\, X^j + (1/c)\omega^k{}_0\, (V^i X_k + v^i x_k) \tag{5.64} $$

$$ x^{i'} = x^i + \omega^i{}_j\, x^j + (1/c)\omega^k{}_0\, (v^i X_k + V^i x_k) \tag{5.65} $$

$$ V^{i'} = \omega^i{}_0\, c + V^i + \omega^i{}_j\, V^j $$
$$ + (1/c)\omega^k{}_0\, (A^i X_k + a^i x_k + V^i V_k + v^i v_k) \tag{5.66} $$

$$ v^{i'} = v^i + \omega^i{}_j\, v^j + (1/c)\omega^k{}_0\, (a^i X_k + A^i x_k + v^i V_k + V^i v_k) \tag{5.67} $$

$$ A^{i'} = A^i + \omega^i{}_j\, A^j + (1/c)\omega^k{}_0\, (\dot{A}^i X_k + \dot{a}^i x_k $$
$$ + 2[A^i V_k + a^i v_k] + V^i A_k + v^i a_k) \tag{5.68} $$

$$ a^{i'} = a^i + \omega^i{}_j\, a^j + (1/c)\omega^k{}_0\, (\dot{a}^i X_k + \dot{A}^i x_k $$
$$ + 2[a^i V_k + A^i v_k] + v^i A_k + V^i a_k) \tag{5.69} $$

The basic assumption of the theory is that the dynamical equations transform according to

$$a^{i'} = a^i + \Delta a^i = f^i(\mathbf{x}', \mathbf{v}', \mathbf{V}') = f^i(\mathbf{x} + \delta\mathbf{x}, \mathbf{v} + \delta\mathbf{v}, \mathbf{V} + \delta\mathbf{V}) \tag{5.70a}$$

$$A^{i'} = A^i + \Delta A^i = F^i(\mathbf{x}', \mathbf{v}', \mathbf{V}') = F^i(\mathbf{x} + \delta\mathbf{x}, \mathbf{v} + \delta\mathbf{v}, \mathbf{V} + \delta\mathbf{V}) \tag{5.70b}$$

where

$$\begin{aligned} \Delta a^i &= a^{i'} - a^i, \qquad & \Delta v^i &= v^{i'} - v^i, \\ \Delta x^i &= x^{i'} - x^i, \qquad & \Delta V^i &= V^{i'} - V^i \end{aligned} \tag{5.71a}$$

and the quantities appearing in (5.71a) are obtained from Eqs. (5.64)–(5.69). Taylor expanding the force functions we obtain to first order

$$\Delta a^i = f^i_{x_j}\,\Delta x^j + f^i_{v_j}\,\Delta v^j + f^i_{V_j}\,\Delta V^j \tag{5.71b}$$

$$\Delta A^i = F^i_{x_j}\,\Delta x^j + F^i_{v_j}\,\Delta v^j + F^i_{V_j}\,\Delta V^j \tag{5.71c}$$

where

$$f^i_{x_j} = \partial f^i/\partial x^j, \qquad f^i_{v_j} = \partial f^i/dv^j, \qquad \text{etc.} \tag{5.72}$$

It follows from (5.71a) and (5.64)–(5.69) that

$$\begin{aligned} &\omega^i{}_j\, a^j + \frac{1}{c}\,\omega_0{}^k\,(\dot{A}^i x_k + 2[a^i V_k + A^i v_k] + v^i A_k + V^i a_k) \\ &\quad = f^i_{x_j}\Big(\omega^j{}_k\, x^k + \frac{1}{c}\,\omega^k{}_0\, V^j x_k\Big) \\ &\qquad + f^i_{v_j}\Big(\omega^j{}_k\, v^k + \frac{1}{c}\,\omega^k{}_0\,[A^j x_k + v^j V_k + V^j v_k]\Big) \\ &\qquad + f^i_{V_j}\Big(\omega_k{}^j\, V^k + \frac{1}{c}\,\omega^k{}_0\,[a^j x_k + V^j V_k + v^j v_k] + \omega^j{}_0\, c\Big) \end{aligned} \tag{5.73a}$$

$$\begin{aligned} &\omega^i{}_j\, A^j + \frac{1}{c}\,\omega^k{}_0\,(\dot{A}^i X_k + 2[A^i V_k + a^i v_k] + V^i A_k + v^i a_k) \\ &\quad = F^i_{x_j}\Big(\omega^j{}_k\, x^k + \frac{1}{c}\,\omega^k{}_0\, V^j x_k\Big) \\ &\qquad + F^i_{v_j}\Big(\omega^j{}_k\, v^k + \frac{1}{c}\,\omega^k{}_0\,[A^j x_k + v^j V_k + V^j v_k]\Big) \\ &\qquad + F^i_{V_j}\Big(\omega^j{}_k\, V^k + \frac{1}{c}\,\omega^k{}_0\,[a^j x_k + V^j V_k + v^j v_k] + \omega^j{}_0\, c\Big) \end{aligned} \tag{5.73b}$$

Some terms have been eliminated by noting that

$$\dot{a}^i = \dot{f}^i = f^i_{x_j} v^j + f^i_{V_j} A^j + f^i_{v_j} a^j \tag{5.74a}$$

$$\dot{A}^i = \dot{F}^i = F^i_{x_j} v^j + F^i_{V_j} A^j + F^i_{v_j} a^j \tag{5.74b}$$

Equations (5.73) should be regarded as conditions imposed on the force functions by the requirement of Lorentz invariance of the dynamical equations when the kinematical quantities transform as indicated in Eqs. (5.64)–(5.69).

Since the $\omega_j{}^i$ and $\omega_0{}^k$ are independent sets of parameters, for (5.73) to be satisfied it is necessary that

$$\omega_j{}^i f^j = \omega_k{}^i \left(f^i_{x_j} x^k + f^i_{v_j} v^k + f^i_{V_j} V^k\right) \tag{5.75a}$$

$$\omega_j{}^i F^j = \omega_k{}^j \left(F^i_{x_j} x^k + F^i_{v_j} v^k + F^i_{V_j} V^k\right) \tag{5.75b}$$

$$\begin{aligned}\dot{F}^i x_k + 2[f^i V_k + F^i v_k] + v^i F_k + V^i f_k = {} & f^i_{x_j} V^j x_k + f^i_{v_j}(F^j x_k + v^j V_k + V^j v_k) \\ & + f^i_{V_j}(f^j x_k + V^j V_k + v^j v_k) + c^2 f^i_{V_k}\end{aligned} \tag{5.76a}$$

$$\begin{aligned}\dot{F}^i X_k + 2[F^i V_k + f^i v_k] + V^i f_k + v^i f_k = {} & F^i_{x_j} V^j x_k + F^i_{v_j}(F^j x_k + v^j V_k + V^j v_k) \\ & + F^i_{V_j}(f^j x_k + V^j V_k + v^j v_k) + c^2 F^i_{V_k}\end{aligned} \tag{5.76b}$$

Equations (5.75)–(5.76) constitute a system of nonlinear partial differential equations which must be satisfied by a force function. The most important question is whether there exists a nontrivial force function which satisfies these equations. If so, then this theory circumvents the no-interaction theorem.

Let us first consider whether action–reaction-type forces are compatible with this formalism. Action–reaction forces are defined by

$$_1f^i = -_2f^i, \qquad F^i = 0, \qquad f^i = 2\,_1f^i \tag{5.77}$$

Inserting $F^i = 0$ into (5.76b), we obtain

$$2f^i v_k + v^i f_k = 0 \tag{5.78}$$

where we have taken our reference frame such that $V^i = 0$ which is possible since $F^i = 0$. Equation (5.78) can be satisfied for $k = i$ only if

$$f^i = 0 \tag{5.79}$$

We, therefore, conclude that action–reaction-type forces are not compatible with this formalism.

Before pursuing the question of the existence of a nontrivial force function, we impose several additional invariance requirements, which will simplify the subsequent discussion. Consider first the *parity transformation* P in which the coordinates are replaced with their negative, i.e.,

$$P\ {}_{\alpha}x^j = -{}_{\alpha}x^j \tag{5.80}$$

Under this operation the other kinematical quantities transform as follows

$$PX^j = -X^j, \qquad PV^j = -V^j, \qquad PA^j = -A^j \tag{5.81a}$$

$$Px^j = -x^j, \qquad Pv^j = -v^j, \qquad Pa^j = -a^j \tag{5.81b}$$

The *time reversal transformation* T corresponds to the replacement of t with $-t$, i.e.,

$$Tt = -t \tag{5.82}$$

Under time reversal the other kinematical quantities transform as follows:

$$TX^j = X^j, \qquad TV^j = -V^j, \qquad TA^j = A^j \tag{5.83a}$$

$$Tx^j = x^j, \qquad Tv^j = -v^j, \qquad Ta^j = a^j \tag{5.83b}$$

Finally, the *particle interchange transformation* C in which

$$C\ {}_1x^j = {}_2x^j, \qquad C\ {}_2x^j = {}_1x^j \tag{5.84}$$

Under particle interchange the kinematical quantities transform as follows:

$$CX^j = X^j, \qquad CV^j = V^j, \qquad CA^j = A^j \tag{5.85a}$$

$$Cx^j = -x^j, \qquad Cv^j = -v^j, \qquad Ca^j = -a^j \tag{5.85b}$$

We now require that the dynamical equations be invariant under any one or combination of the discrete transformations just defined. The particle interchange symmetry is certainly to be expected if the particles are identical, which we henceforth assume. Parity and time reversal are elements of the Poincaré group.

To illustrate the nature of the restrictions imposed by these discrete symmetries consider the application of the transformations PCT to (5.75a). The net affect of PCT on the kinematical quantities is

$$PCTx^j = x^j, \qquad PCTv^j = -v^j, \qquad PCTV^j = V^j, \qquad PCTf^j = f^j \tag{5.86}$$

Therefore PCT operating on both sides of (5.75a) yields

$$\omega^i{}_j f^j = \omega^j{}_k \left(x^k[PCT]f^i_{x_j} - v^k[PCT]f^i_{v_j} + V^k[PCT]f^i_{V_j}\right) \tag{5.87}$$

For (5.84) to equal (5.75a), i.e., for (5.75a) to be invariant under the application of PCT it is necessary that

$$f^i_{x_j}(\mathbf{x}, -\mathbf{v}, \mathbf{V}) = f^i_{x_j}(\mathbf{x}, \mathbf{v}, \mathbf{V}) \tag{5.88a}$$

$$f^i_{v_j}(\mathbf{x}, -\mathbf{v}, \mathbf{V}) = -f^i_{v_j}(\mathbf{x}, \mathbf{v}, \mathbf{V}) \tag{5.88b}$$

$$f^i_{V_j}(\mathbf{x}, -\mathbf{v}, \mathbf{V}) = f^i_{V_j}(\mathbf{x}, \mathbf{v}, \mathbf{V}) \tag{5.88c}$$

In other words invariance requires that $f^i_{x_j}$ be an even function of $\mathbf{v}$. Application of PC, PT, and TC to (5.75a) leads to the following restrictions on f^i and its derivatives (the restrictions on f^i follow from the fact that $a^i = f^i$):

$$f^i(\mathbf{x}) = -f^i(-\mathbf{x}), \qquad f^i(\mathbf{v}) = f^i(-\mathbf{v}), \qquad f^i(\mathbf{V}) = f^i(-\mathbf{V}) \tag{5.89a}$$

$$f^i_{x_j}(\mathbf{x}) = f^i_{x_j}(-\mathbf{x}), \qquad f^i_{x_j}(\mathbf{v}) = f^i_{x_j}(-\mathbf{v}), \qquad f^i_{x_j}(\mathbf{V}) = f^i_{x_j}(-\mathbf{V}) \tag{5.89b}$$

$$f^i_{v_j}(\mathbf{x}) = -f^i_{v_j}(-\mathbf{x}), \qquad f^i_{v_j}(\mathbf{v}) = -f^i_{v_j}(-\mathbf{v}), \qquad f^i_{v_j}(\mathbf{V}) = f^i_{v_j}(-\mathbf{V}) \tag{5.89c}$$

$$f^i_{V_j}(\mathbf{x}) = -f^i_{V_j}(-\mathbf{x}), \qquad f^i_{V_j}(\mathbf{v}) = f^i_{V_j}(-\mathbf{v}), \qquad f^i_{V_j}(\mathbf{V}) = -f^i_{V_j}(-\mathbf{V}) \tag{5.89d}$$

A similar application of the discrete transformations to (5.75b) yields

$$F^i(\mathbf{x}) = -F^i(-\mathbf{x}), \qquad F^i(\mathbf{v}) = -F^i(-\mathbf{v}), \qquad F^i(\mathbf{V}) = -F^i(-\mathbf{V}) \tag{5.90a}$$

$$F^i_{x_j}(\mathbf{x}) = F^i_{x_j}(-\mathbf{x}), \qquad F^i_{x_j}(\mathbf{v}) = -F^i_{x_j}(-\mathbf{v}), \qquad F^i_{x_j}(\mathbf{V}) = -F^i_{x_j}(-\mathbf{V}) \tag{5.90b}$$

$$F^i_{v_j}(\mathbf{x}) = -F^i_{v_j}(-\mathbf{x}), \qquad F^i_{v_j}(\mathbf{v}) = F^i_{v_j}(-\mathbf{v}), \qquad F^i_{v_j}(\mathbf{V}) = -F^i_{v_j}(-\mathbf{V}) \tag{5.90c}$$

$$F^i_{V_j}(\mathbf{x}) = -F^i_{V_j}(-\mathbf{x}), \qquad F^i_{V_j}(\mathbf{v}) = -F^i_{V_j}(-\mathbf{v}), \qquad F^i_{V_j}(\mathbf{V}) = F^i_{V_j}(-\mathbf{V}) \tag{5.90d}$$

Returning now to the question of the existence of a nontrivial force function, we assume that F^i and f^i may be expanded in a power series in V^j. This is not as general as it could be, but we only wish to demonstrate nontrivial forces exist, and if they exist in this case then certainly they exist for the more general case. According to (5.90a) and (5.89a) the series expansion for F^i must be odd in V^j, whereas f^i must be even in V^j; thus

$$F^i = F^i{}_j\,(\mathbf{x}, \mathbf{v})V^j + F^i_{jkl}(\mathbf{x}, \mathbf{v})V^jV^kV^l + \cdots \tag{5.91a}$$

$$f^i = f_0{}^i(\mathbf{x}, \mathbf{v}) + f^i_{jk}(\mathbf{x}, \mathbf{v})V^jV^k + \cdots \tag{5.91b}$$

Inserting (5.91a) and (5.91b) into (5.76b) and equating the coefficients of like powers of V we obtain for the zeroth power

$$F^i{}_j\,(f_0{}^j x_k + v^j v_k + c^2 g^j{}_k) = 2 f_0{}^i v_k + v^i f_{0k} \tag{5.92}$$

This set of equations can be solved to yield the coefficients $F^i{}_j$ in terms of $f_0{}^j$ provided that

$$(f_0{}^j x_k + v^j v_k + c^2 g^j{}_k) \neq 0 \tag{5.93}$$

Repeating this process for higher powers of V, we can determine all the coefficients of the expansions appearing in (5.91a) and (5.91b) in terms of the lowest order term $f_0{}^i$.

The $f_0{}^i$ are not entirely arbitrary since insertion of (5.91b) into (5.89a) leads to the requirement that

$$f_0{}^i(\mathbf{x}) = -f_0{}^i(-\mathbf{x}), \qquad f_0{}^i(\mathbf{v}) = f_0{}^i(-\mathbf{v}) \tag{5.94}$$

and inserting (5.91b) into (5.75a) yields

$$\omega^i{}_j f_0{}^j = \omega^j{}_k\,(f_0{}^i x^k + f_0{}^i v^k) \tag{5.95}$$

It is easy to show that

$$f_0{}^i = x^i R(r) u(v) \tag{5.96}$$

where

$$r = [(x^1)^2 + (x^2)^2 + (x^3)^2]^{1/2}, \qquad v = [(v^1)^2 + (v^2)^2 + (v^3)^2]^{1/2} \tag{5.97}$$

is a nontrivial force function that is compatible with all the restrictions. Of course this is not the only possible kind of force but the fact that forces of this type are consistent with Lorentz invariance demonstrates that the Currie formalism does indeed circumvent the No-Interaction Theorem to provide a nontrivial dynamics of multiparticle systems compatible with Einstein relativity.

It is also worth noting that the relativity associated with either a Euclidean or Einstein dynamics is not sufficient to determine the nature of the forces that may exist between particles, although it does impose some limitations on them. Only experiment can determine the nature of the forces that exist in nature. Examples of relativistic Newtonian type forces are presented by Wray [6] and Currie and Jordan [7].

Currie's formalism circumvents the No-Interaction Theorem by violating Postulate A. That this is so is readily seen from (5.77) and (5.79) where it is shown that to have an interaction between the particles, i.e., $f^i \neq 0$, it is necessary that

$$ {}_1f^i + {}_2f^i \neq 0 \tag{5.98} $$

But this means that

$$ \frac{d}{dt}({}_1v^i + {}_2v^i) \neq 0 \tag{5.99} $$

i.e., that Postulate A is not satisfied. Actually Eqs. (5.77) and (5.79) constitute a statement of the No-Interaction Theorem within the Currie formalism, and has the advantage of being independent of a postulate concerning the asymptotic properties of the motion.

Hill and Kerner [5] have presented a technique for obtaining the Hamilton equations of motion in a formalism such as Currie's, but little, if any actual progress has been made in the application of the theory to actual physical problems or to the development of a quantum version of the theory.

Both of the theories presented have circumvented the No-Interaction Theorem. The reader may consult the literature [6–15] for further developments of these theories and several other theories that have been advanced to handle the relativistic multiparticle system.

5.2. WIGNER–VAN DAM THEORY (TWO-BODY MOTION WHEN $m_2 \gg m_1$)

Since in solving (5.39) for ${}_2\ddot{x}^\mu$ the mass m_2 appears in the denominator, we assume ${}_2\ddot{x}^\mu = 0$ from which it follows that the four-velocities ${}_2\dot{x}^\mu$ are independent of the proper time s_2. Therefore we may choose a Lorentz observer which moves with particle 2 located at its origin. Then

$$ {}_2\dot{x}^\mu = (d_2\dot{x}/ds_2)(d\,{}_2x^\mu/d\,{}_2x^0) = \gamma_2 g^{\mu 0} = g^{\mu 0} \tag{5.100} $$

where

$$ \left(\frac{d\,{}_2x^0}{ds_2}\right) = \gamma_2 \equiv \frac{1}{(1 - {}_2v^2/c^2)^{1/2}} = 1 \tag{5.101} $$

and

$$ {}_2x^k = 0, \qquad {}_2x^0 = ct_2 = s_2 \tag{5.102} $$

To determine the orbit of particle 1 we insert (5.100) and (5.102) in (5.39a) and (5.25). Thus

$$m_1\, {}_1\ddot{x}^k(s_1) = \int_{-\infty}^{\infty} ds_2\, \varphi_{12}(\varrho_{12})\xi_{12}[{}_1x^k(s_1) - {}_2x^k(s_2)]$$

$$= {}_1\dot{x}^0(s_1)\, {}_1x^k(s_1) \int_{-\infty}^{\infty} ds_2\, \varphi_{12}(\varrho_{12}) \tag{5.103}$$

and

$$m_1\, {}_1\ddot{x}^0(s_1) = {}_1\dot{x}^k(s_1)\, {}_1x_k(s_1) \int_{-\infty}^{\infty} ds_2\, \varphi_{12}(\varrho_{12}) \tag{5.104}$$

It follows from (5.102) that ϱ_{12} may be written

$$\varrho_{12} = [({}_1x^0(s_1) - s_2)^2 - ({}_1x^k)^2]^{1/2} = [({}_1x^0(s_1) - s_2)^2 - r^2]^{1/2} \equiv i\varrho \tag{5.105}$$

Since particle 2 is at the spatial origin, r is just the spatial distance between particle 1 and particle 2. From (5.35) we know that φ_{12} is nonzero only when ϱ_{12} is imaginary, and therefore in (5.105) we have introduced ϱ in such a way that φ_{12} is nonzero only when ϱ is real. Using (5.105) we can convert the integrals appearing in (5.103) and (5.104) into integrals over ϱ, i.e.,

$$\int_{-\infty}^{\infty} ds_2\, \varphi_{12}(\varrho) = 2\int_0^r d\varrho\, (\varrho\varphi(\varrho)/(r^2 - \varrho^2)^{1/2}) = I \tag{5.106}$$

For the Wigner–Van Dam theory to be viable, it should yield solutions which in some appropriate nonrelativistic limit correspond to the usual central force-type orbits. With this in mind we write the dynamical equations in terms of the usual orbital parameters r and θ. In an appropriate coordinate system

$${}_1x^1 = r\cos\theta, \qquad {}_1x^2 = r\sin\theta \tag{5.107a}$$

$${}_1\beta^1 = d\,{}_1x^1/d_1\,x^0 = {}_1x^{1\prime} = r'\cos\theta - r\theta'\sin\theta \tag{5.107b}$$

$${}_1\beta^2 = {}_1x^{2\prime} = r'\sin\theta + r\theta'\cos\theta \tag{5.107c}$$

Substituting (5.106) and (5.107) into (5.103) and noting that

$${}_1\dot{x}^k = \frac{d\,{}_1x^0}{ds_1}\,\frac{d\,{}_1x^k}{d\,{}_1x^0} = \gamma'\,{}_1x^{k\prime} \tag{5.108}$$

we obtain

$$m\gamma[r'' - r(\theta')^2] - r(r')^2 I - rI = 0 \tag{5.109a}$$

$$m\gamma[2r'\theta' + r\theta''] - r^2r'\theta' I = 0 \tag{5.109b}$$

To obtain (5.109) we have also used

$$\frac{d\gamma}{dx^0} = \gamma^3\, {}_1\boldsymbol{\beta} \cdot {}_1\boldsymbol{\beta}' = -\frac{{}_1\boldsymbol{\beta} \cdot {}_1\mathbf{x}}{m} I = -\frac{rr'}{m} I \tag{5.110}$$

which follows from (5.104). From (5.109b) it follows that

$$\theta' = \frac{\lambda}{r^2} \exp\left(\int \frac{Ir}{m\gamma}\, dr\right) \tag{5.111}$$

The equation for the orbit can be obtained from (5.109a) by using (5.111) to replace the independent variable ${}_1x^0$ with θ. In addition we secure a somewhat simpler differential equation by replacing r with u where

$$u = 1/r \tag{5.112}$$

Thus the equation for the orbit is

$$\frac{d^2u}{d\theta^2} + u + \frac{I}{m\gamma\lambda^2 u^3} \exp\left(2 \int \frac{I}{m\gamma u^3}\, du\right) = 0 \tag{5.113}$$

Let us for the moment assume that $I = Ku^3$. [It is clear from (5.103) or (5.109) that this corresponds to an inverse square law force.] Equation (5.13) becomes (assuming γ is a constant)

$$\frac{d^2u}{d\theta^2} + u + \frac{k}{m\gamma\lambda^2}\, e^{2uk/m\gamma} = 0 \tag{5.114}$$

Expanding the exponential in a power series and ignoring terms in u to the second and higher powers, (5.114) becomes

$$\frac{d^2u}{d\theta^2} + (1 + \delta)u + \frac{k}{m\gamma\lambda^2} = 0 \tag{5.115}$$

where

$$\delta = 2\left(\frac{k}{m\gamma\lambda}\right)^2 \tag{5.116}$$

It is not difficult to show that (5.115) admits solutions whose orbit is a precessing ellipse with the precession rate being determined by δ.

REFERENCES

1. H. Van Dam and E. P. Wigner, *Phys. Rev.* **142**, 838 (1966).
2. H. Van Dam and E. P. Wigner, *Phys. Rev. B* **138**, 1576 (1965).
3. D. G. Currie, *Phys. Rev.* **142**, 817 (1966).
4. E. H. Kerner, *J. Math. Phys.* (*N. Y.*) **6**, 1218 (1965); *Phys. Rev. Lett.* **16**, 667 (1966).
5. R. N. Hill and E. H. Kerner, *Phys. Rev. Lett.* **17**, 1157 (1966).
6. J. G. Wray, *Phys. Rev.* **1**, 2212 (1970).
7. D. G. Currie and T. F. Jordan, *in* "Lectures in Theoretical Physics." (A. O. Barut and W. E. Brittin, eds.), Vol. XA, p. 91. Gordon Breach, New York, 1968.
8. H. Tetrode, *Z. Phys.* **10**, 317 (1922).
9. A. D. Fokker, *Z. Phys.* **58**, 386 (1929); *Physica* (*Utrecht*) **9**, 33; **12**, 145 (1929).
10. J. A. Wheeler and R. P. Feynman, *Rev. Mod. Phys.* **21**, 424 (1949).
11. C. Garrod, *Phys. Rev.* **167**, 1143 (1968).
12. A. Degasperis, *Phys. Rev. D* **3**, 273 (1971).
13. T. F. Jordan, *Phys. Rev.* **166**, 1308 (1968).
14. P. Havas and J. Stachel, *Phys. Rev.* **185**, 1636 (1969).
15. C. Fronsdal, *Phys. Rev. D* **3**, 1299 (1971).

6

GROUP THEORY

6.1. GROUP THEORY IN CLASSICAL MECHANICS

The impact of group theory and its related algebras on contemporary physics is truly phenomenal. The scientific literature from chemistry to general relativity and elementary particle physics abounds with examples of the power variety and popularity of this mathematical tool. It has become as essential to physics as vector calculus and complex variable theory. Group theory permits us to exploit to the fullest possible extent the symmetry that is so common in nature and lies at the foundation of our physical theories.

Although group theory has not been part of the folklore of classical dynamics, its successes in other branches of physics, e.g., quantum mechanics, have led to an awareness that classical dynamics can also profit handsomely by its use. It is certainly beyond the scope of this book to develop the mathematical theory of groups and to display the full richness of its application to the analysis of physical systems. We present two representative illustrations of its use in classical dynamics in the hope that they will serve to whet the reader's appetite and to induce him to consult the references at the end of this chapter. A very brief review of the fundamentals of group theory required for this chapter is presented in Appendix E.

The first illustration, the factorization of the dynamical matrix, demonstrates a common application of finite groups. The second illustration is concerned with how a canonical formalism may be constructed from a symmetry group.

6.2. FACTORING THE DYNAMICAL MATRIX

While our discussion of small oscillation theory in Chapter 2 is generally applicable to any system oscillating about a stable equilibrium point, the solution of the eigenvalue problem to obtain the normal mode states and frequencies becomes increasingly more tedious (or expensive if a computer is used) as the number of degrees of freedom of the system increases. Since a normal mode analysis is essential to a discussion of the dynamics of both macroscopic and microscopic systems such as molecules and crystals, it is imperative that a device to reduce the labor involved in solving the eigenvalue problem for more complex systems be developed. Group theory, utilizing the symmetry of the system, is just such a device.

We summarize the technique as follows: A physical system having N degrees of freedom is described by a system of differential equations, the dynamical content of which is contained in the N by N dynamical matrix. The dynamical equations are reduced to a readily integrable form if we can find a transformation to new coordinates (the normal mode coordinates) in terms of which the dynamical matrix is reduced to diagonal form (the diagonal elements being directly related to the normal mode frequencies). In Chapter 2, it was shown that the appropriate transformation is determined by solving the eigenvalue problem for the N-dimensional dynamical matrix.

If the system admits a symmetry group, i.e., if the system is invariant under the operations of the elements of a group, the labor involved in diagonalizing the dynamical matrix is often minimized by first reducing the dynamical matrix to *quasi-diagonal* (or *block diagonal*) form as shown in (6.1).

$$A = \begin{bmatrix} a_{11} & a_{12} & a_{13} \cdots & a_{1n} \\ a_{21} & a_{22} & \cdot \cdots & a_{2n} \\ \vdots & & & \\ a_{n1} & a_{n2} & \cdot \cdots & a_{nn} \end{bmatrix} \rightarrow \begin{bmatrix} \boxed{\begin{matrix} a'_{11} & a'_{12} \\ a'_{21} & a'_{22} \end{matrix}} & & & 0 \\ & \ddots & & \\ & & \boxed{a'_{33}} & \\ & & & \ddots \\ 0 & & & \boxed{\begin{matrix} a'_{44} & \cdots & a'_{4n} \\ \vdots & & \vdots \\ a'_{n4} & \cdots & a'_{nn} \end{matrix}} \end{bmatrix} \tag{6.1}$$

The matrix A produces the coupled set of equations

$$\ddot{q}_i = A_{ij}q_j, \qquad i, j = 1, 2, \ldots, N \tag{6.2}$$

but A' produces the equations

$$\begin{bmatrix} \ddot{q}_1' \\ \ddot{q}_2' \\ \vdots \\ \ddot{q}_n' \end{bmatrix} = \begin{bmatrix} a'_{11} & a'_{12} & & & & 0 \\ a'_{21} & a'_{22} & & & & \\ & & a'_{33} & & & \\ & & & a'_{44} & \cdots & a'_{4n} \\ & 0 & & \vdots & & \\ & & & a'_{n4} & \cdots & a'_{nn} \end{bmatrix} \begin{bmatrix} q_1' \\ q_2' \\ q_3' \\ q_4' \\ \vdots \\ q_n' \end{bmatrix} \tag{6.3a}$$

$$\ddot{q}_a' = A'_{ab}q_b', \qquad a, b = 1, 2 \tag{6.3b}$$

$$\ddot{q}_3' = A'_{33}q_3' \tag{6.3c}$$

$$\ddot{q}_c' = A'_{cd}q_d', \qquad c, d = 4, 5, \ldots, N \tag{6.3d}$$

A partial decoupling of the equations has occurred, i.e., block 1 of A' mixes, operates on, only the elements q_1', q_2' of the coordinate matrix, and so forth, and their solution is greatly simplified. Equation (6.3b) could be solved immediately if A'_{33} is a constant, and to solve (6.3a) we merely have to diagonalize a 2×2 matrix, A'_{ab}, and so on. We have in effect "factored" the original matrix into a set of matrices of lower order, each of which may be independently diagonalized. The key to the whole process is the coordinate transformation, represented in (6.1) by the arrow, which converts the original matrix into block diagonal form; the determination of this transformation is a routine step in the process of reducing a representation ol a group into its irreducible parts. So the overall program which we shall study step by step is:

(a) the determination of the symmetry group of the physical system;

(b) the representation of the symmetry group in the configuration space of the system;

(c) the reduction of this representation into its irreducible parts;

(d) the determination of the new coordinate system for which the original matrix is converted into block diagonal form;

(e) the solution of the factored problem.

The subsequent restriction of the discussion to the motion of molecules is for convenience only, and the methods we develop are applicable to all physical systems having a symmetry group. Molecules are characterized by fact that they are often symmetric under the operations of one of the point

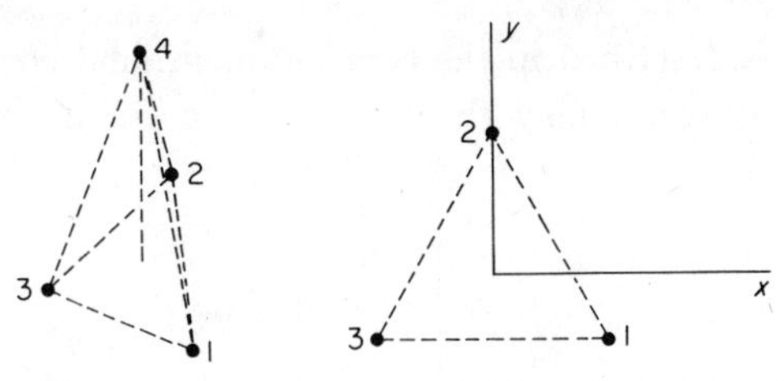

Fig. 6.1. The equilibrium configuration of NH_3 and the Cartesian reference axes.

groups (crystals are usually symmetric under the operations of one of the space groups, etc.). The number of degrees of freedom of a molecular system is in general $3N$, where N is the number of particles in the system. To illustrate the technique, we consider the ammonia molecule NH_3 shown in Fig. 6.1. At equilibrium the hydrogen atoms are at the vertices of an equilateral triangle with the nitrogen atom directly above the midpoint of the triangle. The atoms are numbered as shown with the radius vector $\mathbf{r}_\alpha$ ($\alpha = 1, 2, 3, 4$) locating the positions of the atoms and $\mathbf{r}_{0\alpha}$ locating the equilibrium position of the atoms. The displacement of the atoms from equilibrium is designated by the vector $\mathbf{v}_\alpha = \mathbf{r}_\alpha - \mathbf{r}_{0\alpha}$. At each equilibrium position we establish a Cartesian coordinate system whose axes are parallel to the cartesian system to which $\mathbf{r}_\alpha$ is referred. It was shown in Chapter 2 that for small oscillations the potential energy may be written as a quadratic form of the displacements:

$$V = \tfrac{1}{2} \sum_{\alpha,\beta=1}^{4} \sum_{i,j=1}^{3} (\partial^2 V / \partial v_\alpha{}^i \, \partial v_\beta{}^j)_0 \, v_\alpha{}^i v_\beta{}^j \tag{6.4}$$

If we introduce the mass weighted coordinates

$$\mathbf{u}_\alpha = (m_\alpha)^{1/2} \mathbf{v}_\alpha \qquad \text{(no sum on } \alpha) \tag{6.5}$$

the kinetic energy takes the form

$$T = \tfrac{1}{2} \sum_{\alpha=1}^{4} (\mathbf{u}_\alpha)^2 \tag{6.6}$$

and the potential energy is

$$V = \tfrac{1}{2} \sum_{\alpha,\beta=1}^{4} A_{\alpha\beta}^{ij} u_\alpha{}^i u_\beta^j \tag{6.7}$$

where the dynamical matrix is

$$A_{\alpha\beta}^{ij} = (m_\alpha m_\beta)^{-1/2} (\partial^2 V / \partial v_\alpha{}^i \, \partial v_\beta{}^j)_0 \tag{6.8a}$$

The dynamical equations are

$$\ddot{u}_\beta^j = - \sum_{\alpha=1}^{4} A_{\beta\alpha}^{ji} u_\alpha^i \tag{6.8b}$$

Our system then is described in the twelve-dimensional configuration by a point whose radius vector may be displayed as a column matrix.

$$U = \begin{bmatrix} u_1^{\ 1} \\ u_1^{\ 2} \\ u_1^{\ 3} \\ u_2^{\ 1} \\ \cdot \\ \cdot \\ \cdot \\ u_4^{\ 3} \end{bmatrix} \tag{6.9}$$

Equation (6.9) suggests that the elements of the dynamical matrix be ordered as shown in (6.10).

$$A = \begin{bmatrix} A_{11}^{11} & A_{11}^{12} & A_{11}^{13} & A_{12}^{11} \cdots & A_{14}^{14} \\ A_{11}^{21} & A_{11}^{22} & A_{11}^{23} & A_{12}^{21} \cdots & A_{14}^{24} \\ A_{11}^{31} & & & & \cdot \\ A_{21}^{11} & A_{21}^{12} & & & \cdot \\ \cdot & & & & \cdot \\ \cdot & & & & \cdot \\ \cdot & & & & \cdot \\ A_{41}^{31} & \cdot & \cdot & \cdot \cdots & A_{44}^{33} \end{bmatrix} \tag{6.10}$$

To provide a tangible example, we assume a potential energy that depends only on the square of the distance separating the atoms, i.e.,

$$\begin{aligned} V = {} & \tfrac{1}{2}k_{12}[(\mathbf{v}_1 - \mathbf{v}_2 + \mathbf{r}_{01} - \mathbf{r}_{02})^{1/2} - l_{12}]^2 \\ & + \tfrac{1}{2}k_{13}[(\mathbf{v}_1 - \mathbf{v}_3 + \mathbf{r}_{01} - \mathbf{r}_{03})^{1/2} - l_{13}]^2 \\ & + \tfrac{1}{2}k_{14}[(\mathbf{v}_1 - \mathbf{v}_4 + \mathbf{r}_{01} - \mathbf{r}_{04})^{1/2} - l_{14}]^2 \\ & + \tfrac{1}{2}k_{23}[(\mathbf{v}_2 - \mathbf{v}_3 + \mathbf{r}_{02} - \mathbf{r}_{03})^{1/2} - l_{23}]^2 \\ & + \tfrac{1}{2}k_{24}[(\mathbf{v}_2 - \mathbf{v}_4 + \mathbf{r}_{02} - \mathbf{r}_{04})^{1/2} - l_{24}]^2 \\ & + \tfrac{1}{2}k_{34}[(\mathbf{v}_3 - \mathbf{v}_4 + \mathbf{r}_{03} - \mathbf{r}_{04})^{1/2} - l_{34}]^2 \end{aligned} \tag{6.11}$$

where $l_{\mu\nu}$ is the separation of the μth and νth atoms at equilibrium and the $k_{\mu\nu}$ are constants. Clearly, for NH_3 we must require that

$$k_{12} = k_{13} = k_{23}, \qquad l_{12} = l_{13} = l_{23}, \qquad m_1 = m_2 = m_3 = m \tag{6.12a}$$

$$k_{14} = k_{24} = k_{34}, \qquad l_{14} = l_{24} = l_{34}, \qquad m_4 = M \tag{6.12b}$$

With the coordinate system shown in Fig. 6.1 we find

$$x_{01} = \tfrac{1}{2}l_{12} \equiv a, \qquad x_{02} = x_{04} = 0, \qquad x_{03} = -a \tag{6.13a}$$

$$y_{01} = y_{03} = -a/3^{1/2}, \qquad y_{02} = 2a/3^{1/2} \qquad y_{04} = 0 \tag{6.13b}$$

$$z_{01} = z_{02} = z_{03} = 0, \qquad z_{04} \equiv b, \qquad l_{14} = (\tfrac{4}{3}a^2 + b^2)^{1/2} \tag{6.13c}$$

$$
A = \begin{bmatrix}
\frac{a^2(5\alpha+\beta)}{m} \\
\frac{-a^2(3\alpha+\beta)}{3^{1/2}m} & \frac{a^2(9\alpha+\beta)}{3m} \\
\frac{-ab\beta}{m} & \frac{ab\beta}{3^{1/2}m} & \frac{b^2\beta}{m} \\
\frac{-a^2\alpha}{m} & \frac{3a^2\alpha}{3^{1/2}m} & 0 & \frac{2a^2\alpha}{m} \\
\frac{3a^2\alpha}{3^{1/2}m} & \frac{-3a^2\alpha}{m} & 0 & 0 & \frac{2a^2(9\alpha+2\beta)}{3m} \\
0 & 0 & 0 & 0 & \frac{-2ab\beta}{3^{1/2}m} & \frac{b^2\beta}{m} \\
\frac{-4a^2\alpha}{m} & 0 & 0 & \frac{-a^2\alpha}{m} & \frac{-3a^2\alpha}{3^{1/2}m} & 0 & \frac{a^2(5\alpha+\beta)}{m} \\
0 & 0 & 0 & \frac{-3a^2\alpha}{3^{1/2}m} & \frac{-3a^2\alpha}{m} & 0 & \frac{a^2(3\alpha+\beta)}{3^{1/2}m} & \frac{a^2(9\alpha+\beta)}{3m} \\
0 & 0 & 0 & 0 & 0 & 0 & \frac{ab\beta}{m} & \frac{ab\beta}{3^{1/2}m} & \frac{b^2\beta}{m} \\
\frac{-a^2\beta}{(mM)^{1/2}} & \frac{a^2\beta}{3^{1/2}(mM)^{1/2}} & \frac{ab\beta}{(mM)^{1/2}} & 0 & 0 & 0 & \frac{-a^2\beta}{(mM)^{1/2}} & \frac{-a^2\beta}{3^{1/2}(mM)^{1/2}} & \frac{-ab\beta}{(mM)^{1/2}} & \frac{2a^2\beta}{M} \\
\frac{a^2\beta}{3^{1/2}(mM)^{1/2}} & \frac{-a^2\beta}{3(mM)^{1/2}} & \frac{-ab\beta}{3^{1/2}(mM)^{1/2}} & 0 & \frac{-4a^2\beta}{3(mM)^{1/2}} & \frac{2ab\beta}{3^{1/2}(mM)^{1/2}} & \frac{-a^2\beta}{3^{1/2}(mM)^{1/2}} & \frac{-a^2\beta}{3(mM)^{1/2}} & \frac{-ab\beta}{3^{1/2}(mM)^{1/2}} & 0 & \frac{2a^2\beta}{M} \\
\frac{ab\beta}{(mM)^{1/2}} & \frac{-ab\beta}{3^{1/2}(mM)^{1/2}} & \frac{-b^2\beta}{(mM)^{1/2}} & 0 & \frac{2ab\beta}{3^{1/2}(mM)^{1/2}} & \frac{-b^2\beta}{(mM)^{1/2}} & \frac{-ab\beta}{(mM)^{1/2}} & \frac{-ab\beta}{3^{1/2}(mM)^{1/2}} & \frac{-b^2\beta}{(mM)^{1/2}} & 0 & 0 & \frac{3b^2\beta}{M}
\end{bmatrix} \tag{6.14}
$$

(A is symmetric)

Substituting (6.1), (6.12) and (6.13) into (6.8), we find the dynamical matrix for NH_3 to be (6.14) (p. 197), where

$$\alpha = k_{12}/l_{12}^2, \qquad \beta = k_{14}/l_{14}^2 \tag{6.15}$$

It is the object of our group theory program to convert the matrix in (6.14) into a block diagonal form so that we will not be required to solve a 12×12 determinant to obtain the eigenvalues (normal mode frequencies).

A. Symmetry Group of NH3

To determine the symmetry group, we recall that a symmetry transformation is one that leaves a system invariant. For a molecule a symmetry transformation is one which leaves the geometrical relationships between the equilibrium positions of the atoms unaltered. Thus for NH_3 rotation of the system through an angle of 120° about the z axis (which passes through the nitrogen atom) reproduces the original structure, since the hydrogen atoms are indistinguishable. This is the point group operation designated C_3. Similarly, the operation C_3^2 or rotation through 240° is a symmetry operation. Reflection through any of the three planes shown in Fig. 6.2 also leaves the structure in its equilibrium configuration. There are three such independent reflection operations which we designate as $\sigma_{30°}$, $\sigma_{90°}$, and $\sigma_{150°}$ where the subscript indicates the angle which the intersection of the reflection plane and the triangle makes with the x axis. These operations are the elements of the point group C_{3v}, whose multiplication table is given in Table 6.1.

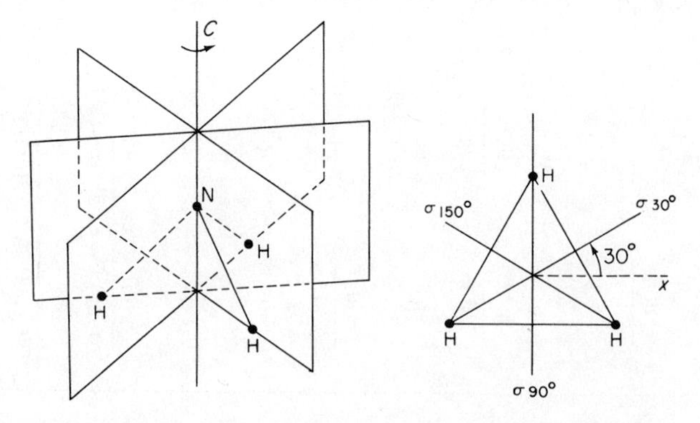

Fig. 6.2. Symmetry operations of the C_{3v} group on NH_3.

TABLE 6.1

Multiplication Table of the C_{3v} Group

E	C_3	C_3^2	$\sigma_{30°}$	$\sigma_{90°}$	$\sigma_{150°}$
C_3	C_3^2	E	σ_{90}	σ_{150}	σ_{30}
C_3^2	E	C_3	σ_{150}	σ_{30}	σ_{90}
$\sigma_{30°}$	σ_{150}	σ_{90}	E	C_3^2	C_3
$\sigma_{90°}$	σ_{30}	σ_{150}	C_3	E	C_3^2
$\sigma_{150°}$	σ_{90}	σ_{30}	C_3^2	C_3	E

B. Representation of the Symmetry Group in Configuration Space

To acquire the representation of the group in the twelve-dimension configuration space of the system, we first obtain the representation of the group in the ordinary three-dimensional space of the molecule. This is facilitated by noting that the C_n operators are simply operators which rotate our radius vector about the Z axis through an angle $2\pi/n$. We have frequently used the matrix operator which transforms the components of a vector because of a counterclockwise rotation of *axes* through an angle φ about the Z axis.

$$R = \begin{bmatrix} \cos\varphi & \sin\varphi & 0 \\ -\sin\varphi & \cos\varphi & 0 \\ 0 & 0 & 1 \end{bmatrix} \tag{6.16}$$

Now a clockwise rotation of the radius vector through an angle φ is equivalent to a counterclockwise rotation of the axes through an angle φ. Thus, we conclude that C_n may be obtained from (6.16) by replacing φ with $-\varphi$, obtaining

$$C_n = \begin{bmatrix} \cos\varphi & -\sin\varphi & 0 \\ \sin\varphi & \cos\varphi & 0 \\ 0 & 0 & 1 \end{bmatrix}, \qquad \varphi = \frac{2\pi}{n} \tag{6.17}$$

Thus for C_{3v} we have from (6.17) with $\varphi = 120°$

$$C_3 = \begin{bmatrix} -1/2 & -3^{1/2}/2 & 0 \\ 3^{1/2}/2 & -1/2 & 0 \\ 0 & 0 & 1 \end{bmatrix}, \qquad C_3^2 = \begin{bmatrix} -1/2 & 3^{1/2}/2 & 0 \\ -3^{1/2}/2 & -1/2 & 0 \\ 0 & 0 & 1 \end{bmatrix} \tag{6.18}$$

To obtain the three-dimensional representation of the reflection operators consider Fig. 6.3. We wish to obtain the operator which converts **r** into **r**′ (i.e., reflection of **r** through the XZ plane). It is clear from Fig. 6.3 that

$$x' - x = 2d \sin \varphi \tag{6.19a}$$

$$y - y' = 2d \cos \varphi \tag{6.19b}$$

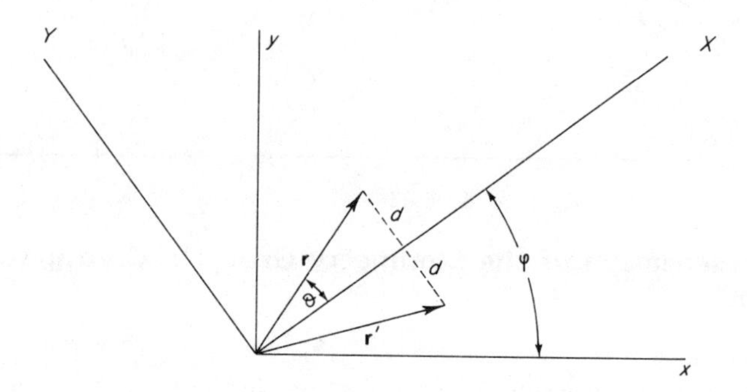

Fig. 6.3. Reflection of the vector **r** through the XZ plane.

Since the XY axes are related to the xy axes by a counterclockwise rotation through an angle φ, it follows from (6.16) and an inspection of Fig. 6.3 that

$$Y = -x \sin \varphi + y \cos \varphi \equiv d \tag{6.20}$$

Substituting (6.20) into (6.19) and solving for x' and y', we find

$$x' = x \cos 2\varphi + y \sin 2\varphi \tag{6.21a}$$

$$y' = x \sin 2\varphi - y \cos 2\varphi \tag{6.21b}$$

or

$$\sigma_\varphi = \begin{bmatrix} \cos 2\varphi & \sin 2\varphi \\ \sin 2\varphi & -\cos 2\varphi \end{bmatrix} \tag{6.22}$$

Thus for C_{3v} we have

$$\sigma_{30} = \begin{bmatrix} 1/2 & 3^{1/2}/2 & 0 \\ 3^{1/2}/2 & -1/2 & 0 \\ 0 & 0 & 1 \end{bmatrix}, \qquad \sigma_{90} = \begin{bmatrix} -1 & 0 & 0 \\ 0 & 1 & 0 \\ 0 & 0 & 1 \end{bmatrix}$$

$$\sigma_{150} = \begin{bmatrix} 1/2 & -3^{1/2}/2 & 0 \\ -3^{1/2}/2 & -1/2 & 0 \\ 0 & 0 & 1 \end{bmatrix}, \qquad E = \begin{bmatrix} 1 & 0 & 0 \\ 0 & 1 & 0 \\ 0 & 0 & 1 \end{bmatrix} \tag{6.23}$$

These operators are represented in the twelve-dimensional configuration space by 12×12 matrices which could be constructed by cleverly inserting the three-dimensional representations as submatrices of the 12×12 matrix in such a way that the 12×12 matrix operates on a configuration space vector such as U of Eq. (6.9) as we know it should (i.e., provides the proper mixing of the components).

A formal procedure for constructing the twelve-dimensional representation follows from the recognition that it is simply the direct product of the operators of the three-dimensional representation and the operators of the permutation group P which have the same effect on the equilibrium positions of the atom as the group operators have. Thus, since C_3 would leave atom 4 unchanged and move atom 3 to position 1, atom 1 to position 2, and atom 2 to position 3, the permutation producing the same change is

$$P_{C_3} = \begin{bmatrix} 0 & 0 & 1 & 0 \\ 1 & 0 & 0 & 0 \\ 0 & 1 & 0 & 0 \\ 0 & 0 & 0 & 1 \end{bmatrix}, \quad \text{i.e.,} \quad P_{C_3} \begin{bmatrix} \text{atom 1} \\ \text{atom 2} \\ \text{atom 3} \\ \text{atom 4} \end{bmatrix} = \begin{bmatrix} \text{atom 3} \\ \text{atom 1} \\ \text{atom 2} \\ \text{atom 4} \end{bmatrix} \tag{6.24}$$

The twelve-dimensional representation of C_3 is just $P_{C_3} \otimes C_3^{(3)}$ [(6.25), p. 202]. Similarly, we find

$$E^{(12)} = \begin{bmatrix} E^{(3)} & & & 0 \\ & E^{(3)} & & \\ & & E^{(3)} & \\ 0 & & & E^{(3)} \end{bmatrix}, \quad C_3^{(12)} = \begin{bmatrix} 0 & C_3^{\,2} & 0 & 0 \\ 0 & 0 & C_3^{\,2} & 0 \\ C_3^{\,2} & 0 & 0 & 0 \\ 0 & 0 & 0 & C_3^{\,2} \end{bmatrix}$$

$$\sigma_{30} = \begin{bmatrix} 0 & \sigma_{30} & 0 & 0 \\ \sigma_{30} & 0 & 0 & 0 \\ 0 & 0 & \sigma_{30} & 0 \\ 0 & 0 & 0 & \sigma_{30} \end{bmatrix}, \quad \sigma_{90} = \begin{bmatrix} 0 & 0 & \sigma_{90} & 0 \\ 0 & \sigma_{90} & 0 & 0 \\ \sigma_{90} & 0 & 0 & 0 \\ 0 & 0 & 0 & 0 \end{bmatrix} \tag{6.26}$$

$$\sigma_{150} = \begin{bmatrix} \sigma_{150} & 0 & 0 & 0 \\ 0 & 0 & \sigma_{150} & 0 \\ 0 & \sigma_{150} & 0 & 0 \\ 0 & 0 & 0 & \sigma_{150} \end{bmatrix}$$

The justification of this procedure is left to the reader, who may consult the literature [1–3] for assistance.

$$C_3^{(12)} = \begin{bmatrix} 0 & 0 & 1 & 0 \\ 1 & 0 & 0 & 0 \\ 0 & 1 & 0 & 0 \\ 0 & 0 & 0 & 1 \end{bmatrix} \otimes C_3^{(3)} = \begin{bmatrix} 0 & 0 & C_3^{(3)} & 0 \\ C_3^{(3)} & 0 & 0 & 0 \\ 0 & C_3^{(3)} & 0 & 0 \\ 0 & 0 & 0 & C_3^{(3)} \end{bmatrix}$$

$$= \begin{bmatrix} 0 & 0 & 0 & 0 & 0 & 0 & -1/2 & -3^{1/2}/2 & 0 & 0 & 0 & 0 \\ 0 & 0 & 0 & 0 & 0 & 0 & 3^{1/2}/2 & -1/2 & 0 & 0 & 0 & 0 \\ 0 & 0 & 0 & 0 & 0 & 0 & 0 & 0 & 1 & 0 & 0 & 0 \\ -1/2 & -3^{1/2}/2 & 0 & 0 & 0 & 0 & 0 & 0 & 0 & 0 & 0 & 0 \\ 3^{1/2}/2 & -1/2 & 0 & 0 & 0 & 0 & 0 & 0 & 0 & 0 & 0 & 0 \\ 0 & 0 & 1 & 0 & 0 & 0 & 0 & 0 & 0 & 0 & 0 & 0 \\ 0 & 0 & 0 & -1/2 & -3^{1/2}/2 & 0 & 0 & 0 & 0 & 0 & 0 & 0 \\ 0 & 0 & 0 & 3^{1/2}/2 & -1/2 & 0 & 0 & 0 & 0 & 0 & 0 & 0 \\ 0 & 0 & 0 & 0 & 0 & 1 & 0 & 0 & 0 & 0 & 0 & 0 \\ 0 & 0 & 0 & 0 & 0 & 0 & 0 & 0 & 0 & -1/2 & -3^{1/2}/2 & 0 \\ 0 & 0 & 0 & 0 & 0 & 0 & 0 & 0 & 0 & 3^{1/2}/2 & -1/2 & 0 \\ 0 & 0 & 0 & 0 & 0 & 0 & 0 & 0 & 0 & 0 & 0 & 1 \end{bmatrix} \tag{6.25}$$

C. Reduction of the Reducible Representation

The twelve-dimensional representation is reducible into the direct sum of irreducible representations. The character table for C_{3v} is presented in Table 6.2. In addition to the characters of the three irreducible representations of the group ($\Gamma_1, \Gamma_2, \Gamma_3$), the table also presents the characters of the reducible twelve-dimensional representation $\Gamma^{(12)}$. A unitary irreducible representation of Γ_3 is also presented.

The character tables of the groups commonly encountered in physical applications are readily available (consult Appendix F and the references at the end of the chapter [1–28]), and they tell us how many irreducible representations there are, their dimension and their character. Thus for C_{3v}, Γ_1, and Γ_2 are one-dimensional representations (the character of the identity E is the dimension of the representation), and Γ_3 is a two-dimensional representation. Since there is no twelve-dimensional irreducible representation, we conclude that our twelve-dimensional representation $\Gamma^{(12)}$ must be reducible into the *direct sum* of the irreducible representations

$$\Gamma^{(12)} = n_1\Gamma_1 \oplus n_2\Gamma_2 \oplus n_3\Gamma_3 \tag{6.27}$$

where n_1, n_2, and n_3 are integers indicating how many times Γ_1, Γ_2, and Γ_3, respectively, appear in the direct sum.

It follows directly from the orthogonality theorem for characters that the coefficients in the direct sum are given by

$$n_j = [\textstyle\sum_R \chi(R)\chi_j(R)]/g \tag{6.28}$$

where g is the order of the group, $\chi(R)$ is the character of the Rth element of the group in the reducible representation, $\chi_j(R)$ is the character of the Rth element of the group in the jth irreducible representation, and the sum is over the elements of the group. Applying (6.28) to the reduction of our twelve-dimensional representation by inserting the information contained in Table 6.2, we obtain

$$n_1 = [(12)(1) + (0)(1) + (0)(1) + (2)(1) + (2)(1) + (2)(1)]/6 = 3$$

Similarly,

$$n_2 = 1, \qquad n_3 = 4$$

Thus

$$\Gamma^{(12)} = 3\Gamma_1 \oplus \Gamma_2 \oplus 4\Gamma_3 \tag{6.29}$$

TABLE 6.2

Character Table for the Group C_{3v}

	E	C_3	C_3^2	σ_{30}	σ_{90}	σ_{150}
Γ_1	1	1	1	1	1	1
Γ_2	1	1	1	−1	−1	−1
Γ_3	2	−1	−1	0	0	0
$\Gamma^{(12)}$	12	0	0	2	2	2
Representation of Γ_3	$\begin{bmatrix} 1 & 0 \\ 0 & 1 \end{bmatrix}$	$\begin{bmatrix} -1/2 & -\sqrt{3}/2 \\ \sqrt{3}/2 & -1/2 \end{bmatrix}$	$\begin{bmatrix} -1/2 & \sqrt{3}/2 \\ -\sqrt{3}/2 & -1/2 \end{bmatrix}$	$\begin{bmatrix} 1/2 & \sqrt{3}/2 \\ \sqrt{3}/2 & -1/2 \end{bmatrix}$	$\begin{bmatrix} -1 & 0 \\ 0 & 1 \end{bmatrix}$	$\begin{bmatrix} 1/2 & -\sqrt{3}/2 \\ -\sqrt{3}/2 & -1/2 \end{bmatrix}$

The information contained in (6.29) is already considerable. It tells us that we can certainly expect our 12×12 matrix to factor into a 3×3 block associated with Γ_1, a 1×1 block associated with Γ_2, and an 8×8 block associated with Γ_3. Sometimes the factoring is more extensive than we just indicated. The best one can hope for is a factoring in which the blocks associated with a given irreducible representation have the same dimension as the representation and appear as many times as the irreducible representation appears in the direct sum. Equation (6.30a) indicates the minimum expectation and (6.30b) the maximum expectation. Our subsequent analysis of NH_3 will show that its dynamical matrix actually reduces to a matrix as shown in (6.30c).

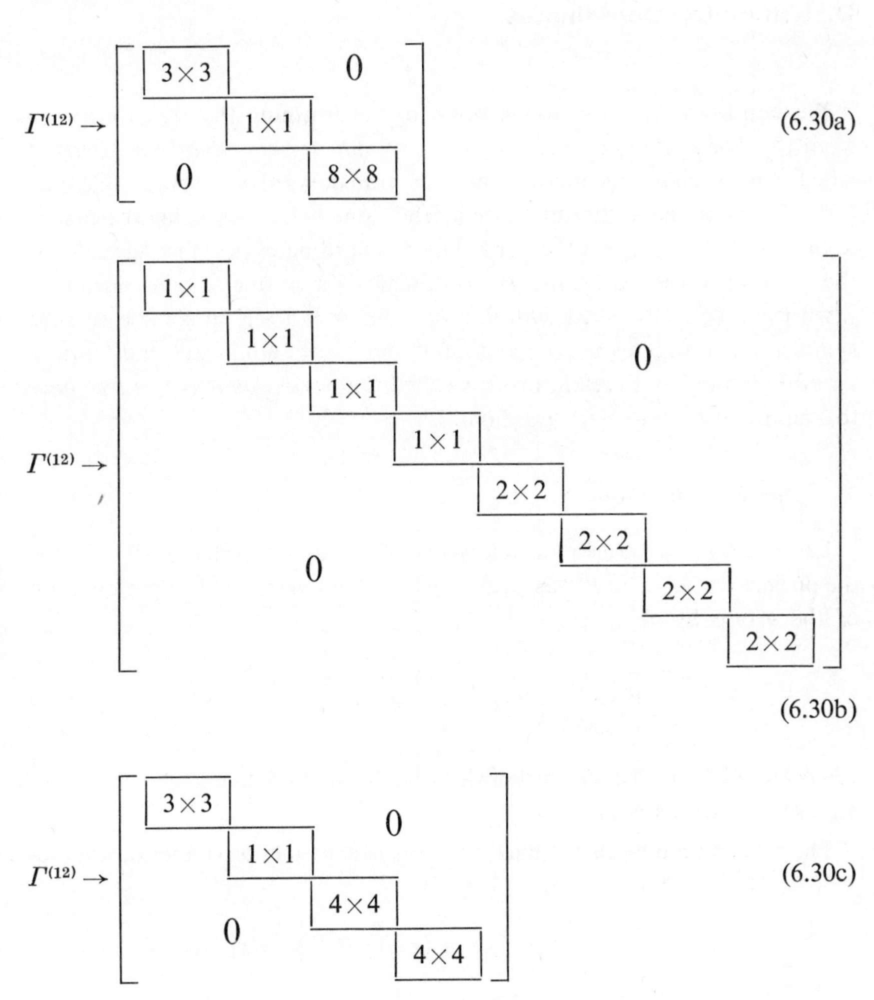

(6.30a)

(6.30b)

(6.30c)

An important insight is obtained if one realizes that (6.29) indicates that the twelve-dimensional representation space may be reduced into a direct sum of spaces of lower dimensionality associated with the irreducible representations Γ_1, Γ_2, Γ_3. It also follows that the space associated with Γ_2 is one dimensional, the space associated with Γ_1 is three dimensional (Γ_1 is one dimensional, but it appears three times in the direct sum), and the space associated with Γ_3 is eight dimensional. It is also possible, but not determined by (6.29), that the eight- and three-dimensional spaces may be further separated into invariant spaces of lower dimension.

D. Symmetry Coordinates

We continue the reduction process by determining the transformation from the original coordinates of (6.9) to *symmetry coordinates* in terms of which the original dynamical matrix is transformed into block diagonal form. The symmetry coordinates play the same role in reducing the matrix to block diagonal form as the normal mode coordinates play in diagonalizing the matrix; in fact, a symmetry coordinate is a normal mode coordinate when the block associated with that coordinate is 1×1. In some cases, the symmetry coordinates are obvious, but more commonly one must utilize an often tedious but certain process called the *projection operator technique* to acquire the symmetry coordinates.

1. Projection Operators

Let $G_1, G_2, \ldots, G_g$ be the elements of group of order g. We define the projection operator $P_{\alpha\beta}^{(a)}$ associated with the ath irreducible representation of the group by

$$P_{\alpha\beta}^{(a)} \equiv (d_a/g) \sum_{k=1}^{g} D_{\alpha\beta}^{(a)k} G_k \qquad (\alpha, \beta = 1, 2, \ldots, n_a) \tag{6.31}$$

where the $D_{\alpha\beta}^{(a)k}$ are the matrices of the ath, n_a-dimensional irreducible unitary representation of the group G.

The projection operator applied to an arbitrary function of the coordinates yields

$$f_{\alpha\beta}^{(a)} \equiv P_{\alpha\beta}^{(a)} f(\mathbf{r}) = (d_a/g) \sum_{k=1}^{g} D_{\alpha\beta}^{(a)k} G_k f(\mathbf{r}) \tag{6.32}$$

Now operating on both sides of (6.32) with an element G_l of the group

$$G_l f_{\alpha\beta}^{(a)} = (d_a/g) \sum_{k=1}^{g} D_{\alpha\beta}^{(a)k} G_l G_k f(\mathbf{r}) \tag{6.33}$$

Now since the group elements satisfy the relation,

$$G_l G_k = G_s \qquad \text{or} \quad G_k = G_l^{-1} G_s \tag{6.34}$$

their representations must satisfy an identical relationship

$$D_{\alpha\beta}^{(a)k} = \sum_{\gamma=1}^{n_a} (D_{\alpha\gamma}^{(a)l})^{-1} D_{\gamma j}^{(a)s} = \sum_{\gamma=1}^{n_a} (D_{\alpha\gamma}^{(a)l})^{\dagger} D_{\gamma\beta}^{(a)s} \tag{6.35}$$

where the last relationship follows from the unitarity of the matrices. Substituting (6.35) into (6.33), we obtain

$$G_l f_{\alpha\beta}^{(a)} = (d_a/g) \sum_{s=1}^{g} \sum_{\gamma=1}^{n_a} (D_{\alpha\gamma}^{(a)l})^{\dagger} D_{\gamma\beta}^{(a)s} G_s f(\mathbf{r}) = \sum_{\gamma=1}^{n_a} \overline{D_{\gamma\alpha}^{(a)l}} f_{\gamma\beta}^{(a)} \tag{6.36}$$

For fixed β let

$$\mathbf{f}_{\beta}^{(a)} \equiv \begin{bmatrix} f_{1\beta}^{(a)} \\ f_{2\beta}^{(a)} \\ \vdots \\ f_{n\beta}^{(a)} \end{bmatrix} \tag{6.37}$$

Equation (6.36) then becomes

$$G_l \mathbf{f}_{\beta}^{(a)} = D^{(a)l\dagger} \mathbf{f}_{\beta}^{(a)} \tag{6.38}$$

Equation (6.38) demonstrates that for given β, the functions $f_{1\beta}^{(a)}, f_{2\beta}^{(a)}, \ldots f_{n_a\beta}^{(a)}$ serve as basis functions for the ath irreducible representation of the group. The operators $P_{\alpha\beta}^{(a)}$ take an arbitrary function and project out of it those portions which transform as partners in a basis for the ath irreducible representation.

Two important properties of projection operators are

$$P_{\alpha\beta}^{(a)} P_{\gamma\delta}^{(b)} = \delta_{\alpha\delta} \delta_{\beta\gamma} \delta_{ab} \tag{6.39}$$

and

$$P_{\alpha\beta}^{(a)} \xi_{\beta}^{(a)} = \xi_{\alpha}^{(a)} \tag{6.40}$$

where $\xi_1^{(a)}, \xi_2^{(a)}, \ldots, \xi_{n_a}^{(a)}$ are basis functions of the irreducible representation. The proof of (6.39) and (6.40) is left to the reader (Problem 6.2). Equation (6.39) is the *orthogonality relation* for projection operators, and (6.40) manifests the *step operator property* of projection operators, viz., that given one basis function all others can be obtained by applications of the projection operator.

From (6.31) and Table 6.2, it follows that the projection operators of the C_{3v} group are

$$P_{11}^{(1)} = \tfrac{1}{6}(E + C_3 + C_3^2 + \sigma_{30} + \sigma_{150} + \sigma_{90}) \tag{6.41}$$

$$P_{11}^{(2)} = \tfrac{1}{6}(E + C_3 + C_3^2 - \sigma_{30} - \sigma_{150} - \sigma_{90}) \tag{6.42}$$

$$P_{11}^{(3)} = \tfrac{1}{3}(E - \tfrac{1}{2}C_3 - \tfrac{1}{2}C_3^2 + \tfrac{1}{2}\sigma_{30} + \tfrac{1}{2}\sigma_{150} - \sigma_{90}) \tag{6.43}$$

$$P_{12}^{(3)} = \tfrac{1}{3}(- \tfrac{1}{2}3^{1/2}C_3 + \tfrac{1}{2}3^{1/2}C_3^2 + \tfrac{1}{2}3^{1/2}\sigma_{30} - \tfrac{1}{2}3^{1/2}\sigma_{150}) \tag{6.44}$$

$$P_{21}^{(3)} = \tfrac{1}{3}(\tfrac{1}{2}3^{1/2}C_3 - \tfrac{1}{2}3^{1/2}C_3^2 + \tfrac{1}{2}3^{1/2}\sigma_{30} - \tfrac{1}{2}3^{1/2}\sigma_{150}) \tag{6.45}$$

$$P_{22}^{(3)} = \tfrac{1}{3}(E - \tfrac{1}{2}C_3 - \tfrac{1}{2}C_3^2 - \tfrac{1}{2}\sigma_{30} - \tfrac{1}{2}_{150} + \sigma_{90}) \tag{6.46}$$

2. Symmetry Coordinates

The projection operators are used to "project out" the symmetry coordinates of the $D^{(a)}$ representation. To accomplish this, one constructs the vectors $P_{ij}^{(a)}U$, where U is the column matrix of the radius vectors representing an arbitrary displacement of the atoms, i.e.,

$$U = \begin{bmatrix} u_1^1 \\ u_1^2 \\ u_1^3 \\ u_2^1 \\ \cdot \\ \cdot \\ \cdot \\ u_4^3 \end{bmatrix}$$

The independent components of these vectors are the linearly independent symmetry coordinates which span the representation spaces associated with the irreducible representations $D^{(a)}$.

To illustrate the procedure, we consider first the Γ_2 irreducible representation of C_{3v}. From (6.29) we see that the one-dimensional representation Γ_2 appears only once in the reduction; and, therefore, we expect that there is only one symmetry coordinate associated with Γ_2, i.e., only one vector is necessary to span the space. Using (6.42) and (6.26), we find

$$P_{11}^{(2)}U = \frac{1}{6}\begin{bmatrix} \frac{1}{2}(u_1{}^1 - 2u_2{}^1 + u_3{}^1 + 3^{1/2}u_1{}^2 - 3^{1/2}u_3{}^2) \\ \frac{1}{2}3^{1/2}(u_1{}^1 + u_3{}^1 + 3^{1/2}u_1{}^2 - 3^{1/2}u_3{}^2 - 2u_2{}^1) \\ 0 \\ -(u_1{}^1 - 2u_2{}^1 + u_3{}^1 + 3^{1/2}u_1{}^2 - 3^{1/2}u_3{}^2) \\ 0 \\ 0 \\ \frac{1}{2}(u_1{}^1 - 2u_2{}^1 + u_3{}^1 + 3^{1/2}u_1{}^2 - 3^{1/2}u_3{}^2) \\ -\frac{1}{2}3^{1/2}(u_1{}^1 + u_3{}^1 + 3^{1/2}u_1{}^2 - 3^{1/2}u_3{}^2 - 2u_2{}^1) \\ 0 \\ 0 \\ 0 \\ 0 \end{bmatrix} \tag{6.47}$$

There appears in the column matrix but one independent component

$$u_1{}^1 - 2u_2{}^1 + u_3{}^1 + 3^{1/2}u_1{}^2 - 3^{1/2}u_3{}^2 \tag{6.48}$$

and we conclude that the symmetry coordinate which spans the representation space is

$$\begin{bmatrix} 1 \\ 3^{1/2} \\ 0 \\ -2 \\ 0 \\ 0 \\ 1 \\ -3^{1/2} \\ 0 \\ 0 \\ 0 \\ 0 \end{bmatrix} \xrightarrow{\text{normalizing}} \frac{1}{2(3)^{1/2}}\begin{bmatrix} 1 \\ 3^{1/2} \\ 0 \\ -2 \\ 0 \\ 0 \\ 1 \\ -3^{1/2} \\ 0 \\ 0 \\ 0 \\ 0 \end{bmatrix} = Q_1^{(2)} \tag{6.49}$$

Here $Q_1^{(2)}$ is the normalized symmetry coordinate which serves as the basis in the representation space associated with Γ_2.

Similarly, one finds for the Γ_1 representation

$$6P_{11}^{(1)}U = \begin{bmatrix} \frac{1}{2}3^{1/2}(3^{1/2}u_1{}^1 - 3^{1/2}u_3{}^1 - u_1{}^2 + 2u_2{}^2 - u_3{}^2) \\ -\frac{1}{2}(3^{1/2}u_1{}^1 - 3^{1/2}u_3{}^1 - u_1{}^2 + 2u_2{}^2 - u_3{}^2) \\ 2(u_1{}^3 + u_2{}^3 + u_3{}^3) \\ 0 \\ 3^{1/2}u_1{}^1 - 3^{1/2}u_3{}^1 - u_1{}^2 + 2u_2{}^2 - u_3{}^2 \\ 2(u_1{}^3 + u_2{}^3 + u_3{}^3) \\ -\frac{1}{2}3^{1/2}(3^{1/2}u_1{}^1 - 3^{1/2}u_3{}^1 - u_1{}^2 + 2u_2{}^2 - u_3{}^2) \\ -\frac{1}{2}(3^{1/2}u_1{}^1 - 3^{1/2}u_3{}^1 - u_1{}^2 + 2u_2{}^2 - u_3{}^2) \\ 2(u_1{}^3 + u_2{}^3 + u_3{}^3) \\ 0 \\ 0 \\ 6u_4{}^3 \end{bmatrix} \tag{6.50}$$

This vector has three independent components:

$$u_4{}^3, \quad u_1{}^3 + u_2{}^3 + u_4{}^3, \quad 3^{1/2}u_1{}^1 - 3^{1/2}u_3{}^1 - u_1{}^2 + 2u_2{}^2 - u_3{}^2$$

and the associated normalized symmetry coordinates are

$$Q_1^{(1)} = \begin{bmatrix} 0 \\ 0 \\ 0 \\ 0 \\ 0 \\ 0 \\ 0 \\ 0 \\ 0 \\ 0 \\ 0 \\ 1 \end{bmatrix}, \quad Q_2^{(1)} = \frac{1}{3^{1/2}} \begin{bmatrix} 0 \\ 0 \\ 1 \\ 0 \\ 0 \\ 1 \\ 0 \\ 0 \\ 1 \\ 0 \\ 0 \\ 0 \end{bmatrix}, \quad Q_3^{(1)} = \frac{1}{2(3)^{1/2}} \begin{bmatrix} 3^{1/2} \\ -1 \\ 0 \\ 0 \\ 2 \\ 0 \\ -3^{1/2} \\ -1 \\ 0 \\ 0 \\ 0 \\ 0 \end{bmatrix} \tag{6.51}$$

The $Q_1^{(1)}, Q_2^{(1)}, Q_3^{(1)}$ are the symmetry coordinates which serve as the basis spanning the three-dimensional space associated with the $3\Gamma_1$ appearing in the reduction of the twelve-dimensional representation space [see (6.29)]. The column matrices give the components of these symmetry vectors relative to the original coordinate system of the twelve-dimensional configuration space.

The work involved in obtaining the symmetry coordinates associated with Γ_3 is more tedious because there are four projection operators and, therefore, four vectors which must be constructed, each one of which contributes several, but not all, of the eight remaining symmetry coordinates associated with the $4\Gamma_3$ appearing in (6.29) [see (6.52), p. 212].

The symmetry displacements of NH_3 are shown in Fig. 6.4. Thus, $Q_2^{(3)}$ is simply the displacement of the nitrogen atom along the x axis; $Q_1^{(1)}$ is the displacement of the nitrogen atom along the z axis; $Q_1^{(2)}$ corresponds to a rotation of the hydrogen atoms about the z axis as shown in Fig. 6.4d; and so on. Displacements normal to the page are denoted by a circle around the dot representing the atom with the relative magnitude of the displace-

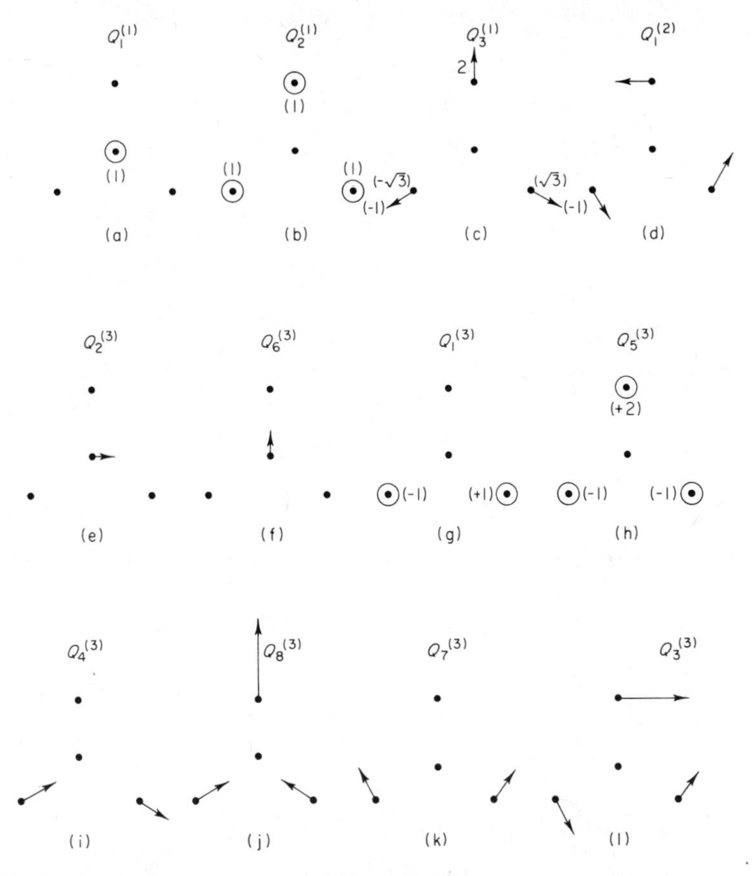

Fig. 6.4. The symmetry displacements of the NH_3 molecule displacements normal to the page are indicated by circles about the atom. The relative magnitude of the displacement along an axis is indicated by the number in parentheses.

$$Q_1^{(3)} = \frac{1}{2^{1/2}}\begin{bmatrix} 0 \\ 0 \\ 1 \\ 0 \\ 0 \\ 0 \\ 0 \\ 0 \\ -1 \\ 0 \\ 0 \\ 0 \end{bmatrix}, \quad Q_2^{(3)} = \begin{bmatrix} 0 \\ 0 \\ 0 \\ 0 \\ 0 \\ 0 \\ 0 \\ 0 \\ 0 \\ 1 \\ 0 \\ 0 \end{bmatrix}, \quad Q_3^{(3)} = \frac{1}{2\,(6^{1/2})}\begin{bmatrix} 1 \\ 3^{1/2} \\ 0 \\ 4 \\ 0 \\ 0 \\ 1 \\ -3^{1/2} \\ 0 \\ 0 \\ 0 \\ 0 \end{bmatrix}, \quad Q_4^{(3)} = \frac{1}{2\,(2^{1/2})}\begin{bmatrix} 3^{1/2} \\ -1 \\ 0 \\ 0 \\ 0 \\ 0 \\ 3^{1/2} \\ 1 \\ 0 \\ 0 \\ 0 \\ 0 \end{bmatrix}$$

$$Q_5^{(3)} = \frac{1}{6^{1/2}}\begin{bmatrix} 0 \\ 0 \\ -1 \\ 0 \\ 0 \\ 2 \\ 0 \\ 0 \\ -1 \\ 0 \\ 0 \\ 0 \end{bmatrix}, \quad Q_6^{(3)} = \begin{bmatrix} 0 \\ 0 \\ 0 \\ 0 \\ 0 \\ 0 \\ 0 \\ 0 \\ 0 \\ 0 \\ 1 \\ 0 \end{bmatrix}, \quad Q_7^{(3)} = \frac{1}{2\,(2^{1/2})}\begin{bmatrix} 1 \\ 3^{1/2} \\ 0 \\ 0 \\ 0 \\ 0 \\ -1 \\ 3^{1/2} \\ 0 \\ 0 \\ 0 \\ 0 \end{bmatrix}, \quad Q_8^{(3)} = \frac{1}{2\,(6^{1/2})}\begin{bmatrix} -3^{1/2} \\ 1 \\ 0 \\ 0 \\ 4 \\ 0 \\ 3^{1/2} \\ 1 \\ 0 \\ 0 \\ 0 \\ 0 \end{bmatrix} \tag{6.52}$$

ment indicated beside the circle (a + indicates displacement out of the page toward the reader; a – indicates displacement out of the page away from the reader).

It must be emphasized that the displacements represented in Fig. 6.4 are symmetry displacements, not normal mode displacements. It does happen that Fig. 6.4d represents also a normal mode displacement because the symmetry coordinate associated with a one dimensional irreducible representation that appears only once in the reduction of a reducible representation is also a normal mode coordinate.

With the symmetry coordinates, we may now reduce our original dynamical matrix [Eq. (6.14)] to block diagonal form. Thus, our original dynamical equations were

$$\ddot{U} = -AU$$

Operating on both sides by the unitary matrix $S^{-1} = S^{\dagger}$, we may write

$$S^{\dagger}\ddot{U} = -(S^{\dagger}AS)S^{\dagger}U \tag{6.53}$$

or

$$\ddot{U}' = -A'U' \tag{6.54}$$

where S is defined in such a way as to make A' have a block diagonal form. To accomplish this, we construct S from the symmetry coordinates in the following way: The twelve columns of S are simply the twelve symmetry coordinates. Thus, we have (6.55) (p. 214) and A' is as shown in (6.56) (p. 215). The quantities appearing in (6.56) are defined in Eqs. (6.12)–(6.15).

E. Normal Modes

The normal modes of the NH_3 system are now obtained from the eigenvalues and eigenvectors of A'. From (6.56) it is clear that the two 4×4 submatrices are identical and, therefore, we need only obtain the eigenvalues for the 3×3 submatrix, the 1×1 submatrix, and one 4×4 submatrix.

It has already been noted that the symmetry coordinate $Q_1^{(2)}$ associated with the 1×1 submatrix is also a normal mode coordinate, and from (6.56) it is clear that its eigenvalue is zero. Figure 6.4d illustrates the displacement characteristic of this normal mode, viz., rotation of the system as a whole about the Z axis. As this is not a vibratory motion, it is not surprising that the eigenfrequency is zero.

$$
S = \begin{bmatrix}
0 & 0 & \frac{1}{2} & \frac{1}{2(3^{1/2})} & 0 & 0 & \frac{1}{2(6^{1/2})} & \frac{3^{1/2}}{2(2^{1/2})} & 0 & 0 & \frac{1}{2(2^{1/2})} & \frac{-1}{2(2^{1/2})} \\
0 & 0 & \frac{-1}{2(3^{1/2})} & \frac{1}{2} & 0 & 0 & \frac{1}{2(2^{1/2})} & \frac{-1}{2(2^{1/2})} & 0 & 0 & \frac{3^{1/2}}{2(2^{1/2})} & \frac{1}{2(6^{1/2})} \\
0 & \frac{1}{3^{1/2}} & 0 & 0 & \frac{1}{2^{1/2}} & 0 & 0 & 0 & \frac{-1}{6^{1/2}} & 0 & 0 & 0 \\
0 & 0 & 0 & \frac{-1}{3^{1/2}} & 0 & 0 & \frac{2}{6^{1/2}} & 0 & 0 & 0 & 0 & 0 \\
0 & 0 & \frac{1}{3^{1/2}} & 0 & 0 & 0 & 0 & 0 & 0 & 0 & 0 & \frac{2}{6^{1/2}} \\
0 & \frac{1}{3^{1/2}} & 0 & 0 & 0 & 0 & 0 & 0 & \frac{2}{6^{1/2}} & 0 & 0 & 0 \\
0 & 0 & \frac{-1}{2} & \frac{1}{2(3^{1/2})} & 0 & 0 & \frac{1}{2(6^{1/2})} & \frac{3^{1/2}}{2(2^{1/2})} & 0 & 0 & \frac{-1}{2(2^{1/2})} & \frac{1}{2(2^{1/2})} \\
0 & 0 & \frac{-1}{2(3^{1/2})} & \frac{-1}{2} & 0 & 0 & \frac{-1}{2(2^{1/2})} & \frac{1}{2(2^{1/2})} & 0 & 0 & \frac{3^{1/2}}{2(2^{1/2})} & \frac{1}{2(6^{1/2})} \\
0 & \frac{1}{3^{1/2}} & 0 & 0 & \frac{-1}{2^{1/2}} & 0 & 0 & 0 & \frac{-1}{6^{1/2}} & 0 & 0 & 0 \\
0 & 0 & 0 & 0 & 0 & 1 & 0 & 0 & 0 & 0 & 0 & 0 \\
0 & 0 & 0 & 0 & 0 & 0 & 0 & 0 & 0 & 1 & 0 & 0 \\
1 & 0 & 0 & 0 & 0 & 0 & 0 & 0 & 0 & 0 & 0 & 0
\end{bmatrix} \tag{6.55}
$$

$$S^{\dagger}AS = A' =$$

$$\left[\begin{array}{ccc|c|cccc|cccc}
\frac{3b^2\beta}{M} & \frac{-3b^2\beta}{(3mM)^{1/2}} & \frac{2ab\beta}{(mM)^{1/2}} & & & & & & & & & \\
\frac{-3b^2\beta}{(3mM)^{1/2}} & \frac{b^2\beta}{m} & \frac{-2ab\beta}{m3^{1/2}} & & & & & & & & 0 & \\
\frac{2ab\beta}{(mM)^{1/2}} & \frac{-2ab\beta}{3^{1/2}m} & \frac{4a^2(9\alpha+\beta)}{3m} & & & & & & & & & \\
\hline
 & & & 0 & & & & & & & & \\
\hline
 & & & & \frac{b^2\beta}{m} & \frac{2^{1/2}ab\beta}{(mM)^{1/2}} & 0 & \frac{-2ab\beta}{3^{1/2}m} & & & & \\
 & & & & \frac{2^{1/2}ab\beta}{(mM)^{1/2}} & \frac{2a^2\beta}{M} & 0 & \frac{-4a^2\beta}{6^{1/2}(mM)^{1/2}} & & & & \\
 & & & & 0 & 0 & \frac{3a^2\alpha}{m} & \frac{-3a^2\alpha}{m} & & & & \\
 & & & & \frac{-2ab\beta}{3^{1/2}m} & \frac{-4a^2\beta}{6^{1/2}(mM)^{1/2}} & \frac{-3a^2\beta}{m} & \frac{a^2(9\alpha+4\beta)}{3m} & & & & \\
\hline
 & & & & & & & & \frac{b^2\beta}{m} & \frac{2^{1/2}ab\beta}{(mM)^{1/2}} & 0 & \frac{-2ab\beta}{3^{1/2}m} \\
 & 0 & & & & & & & \frac{2^{1/2}ab\beta}{(mM)^{1/2}} & \frac{2a^2\beta}{M} & 0 & \frac{-4a^2\beta}{6^{1/2}(mM)^{1/2}} \\
 & & & & & & & & 0 & 0 & \frac{3a^2\alpha}{m} & \frac{-3a^2\alpha}{m} \\
 & & & & & & & & \frac{-2ab\beta}{3^{1/2}m} & \frac{-4a^2\beta}{6^{1/2}(mM)^{1/2}} & \frac{-3a^2\alpha}{m} & \frac{a^2(9\alpha+4\beta)}{3m}
\end{array}\right] \tag{6.56}$$

1. Zero-Frequency Modes

The appearance of a zero-frequency mode is not a peculiarity of NH_3. In fact, the existence of zero-frequency modes follows from a symmetry of the system as a whole, viz., its translational and rotational symmetry.

The translational symmetry follows from the fact that the potential energy of the system depends only on the distance of the atoms from one another and not on their position in space. Thus, if we displace each by the same amount in the same direction, the distances between the atoms remain unchanged. Thus

$$V(\mathbf{r}_\alpha + \mathbf{R}) = V(\mathbf{r}_\alpha) \tag{6.57}$$

After the usual Taylor expansion it follows from (6.7) that we may write (6.27) as

$$(U + U_R)^T A(U + U_R) = U^T A U \tag{6.58}$$

where U is defined in (6.9), A is the dynamical matrix, and U_R is the matrix representation of $\mathbf{R}$ in the twelve-dimensional configuration space. Expanding the left side of (6.58) yields

$$U_R{}^T A U + U^T A U_R + U_R{}^T A U_R = 0 \tag{6.59}$$

Since the displacements represented by U_R are arbitrary, we choose them to be sufficiently small that we may neglect the third term in (6.59). Then, since

$$U_R{}^T A U = U^T A U_R$$

(6.59) becomes

$$U^T A U_R = 0 \tag{6.60}$$

Since this must hold for all U^T, we have

$$A U_R = 0 \tag{6.61}$$

Now U_R contains three arbitrary elements (corresponding to the three components of $\mathbf{R}$) and, therefore, A has three linearly independent eigenvectors belonging to the eigenvalue zero.

Proceeding in a similar fashion for rotational symmetry, it follows that there are also three linearly independent eigenvectors belonging to the eigenvalue zero. Thus, if a system as a whole has rotational and translational symmetry, six of the normal mode frequencies are zero, and the maximum number of nonzero vibratory states is $3N - 6$, where N is the number of particles in the system. In NH_3 we expect a maximum of six vibratory states. It is possible to remove the nonvibratory coordinates from the problem at the outset, and a further discussion of this may be found in the literature [1–3]. Clearly, the advantage would be a dynamical matrix of $3N - 6$ dimensions rather than $3N$ dimensions.

2. Vibratory Modes of NH_3

To obtain the eigenfrequencies we need only find the eigenvalues of the submatrices appearing in (6.56). Table 6.3 presents the eigenvalues and eigenfrequencies of NH_3. The physical parameters for NH_3 appear in (6.62). We note that there are four rather than six vibratory modes because the two 4×4 submatrices are identical. It is left to the reader to determine the

TABLE 6.3
Eigenvalues and Eigenfrequencies of NH_3

Submatrix	Eigenvalues, λ^a ($\times 10^{28}$)	Eigenfrequencies, ν^a (10^{13} Hz)
3×3	0	0
	26.3	8.16
	1.06	1.67
1×1	0	0
4×4	0	0
	0	0
	13.4	5.88
	1.96	2.23

[a] $\nu = \sqrt{\lambda}/2\pi$.

normal mode coordinates. It is important to realize that the entire factorization process and subsequent solution of the eigenvalue problem for the reduced dynamical matrix can be done on the computer.

$$k_{12} = 1.3\times10^5 \quad \text{dyn cm}^{-1}, \qquad k_{14} = 0.558\times10^5 \quad \text{dyn cm}^{-1}$$
$$a = 0.5\times10^{-10} \quad \text{m}, \qquad b = 0.24\times10^{-10} \quad \text{m} \tag{6.62}$$

6.3. CONSTRUCTING A CANONICAL FORMALISM FROM A SYMMETRY GROUP

A. Introduction

Most groups occur in physics as transformation groups, i.e., as groups whose elements are realized by geometrical or physical transformations. Thus in Chapter 2, the set of relativity transformations formed a group. It is important to recognize that the concrete transformations are but *realizations* of an abstract group whose structure is completely characterized by the composition law of the abstract group elements.

The Lie algebras associated with Lie groups are most important because mathematically they are easier to study, and physically their elements correspond to conserved quantities, whereas the group elements, themselves, correspond to transformations. In this section we will construct a canonical formalism that has the Galilean group as a symmetry group.

In order to clarify the motivation for our subsequent discussion, we first discuss quite generally the relationship of invariance and dynamics [22]. First, let us consider a point P of a space whose variables are sufficient for the complete description of a physical system by an observer A at a particular time t. We assume that the point at the time t, $P(t)$, is related to the point at the time $t = 0$, $P(0)$, by means of a suitable transformation

$$P(t) = U_t P(0) \tag{6.63}$$

In other words, (6.63) expresses the time evolution of the system, and U_t specifies the dynamics of the system.

A second observer A' capable of performing the same kinds of measurements on the physical system as A would likewise be able to describe the time evolution of the system by a function $P'(t')$ in the same space to which $P(t)$ belongs and satisfying an equation similar to (6.63). We assume that a one-to-one correspondence can be established between $P(t)$ and $P'(t')$, and, in particular, that this correspondence can be established when t' is numerically equal to t. Then

$$P'(t) = L_t P(t) \tag{6.64}$$

where the transformation L_t will, in general, depend on the time itself. We define the two observers to be equivalent if the time evolution is described by both observers in terms of the same operator U_t, i.e.,

$$P'(t) = U_t P'(0) \tag{6.65}$$

Substituting (6.65) and (6.63) into (6.64) we find that

$$U_t P'(0) = L_t U_t P(0) \tag{6.66}$$

Now using (6.64) with $t = 0$, (6.66) becomes

$$U_t L_0 = L_t U_t \tag{6.67}$$

or

$$L_0 = U_t L_0 U_t^{-1} \tag{6.68}$$

This equation expresses quite generally the equivalence of the observers A and A'.

Now consider a class of observers for whom the set of all space–time coordinate transformations connecting the observers forms a group G. The physical laws are invariant with respect to G, i.e., G is a symmetry group, provided that:

(1) Any two observers of the class are equivalent as defined above;

(2) The transformations L_t connecting any two observers A and A' belonging to the class, depend only on the relation connecting A and A' and not on the particular choice of A. (This assures that the set of transformations L_t will be homomorphic to G.)

If the group G includes the time translation which connects two observers whose only difference is their choice of the origin of time, i.e.,

$$t' = t - \tau \tag{6.69}$$

then for the transformation of (6.69), (6.64) becomes

$$P'(0) = E_0(\tau) P(0) \tag{6.70}$$

where $E_0(\tau)$ is the particular L_t corresponding to a time translation. But in this case

$$P'(0) = P(\tau) \tag{6.71}$$

Therefore,

$$E_0(\tau) = U_t \tag{6.72}$$

i.e., the time translation operator coincides with the time evolution operator. Equation (6.68) can be written

$$L_t = E_0(t)L_0E_0^{-1}(t) \tag{6.73}$$

From (6.72) and (6.73) it is clear that the time evolution and the transformation properties of the complete description given by different observers at any time t are completely determined once the transformations are given. Thus the problem of constructing the most general dynamical theory invariant under G itself is reduced to that of constructing its possible realizations. We limit our considerations to canonical realizations.

B. Canonical Realizations of the Galilean Group

The Galilean group is a ten-parameter group whose parameters may be chosen such that the generators **P**, H, **J**, and **K** are, respectively, the generators of spatial displacement, temporal displacement, spatial rotations, and pure Galilean transformations (i.e., transformations to reference frames moving with a constant velocity relative to the original frame). In Chapter 1, we found that the Galilean transformations were

$$t' = t + \tau, \qquad x_i' = \omega_{ij}x_j + v_it + \beta_i \tag{6.74}$$

where ω_{ij} is the 3×3 rotation matrix, v_i is the relative velocity of the primed and unprimed frames, β_i is the spatial displacement parameter, and τ is the temporal displacement parameter. Denoting an abstract group element of the Galilean group by $(\tau, \boldsymbol{\beta}, \mathbf{v}, \omega)$, it follows from two successive applications of (6.74) on some space–time point that the group composition law is

$$(\tau', \boldsymbol{\beta}, \mathbf{v}, \omega')(\tau_\mu, \boldsymbol{\beta}, \mathbf{v}, \omega) = (\tau' + \tau, \boldsymbol{\beta}' + \omega'\boldsymbol{\beta} + \tau\mathbf{v}', \mathbf{v}' + \omega'\mathbf{v}, \omega'\omega) \tag{6.75}$$

and the identity element is (0, 0, 0, 1). The inverse is

$$(\tau, \boldsymbol{\beta}, \mathbf{v}, \omega)^{-1} = (-\tau, \omega^{-1}(\boldsymbol{\beta} - \tau\mathbf{v}), -\omega^{-1}\mathbf{v}, \omega^{-1}) \tag{6.76}$$

The Galilean Lie algebra may be obtained directly from the abstract group composition law (6.75); however, it is simpler to use a faithful representation to construct the elements of the Lie algebra and their Lie brackets.

Two representations which are valuable for this purpose are the five-dimensional representation and the unitary infinite-dimensional representation operating on the representation space of square integrable functions of $\mathbf{r}$ and t. The five-dimensional matrix representation has the form

$$G = \begin{bmatrix} \Omega & \mathbf{v} & \boldsymbol{\beta} \\ 0 & 1 & \tau \\ 0 & 0 & 1 \end{bmatrix} \tag{6.77}$$

where Ω is a submatrix, the 3×3 rotation matrix, and $\mathbf{v}$ and $\boldsymbol{\beta}$ are the column matrices

$$\begin{bmatrix} v_1 \\ v_2 \\ v_3 \end{bmatrix} \quad \text{and} \quad \begin{bmatrix} \beta_1 \\ \beta_2 \\ \beta_3 \end{bmatrix}$$

corresponding to pure Galilean transformations and spatial translations. Applying (E.54) to (6.77), one obtains a five-dimensional matrix representation of the ten elements of the Lie algebra. By forming the commutator of these matrices, the Lie brackets and the structure constants are obtained. The details are left to the reader as an exercise. (See Problem 6.12).

The properties of the infinite-dimensional unitary representation are displayed in the following equation:

$$U(\tau, \boldsymbol{\beta}, \mathbf{v}, \Omega) f(\mathbf{r}, t) = f(\Omega^{-1}(\mathbf{r} - \mathbf{v}t - \boldsymbol{\beta} + \tau\mathbf{v}), t - \tau) \tag{6.78}$$

A realization of the generators is obtained by performing a Taylor expansion of the function on the right side of (6.78) about the identity transformation and comparing the resulting expansion with Eq. (E.53). In this realization, the elements are differential operators. One finds

$$H = \partial/\partial t \tag{6.79a}$$

$$\mathbf{P} = \nabla_r \tag{6.79b}$$

$$\mathbf{J} = \mathbf{r}\times\nabla_r \tag{6.79c}$$

$$\mathbf{K} = t\nabla_r \tag{6.79d}$$

The Lie brackets are obtained by using the commutation properties of the differential operators appearing in (6.79). (See Problems 6.14 and 6.15).

In either case, one finds that the Lie brackets of the Galilean Lie algebra are

$$[J_i, J_k] = \varepsilon_{ijk} J_k \tag{6.80a}$$

$$[J_i, K_j] = \varepsilon_{ijk} K_k \tag{6.80b}$$

$$[J_i, P_j] = \varepsilon_{ijk} P_k \tag{6.80c}$$

$$[J_i, H] = 0 \tag{6.80d}$$

$$[K_i, K_j] = 0 \tag{6.80e}$$

$$[K_i, P_j] = 0 \tag{6.80f}$$

$$[K_i, H] = P_i \tag{6.80g}$$

$$[P_i, P_j] = 0 \tag{6.80h}$$

$$[P_i, H] = 0 \tag{6.80i}$$

where ε_{ijk} is the Levi–Civita tensor.

To apply algebraic techniques to physical systems it is often necessary to use *extensions* of the Lie group or Lie algebra. The mathematical ideas involved in group and algebra extensions are developed in the area of mathematics known as *cohomology theory*. We shall not develop these ideas here, but the reader may consult the following works for an introduction to the subject and further references: Hamermesh [4, Chapter 12], Loebl [6, Vol. II, p. 235] Lurcat [21, L. Michel's article], Bargmann [20]. For our purposes it is sufficient to note that an extension to a Lie algebra consists in adding a nontrivial *neutral element* to our algebra. A neutral element is an element that commutes with all of the other elements of the algebra. Thus, if we add to an algebra having the brackets

$$[a_i, a_j] = C^k_{ij} a_k, \qquad i, j, k = 1, 2, \ldots, n \tag{6.81}$$

a neutral element a_0 such that

$$[a_i, a_j] = C^k_{ij} a_k + \beta_{ij} a_0, \qquad [a_i, a_0] = 0, \qquad i, j, k = 1, 2, \ldots, n \tag{6.82}$$

then the algebra of (6.82) is said to be an *extension* of the algebra of (6.70). It can be shown that for the Galilean algebra the only nontrivial extension is one that replaces (6.80f) with

$$[K_i, P_j] = \delta_{ij} mI \tag{6.83}$$

We wish to obtain a canonical realization of the extended Galilean algebra and group. A canonical transformation is a transformation in phase space which leaves the Poisson brackets of the fundamental dynamical variables invariant. A canonical realization of a group is the set of canonical transformations having the same composition rule as the group. It is important to note that the Lie brackets have four important properties:

$$[a, b] = -[b, a], \qquad \text{antisymmetry} \tag{6.84a}$$

$$[\alpha a + \beta b, c] = \alpha[a, c] + \beta[b, c], \qquad \text{linearity} \tag{6.84b}$$

$$[a, [b, c]] + [c, [a, b]] + [b, [c, a]] = 0, \qquad \text{Jacobi identity} \tag{6.84c}$$

$$[a, b] = c, \qquad \text{closure} \tag{6.84d}$$

where a, b, and c are elements of the Lie algebra and α and β are arbitrary numbers. Properties (6.84a)–(6.84c) are also satisfied by the Poisson bracket and thus in obtaining a canonical realization of the algebra we will establish the Poisson bracket as the realization of the Lie bracket. The generators of the algebra are realized as functions of the dynamical variables of phase space. An excellent methodical approach to the acquisition of the specific form of these realizations is presented by Pauri and Prosperi [22]. Here, we simply employ some reasonable guesses as to the proper functional form of the realization and verify that these guesses satisfy the bracket relations of the algebra.

Consider now the phase space of a system having three degrees of freedom, i.e., $\mathbf{q} : (q_1, q_2, q_3)$ and $\mathbf{p} : (p_1, p_2, p_3)$. We note from (6.69a) that the generators J_i satisfy the same bracket relations as the angular momentum; and, therefore, we try to realize the generators J_i by the following functional form:

$$J_i = \varepsilon_{ijk} q_j p_k, \qquad i, j, k = 1, 2, 3 \tag{6.85}$$

Identifying the bracket of (6.80a) as a Poisson bracket, it follows immediately that (6.85) satisfies (6.80a). Realization of the generators P_i by the momentum variables

$$P_i = p_i \tag{6.86}$$

is consistent with (6.80h) and (6.80c).

Since H commutes with all of the J_i, we take H to be a scalar function of the dynamical variables

$$H \rightarrow h = h(q_i, p_i) \tag{6.87}$$

Substituting (6.87) into (6.80i), we find

$$[P_i, H] \to [p_i, h]_P = \frac{\partial p_i}{\partial q_j}\frac{\partial h}{\partial p_j} - \frac{\partial p_i}{\partial p_j}\frac{\partial h}{\partial q_j} = -\frac{\partial h}{\partial q_i} = 0 \qquad (6.88)$$

Thus from (6.88) it follows that h must be a scalar function of p_i alone. We choose the simplest such function

$$H \to h = \alpha p_i p_i \qquad (6.89)$$

where α is a constant.

The bracket relations of K_j with J_i suggest that the K_j are vector functions of the dynamical variables. We realize K_j by

$$K_j = k_j(q, p) \qquad (6.90)$$

Substituting (6.90) into (6.80g), we find

$$\begin{aligned}[K_i, H] \to [k_i, h]_P &= \alpha \frac{\partial k_i}{\partial q_j}\frac{\partial (p_l p_l)}{\partial p_j} - \alpha \frac{\partial k_i}{\partial p_j}\frac{\partial (p_l p_l)}{\partial q_j} \\ &= 2\alpha p_j \frac{\partial k_i}{\partial q_j} = p_i \end{aligned} \qquad (6.91)$$

Therefore,

$$\partial k_i/\partial q_j = (1/2\alpha)\, \delta_{ij} \qquad (6.92)$$

or

$$k_i = (1/2\alpha) q_i + f_i(p) \qquad (6.93)$$

Thus we have determined the q dependence of k_i which is consistent with (6.80g). It remains to determine the specific form of $f_i(p)$. From (6.80e) it follows that

$$\partial f_i/\partial p_j = \partial f_j/\partial p_i \qquad (6.94)$$

Therefore, there exists a function $F(p)$ such that

$$f_i = \partial F(p)/\partial p_i$$

and

$$k_i = (1/2\alpha) q_i + \partial F(p)/\partial p_i \qquad (6.95)$$

Only (6.80f) remains to be checked. Substituting (6.95) into (6.80f), we find

$$[K_i, P_j] \to [k_i, p_j]_P = \frac{\partial k_i}{\partial q_l}\frac{\partial p_i}{\partial p_l} - \frac{\partial k_i}{\partial p_l}\frac{\partial p_j}{\partial q_l}$$

$$= \frac{1}{2\alpha}\,\delta_{il}\,\delta_{jl} = \frac{1}{2\alpha}\,\delta_{ij} \neq 0 \qquad (6.96)$$

Thus we see that our realization does not satisfy (6.80f) because (6.96) does not go to zero for any finite value of α. Remember, however, that there is a nontrivial extension of the Galilean algebra in which (6.80f) is replaced by (6.83). Using (6.83) instead of (6.80f), (6.96) becomes

$$(1/2\alpha)\,\delta_{ij} = \delta_{ij} m \qquad (6.97)$$

and our realization is consistent with all the bracket relations if we use the extended Lie algebra and set $m = 1/2\alpha$. Thus the simplest nontrivial canonical realization of the extended Galilean algebra is

$$J_i = \varepsilon_{ijk} q_j p_k, \qquad P_i = p_i, \qquad K_i = m q_i, \qquad H = p_i p_i/2m \qquad (6.98)$$

Since H is the generator of a time translation, it is reasonable to consider it to be the Hamiltonian of our system, and we recognize that our realization corresponds to a free particle of mass m. More general free particle realizations exist, e.g., those corresponding to particles of mass m having an intrinsic angular momentum or spin; these more general realizations are discussed by Pauri and Prosperi [22] and Loebl [6, Vol. II].

For a system containing two masses, m_1 and m_2, the following realization, an obvious extension of the realization of (6.98), is compatible with the Lie brackets;

$$J_i = \varepsilon_{ijk} \sum_{n=1}^{2} q_{nj} p_{nj}, \qquad P_i = \sum_{n=1}^{2} p_{ni}, \qquad K_i = \sum_{n=1}^{2} m_n q_{ni} \qquad (6.99)$$

Now, however, (6.80i) no longer requires that $\partial h/\partial q_i = 0$, but only the much weaker condition that

$$\partial h/\partial q_{1i} = -\partial h/\partial q_{2i} \qquad (6.100)$$

which is satisfied by

$$H = \sum_{n=1}^{2} \frac{p_{ni} p_{ni}}{2m_1} + V(q_{2j} - q_{1j}) \qquad (6.101)$$

The realization defined in (6.99) and (6.101) is that of a two-particle system whose interaction potential energy is $V(q_{2j} - q_{1j})$. The significant feature is that the nature of the interaction is not arbitrary but depends on the relative position of the particles. This restriction of the form of the interaction by a symmetry group is a characteristic feature of symmetry groups; and, if the procedure outlined in this chapter were applied to the Poincaré group, we would again acquire the No-Interaction Theorem of the previous chapter.

Having the canonical realization of the elements of the Lie algebra, one can obtain the corresponding realization of the elements of the Lie group by exponentiation, i.e., by employing Eq. (E.45). Thus for a time translation, we may write

$$(\tau, 0, 0, 1) = e^{\tau H} \tag{6.102}$$

where τ is the infinitesimal time translation parameter and H is the canonical realization of the time translation generator, e.g., (6.101) or (6.98).

Now the canonical realizations of the group operate on functions defined on phase space. Thus

$$f(q(\tau), p(\tau)) = (\tau, 0, 0, 1) f(q(0), p(0)) = e^{\tau H} f \tag{6.103}$$

But the exponential is defined in terms of its power series expansion, and the composition law between the generators and phase space functions is the Poisson bracket. Therefore, (6.103) may be written

$$f(q(\tau), p(\tau)) = f + \tau[H, f] + \tfrac{1}{2}\tau^2[H, [H, f]] + \cdots \tag{6.104}$$

Moreover, the change in f along the one-parameter trajectory defined by H is for infinitesimal τ

$$df/d\tau = [H, f] \tag{6.105}$$

or for any generator I,

$$df/d\varepsilon = [I, f] \tag{6.106}$$

In particular, if $[I, f] = 0$, the function f is a constant along the trajectory. We thus have recovered the connection between conservation laws and symmetry transformations. By reference to the bracket relations contained in (6.80), we see that J_i and P_i are constants of the motion. In the realizations of (6.98) and (6.99), J_i and P_i are the total angular momentum and the total linear momentum of the system. Hence a complete dynamical formalism has been constructed with (6.105) providing the time evolution of functions of dynamical variables, and (6.104) providing the series solution of (6.105).

PROBLEMS

6.1 Verify that the twelve-dimensional representations in equation (6.26) are correct by applying them to a vector in the twelve-dimensional representation of the NH_3 system and showing that the mathematical result is what one would expect.

6.2 Prove (6.39) and (6.40).

6.3 Find the projection operators of the group C_{4v}.

6.4 Find the projection operators of the group Δ_2.

6.5 Show that the water molecule has Δ_2 symmetry.

6.6 Factor the dynamical matrix of the water molecule and find the normal mode frequencies. The angle between the two OH bonds is 105°, and the equilibrium length of the OH bond is 0.96 Å. The force constants are: $k_{\mathrm{OH}} = 7.76 \times 10^5$ dyne·cm^{-1}, $k_{\mathrm{HH}} = 6.36 \times 10^5$ dyne·cm^{-1}.

6.7 Show how one can obtain a fairly good estimate of the physical parameters necessary for a normal mode calculation for a molecule (i.e., force constants $k_{\alpha\beta}$ and atomic separations $k_{\alpha\beta}$ from information contained in the "Handbook of Chemistry and Physics").

6.8 Determine the symmetry of BF_3.

6.9 Using the information contained in Appendix E and the results of Problems 6.7 and 6.8, find the normal mode frequencies of BF_3.

6.10 Factor the dynamical matrix of a planar molecule of the type XY_3, where the Y atoms form an equilateral triangle and X is at the midpoint.

6.11 Find the normal mode coordinates of NH_3.

6.12 Using Eqs. (E.54) and (6.66) find the five-dimensional representation of the elements of the Galilean Lie algebra.

6.13 Using the results of the previous problem and Eq. (E.55) find the Lie brackets of the Galilean Lie algebra.

6.14 Show that the operators defined in (6.68) form a realization of the elements of the Galilean Lie algebra by Taylor expanding the right side of (6.67) about the identity transformation and comparing the results with (E.53).

6.15 Using (6.68) and (E.55) find the Lie brackets of the Galilean Lie algebra.

6.16 Proceeding as in Problems 6.12–6.15, find realizations of the generators of the Poincaré group and the Lie brackets of the Poincaré Lie algebra.

REFERENCES

1. J. S. Lomont, "Applications of Finite Groups." Academic Press, New York, 1959.
2. A. Nussbaum, "Applied Group Theory." Prentice-Hall, Englewood Cliffs, New Jersey, 1971.
3. M. I. Petrashen and E. D. Trifonov, "Applications of Group Theory in Quantum Mechanics." MIT Press, Cambridge, Massachusetts, 1969.
4. M. Hamermesh, "Group Theory." Addison-Wesley, Reading, Massachusetts, 1964.
5. G. Lyubarskii, "The Application of Group Theory in Physics." Pergamon, Oxford, 1960.
6. E. M. Loebl, ed., "Group Theory and Its Applications." Academic Press, New York, 1968.
7. E. B. Wilson, J. C. Decius, and P. C. Cross, "Molecular Vibrations." McGraw-Hill, New York, 1955.
8. R. McWeeny, "Symmetry - An Introduction to Group Theory." Pergamon, Oxford, 1963.
9. R. Hochstrasser, "Molecular Aspects of Symmetry." Benjamin, New York, 1966.
10. P. T. Landsberg, ed. "Solid State Theory." Wiley, New York, 1969.
11. F. A. Cotton, "Chemical Applications of Group Theory." Wiley (Interscience), New York, 1963.
12. G. Herzberg, "Molecular Structure and Molecular Spectra, II. Infrared and Raman Structure of Polyatomic Molecules." Van Nostrand-Reinhold, Princeton, New Jersey, 1945.
13. S. J. Cyvin, "Molecular Vibrations and Mean Square Amplitudes." Elsevier, Amsterdam, 1968.
14. J. W. Leech and D. W. Newman, "How to Use Group Theory." Methuen, London, 1969.
15. E. P. Wigner, "Group Theory and Its Application to the Quantum Mechanics of Atomic Spectra." Academic Press, New York, 1959.
16. C. Chevally, "Theory of Lie Groups." Princeton Univ. Press, Princeton, New Jersey, 1946.

17. L. Pontrjagin, "Topological Groups." Princeton Univ. Press, Princeton, New Jersey, 1939.
18. P. M. Cohn, "Lie Groups." Cambridge Univ. Press, London and New York, 1957.
19. G. Reach, *Nuovo Cimento Suppl.* **14**, 67 (1959).
20. V. Bargmann, *Ann. Math.* **59**, 1 (1954).
21. F. Lurcat ed., "Applications of Mathematics to Problems in Theoretical Physics." Gordon & Breach, New York, 1967.
22. M. Pauri and G. M. Prosperi, *J. Math. Phys.* (*N. Y.*) **7**, 366 (1966); **8**, 2256 (1967); **9**, 1146 (1968).
23. E. C. G. Sudarshan, "Principles of Classical Mechanics." Rep. NYO - 10250. Univ. of Rochester, Rochester, New York, 1963.
24. J. M. Levy-Leblond, *Ann. Phys.* (*N. Y.*) **57**, 481 (1970).
25. N. Mukunda, *J. Math. Phys.* (*N. Y.*) **8**, 1069 (1967).
26. D. G. Currie, T. F. Jordan, and E. C. G. Sudarshan, *Rev. Mod. Phys.* **35**, 350 (1963).
27. R. Hermann, "Lie Groups for Physicists." Benjamin, New York, 1966.
28. R. Weast, ed., "Handbook of Chemistry and Physics." Chemical Rubber Co. Cleveland, Ohio, 1972.

Appendix A

SOME USEFUL MATHEMATICS

A number of methematical concepts that bear directly or indirectly on the material presented in the main text are reviewed here. It should be noted at the outset that when there is a disparity between the mathematician's and the physicist's use of a term, the physicist's definition will be presented.

Consider two arbitrary sets of elements $y = \{y_1, y_2, \ldots\}$ and $x = \{x_1, x_2, \ldots\}$. A *function* F is said to be defined on x if to each element of x there corresponds one and only one element of y. We normally designate this relationship by writing $y = F(x)$ or simply $y = Fx$. F may equivalently be called a *transformation*, a *mapping*, or an *operator*. Many authors will call F a function only if the elements of both x and y are numbers; whereas if the elements of y are numbers and the elements of x are functions, then F is usually called a *functional.* The set x is called the *domain* and the set y the *range* of F. The transformation F is *isomorphic,* i.e., one-to-one, if to each element of y there is but one element of x such that $Fx = y$. When F is isomorphic, an *inverse* transformation F^{-1} exists defined by $x = F^{-1}y$.

The set of elements $\{x, y, \ldots\}$ is a *vector space* if the following axioms are satisfied.

Axiom 1. To every pair of elements x and y there corresponds an element $x + y$, called the *sum* of x and y, with the properties

(a) $x + y = y + x$

(b) $x + (y + z) = (x + y) + z$

(c) There exists a unique vector 0 such that $x + 0 = x$ for every element of the set.

(d) To every element x of the set there exists a unique element labeled $-x$ such that $x + (-x) = 0$.

Axiom 2. To every element x and every number α, there corresponds a vector αx such that

(a) $\alpha(\beta x) = (\alpha\beta)x$
(b) $(\alpha + \beta)x = \alpha x + \beta x$
(c) $\alpha(x + y) = \alpha x + \alpha y$
(d) $1x = x$

The elements of the vector space are called *vectors*. The vector space is either *real* or *complex* depending on whether α is a real or complex number.

A vector x is a linear combination of the vectors $x_1, \ldots, x_k$ if there exist numbers $\alpha_1, \ldots, \alpha_k$ such that $x = \sum_{i=1}^{N} \alpha_i x_i$. Vectors $x_1, \ldots, x_k$ are dependent if there exist numbers $\alpha_1, \ldots, \alpha_k$, not all zero, such that

$$\alpha_1 x_1 + \alpha_2 x_2 + \cdots + \alpha_k x_k = 0 \tag{A.1}$$

If (A. 1) is valid only for $\alpha_1 = \cdots = \alpha_k = 0$, the vectors are *independent*.

A vector space is *n dimensional* if it has a set of n independent vectors, but every set of $n + 1$ vectors is a dependent set. If for every positive integer k we can find k independent vectors, the space is ∞ *dimensional.* A finite set of vectors $e_1, \ldots, e_n$ is said to be a basis of a vector space, if

(a) the vectors $e_1, \ldots, e_n$ are independent.
(b) every vector x in the space can be written as a linear combination of the basis vectors; i.e.,

$$x = \alpha_1 e_1 + \cdots + \alpha_n e_n \tag{A.2}$$

The numbers $\alpha_1, \alpha_2, \ldots, \alpha_n$ are called the *components* of the vector x relative to the basis $e_1, \ldots, e_n$. It can be shown that any set of n independent vectors may be used as a basis of an n-dimensional space.

Of particular importance in physics are *linear* transformations. The transformation of the vector x into the vector x' may be written as

$$x'_j = a_{ji} x_i \tag{A.3}$$

where the x_i and x'_j are the components of x and x', and the a_{ij} are numbers that define the transformation and therefore the relationship between the components of the vectors. The Galilean and Lorentz transformations are linear transformations.

It is often convenient to write (A.3) in matrix form. Components of terms having two subscripts are conventionally placed in the matrix by allowing the first subscript to designate the row and the second index the column in which the component appears. Thus a_{21} is placed in the second row first column. Terms with a single subscript have their components arranged in a single column or a single row, e.g.,

$$x_i = \begin{bmatrix} x_1 \\ x_2 \\ \vdots \\ x_n \end{bmatrix} \quad \text{or} \quad x_i = (x_1 \quad x_2 \cdots x_n) \tag{A.4}$$

Thus (A.3) may be written in matrix form

$$X' = AX \tag{A.5}$$

or

$$\begin{bmatrix} x_1' \\ x_2' \\ \vdots \\ x_n' \end{bmatrix} = \begin{bmatrix} a_{11} & a_{12} & \cdots & a_{1n} \\ a_{21} & a_{22} & \cdots & a_{2n} \\ \vdots & & & \vdots \\ a_{n1} & a_{n2} & \cdots & a_{nn} \end{bmatrix} \begin{bmatrix} x_1 \\ x_2 \\ \vdots \\ x_n \end{bmatrix}$$

It is often necessary to be able to convert expressions written in index notation into equivalent matrix expressions. Thus the multiplication of the matrices A and B is written in index notation as

$$a_{ij}b_{jk} = c_{ik} \quad \sim \quad AB = C \tag{A.6}$$

where we are using the summation convention.

Another common form is the multiplication of the transpose of matrix and another matrix A^TB. In index notation, this becomes

$$d_{jk} = a_{ij}b_{ik} \quad \sim \quad D = A^TB \tag{A.7}$$

Similarly,

$$l_{ik} = a_{ij}b_{ki} \quad \sim \quad E = AB^T \tag{A.8}$$

A set of elements $S = \{x, y, z, \ldots,\}$ is a *group* if a law of *binary* composition (i.e., the composition involves a pair of elements) is defined having the following properties:

1. Closure $x \cdot y = z$, where x, y, and $z \in S$
2. Associative $x \cdot (y \cdot z) = (x \cdot y) \cdot z$
3. Identity There exists an element I, the identity element, such that $x \cdot I = I \cdot x = x$ for any $x \in S$
4. Inverse To each element x there corresponds an element $x^{-1} \in S$ such that $x^{-1} \cdot x = x \cdot x^{-1} = I$ x^{-1} is called the inverse of x.

The set of all real integers $\{\ldots, -3, -2, -1, 0, 1, 2, 3, \ldots\}$ is a group if the law of composition is the ordinary addition of numbers. (-1) is the inverse of (1), 0 is the identity, the addition of any two integers yields an integer, and ordinary addition is associative. Note that the set $\{-2, -1, 0, 1, 2\}$ is not a group under ordinary addition because the closure property is not satisfied, i.e., $1 + 2 = 3$, and 3 is not a member of the set. Two of the more important groups used in physics are the Galilean group, defined by (1.23), and the Lorentz group, defined by (1.27).

Consider a group having elements $a, b, c, \ldots$ and a set of transformations $A, B, C, \ldots$. If we can establish a correspondence $a \leftrightarrow A$, $b \leftrightarrow B$, etc., such that to the product (ab) there corresponds the product (AB), i.e., $(ab) \leftrightarrow (AB)$, then the set of transformations is said to form a *representation* of the group, and the vector space on which the transformations act is called the *representation space* of the group.

If the representation space is finite dimensional, the transformation elements (operators) may be expressed in matrix form. An object which transforms according to the following prescription:

$$T^{\alpha\beta\cdots\delta'} = A^{\alpha}{}_{a}A^{\beta}{}_{b} \cdots A^{\delta}{}_{d}T^{ab\cdots d} \tag{A.9}$$

is called a *tensor* defined with respect to the transformation group $A, B, \ldots$. The *rank* of the tensor is the number of indices appearing on the tensor; e.g., $T^{\alpha\beta\gamma}$ is a tensor of the third rank, and $T^{\alpha\beta\gamma\delta\epsilon}$ is a tensor of the fifth rank. A tensor of the first rank, T^{α}, is called a vector; a tensor of the zeroth rank, T, is called a scalar. It is important to note that a tensor is defined relative to a given group; and, therefore, an object that is a tensor relative to one group is not necessarily a tensor relative to another group. For example, a tensor defined on the Galilean group is not a tensor for the Lorentz group.

Tensor analysis is an extensive subject, especially the calculus of tensors. Fortunately, in this text only tensors defined on a Euclidean or a Lorentz space are employed; and it happens that for these tensors the complexities encountered in general tensor analysis are absent, and they can, for the most part, be handled in much the same way as the vectors of Newtonian mechanics. The few additional features are discussed in the remaining portion of this Appendix.

An important property of vector spaces used in physics is the *inner product* of two vectors comprising the space. The inner product of two vectors, x and y, is represented by $\langle x, y\rangle$ and is a function of the ordered pairs x, y with the properties

(a) $\langle x, \alpha y\rangle = \alpha\langle x, y\rangle$ (α is a number)
(b) $\langle x, y + z\rangle = \langle x, y\rangle + \langle x, z\rangle$
(c) $\langle x, y\rangle = \langle y, x\rangle^*$ ($*$ denotes complex conjugate)

It follows from (a) and (c) that $\langle \alpha x, y \rangle = \alpha^* \langle x, y \rangle$. The inner product of a vector with itself, $\langle x, x \rangle$, is called the *norm* of the vector x.

The inner product is simply the generalization of the scalar or dot product of the vectors used in elementary mechanics. In classical mechanics, the two vector spaces most commonly used are the Euclidean space and the Lorentz space. In an n-dimensional Euclidean space, the inner product is defined as

$$\langle x, y \rangle \equiv x_1 y_1 + x_2 y_2 + \cdots + x_n y_n \tag{A.10}$$

where $x_1, x_2 \ldots$ are the components of the vector x and $y_1, y_2, \ldots$ are the components of the vector y.

In the four-dimensional Lorentz space, the inner product is defined as

$$\langle x, y \rangle \equiv x_0 y_0 - x_1 y_1 - x_2 y_2 - x_3 y_3 \tag{A.11}$$

There is an important difference between the norm in a Euclidean space and the norm in a Lorentz space, viz., the Euclidean norm is positive definite (i.e., $\langle x, x \rangle \geq 0$ with the equality holding only if $x = 0$), whereas the Lorentz norm is *indefinite* (i.e., $\langle x, x \rangle$ may be positive, negative, or zero). This difference has profound physical consequences, some of which are discussed in Chapter 1.

In finite-dimensional vector spaces, the inner product is often written as

$$\langle x, y \rangle = x_\mu g^{\mu\nu} y_\nu \tag{A.12}$$

where $g^{\mu\nu}$ is the *contravariant metric tensor*. Comparing (A.12) with (A.10) and (A.11), we conclude

Euclidean space

$$g^{\mu\nu} = \begin{bmatrix} 1 & 0 & 0 & 0 \\ 0 & 1 & 0 & 0 \\ 0 & 0 & 1 & 0 \\ 0 & 0 & 0 & 1 \end{bmatrix} \tag{A.13}$$

Lorentz space

$$g^{\mu\nu} = \begin{bmatrix} 1 & 0 & 0 & 0 \\ 0 & -1 & 0 & 0 \\ 0 & 0 & -1 & 0 \\ 0 & 0 & 0 & -1 \end{bmatrix} \tag{A.14}$$

We define a *contravariant* vector y^μ as

$$y^\mu = g^{\mu\nu} y_\nu \tag{A.15}$$

and (A.12) may be written

$$\langle x, y\rangle = x_\mu y^\mu = x_0 y^0 + x_1 y^1 + x_2 y^2 + x_3 y^3 \tag{A.16}$$

A vector with components, y_μ, is called a *covariant* vector. Thus the superscript position is the "contravariant" position, and the subscript position in the "*covariant*" position. The relationship between the covariant and contravariant components of a vector is determined by the metric tensor, i.e., by (A.15).

The contravariant metric tensor may be thought of as an operator which raises a subscript (a covariant index) to the superscript position (contravariant position). We can introduce a *covariant metric tensor* which converts a contravariant index into a covariant index, i.e.,

$$y_\mu = g_{\mu\nu} y^\nu \tag{A.17}$$

Substituting (A.15) into (A.17), we obtain

$$y_\mu = g_{\mu\nu} g^{\nu\lambda} y_\lambda \equiv g_\mu{}^\lambda y_\lambda \tag{A.18}$$

where we have introduced the *mixed metric tensor* $g_\mu{}^\lambda$. In other words, the $g_{\mu\nu}$ converts the contravariant index ν of $g^{\nu\lambda}$ into a covariant μ. It is immediately obvious from (A. 18) that $g_\mu{}^\lambda$ must be the identity matrix; and, therefore, it also follows that the covariant metric tensor is the inverse of the contravariant metric tensor. Thus,

Euclidean space

$$g_{\mu\nu} = \begin{bmatrix} 1 & & & 0 \\ & 1 & & \\ & & 1 & \\ 0 & & & 1 \end{bmatrix} \tag{A.19}$$

Lorentz space

$$g_{\mu\nu} = \begin{bmatrix} 1 & & & 0 \\ & -1 & & \\ & & -1 & \\ 0 & & & -1 \end{bmatrix} \tag{A.20}$$

The mixed metric tensor plays the same role as the Kronecker delta δ_{ij}, and it can be used in the same way. With the aid of the various forms of the metric tensor introduced here, it is possible to raise or lower tensor indices to suit our convenience. It is clear that in Euclidean space the covariant and contravariant components are equal; however, in Lorentz space it follows from (A.17) and (A.20) that

$$x^0 = x_0, \qquad x^k = -x_k \tag{A.21}$$

When employing the tensor notation, it is common to denote the tensor character of the differential operator by writing

$$\partial^\mu \equiv \frac{\partial}{\partial x_\mu}, \qquad \partial_\mu \equiv \frac{\partial}{\partial x^\mu} \tag{A.22}$$

Thus in a three-dimensional Euclidean space, the "Laplace operator" is

$$\nabla^2 \rightarrow \partial^i \partial_i = \partial_i \partial_i \tag{A.23}$$

whereas in Lorentz space the "D'Alembertian operator" is

$$\square \rightarrow \partial^\mu \partial_\mu = \partial^0 \partial_0 + \partial^i \partial_i = (\partial^0)^2 - (\partial^1)^2 - (\partial^2)^2 - (\partial^3)^2 \tag{A.24}$$

The D'Alembertian is simply the scalar product of the four vector ∂^μ with itself.

Appendix B

DERIVATION OF LAGRANGE'S EQUATIONS FROM NEWTON'S LAWS

In this Appendix we derive the Lagrange equations of motion directly from the Newtonian equations

$$ {}_\alpha\mathbf{F}^{(T)} = {}_\alpha\dot{\mathbf{p}} \tag{B.1} $$

A major difficulty in the application of (B. 1) to complex systems is that ${}_\alpha\mathbf{F}^{(T)}$ includes both the applied forces and the forces of constraint; however, the forces of constraint are rarely known a priori and it would be advantageous to be able to determine the motion of the system without the need of introducing the constraint forces. With this in mind we distinguish between applied forces ${}_\alpha\mathbf{F}$, and constraint forces ${}_\alpha\mathbf{f}$.

$$ {}_\alpha\mathbf{F}^{(T)} = {}_\alpha\mathbf{F} + {}_\alpha\mathbf{f} \tag{B.2} $$

We now subject our system to a *virtual displacement* which is defined as a change in the configuration of the system produced by a change $\delta\,{}_\alpha\mathbf{r}$ in the configuration coordinates consistent with the forces and constraints imposed on the system at a given instant. The virtual displacement is performed at an instant of time, i.e., there is no time elapse in the process. The virtual work associated with the virtual displacement can be obtained by taking the scalar product of (B.1) with $\delta\,{}_\alpha\mathbf{r}$:

$$ \sum_\alpha [({}_\alpha\mathbf{F} - {}_\alpha\dot{\mathbf{p}}) \cdot \delta\,{}_\alpha\mathbf{r} + \mathbf{f}_\alpha \cdot \delta\,{}_\alpha\mathbf{r}] = 0 \tag{B.3} $$

We now limit our subsequent discussion to systems for which

$$ \sum_\alpha {}_\alpha\mathbf{f} \cdot \delta\,{}_\alpha\mathbf{r} = 0 \tag{B.4} $$

Thus (B.3) becomes

$$ \sum ({}_\alpha\mathbf{F} - {}_\alpha\dot{\mathbf{p}}) \cdot \delta\,{}_\alpha\mathbf{r} = 0 \tag{B.5} $$

The forces of constraint have now been eliminated from the formalism and (B. 5) can be used to obtain the equations of motion if we may set the coefficients of the $\delta\ {}_\alpha\mathbf{r}$ equal to zero. But we cannot in general do this since the $\delta\ {}_\alpha\mathbf{r}$ are not necessarily independent. To remedy this situation we introduce a new set of coordinates, the generalized coordinates q_a, which are independent. The number of coordinates q_a necessary to define the configuration of the system is just equal to N, the number of degrees of freedom of the system. The equations of transformation connecting the Cartesian coordinates and the generalized coordinates are

$$ {}_\alpha\mathbf{r} = {}_\alpha\mathbf{r}(q_1, q_2, \ldots, q_N, t) \tag{B.6}$$

Thus the velocity of the αth particle is

$$ {}_\alpha\dot{\mathbf{v}} = {}_\alpha\dot{\mathbf{r}} = \sum_{a=1}^{N} \frac{\partial\ {}_\alpha\mathbf{r}}{\partial q_a}\dot{q}_a + \frac{\partial\ {}_\alpha\mathbf{r}}{\partial t}$$

and

$$\delta\ {}_\alpha\mathbf{r} = \sum_a \frac{\partial\ {}_\alpha\mathbf{r}}{\partial q_a}\,\delta q_a \tag{B.7}$$

We may write the virtual work of the applied forces as

$$\sum_\alpha {}_\alpha\mathbf{F}\cdot\delta\ {}_\alpha\mathbf{r} = \sum_\alpha\sum_a {}_\alpha\mathbf{F}\cdot\frac{\partial\ {}_\alpha\mathbf{r}}{\partial q_a}\,\delta q_a \equiv \sum_a Q_a\,\delta q_a \tag{B.8}$$

where Q_a, the *generalized force*, is defined as

$$Q_a \equiv \sum_\alpha {}_\alpha\mathbf{F}\cdot\frac{\partial\ {}_\alpha\mathbf{r}}{\partial q_a} \tag{B.9}$$

Since the generalized coordinates need not have the dimensions of length, e.g., q may be an angle, the generalized force need not have the dimensions of force.

Using (B. 7) and the usual Newtonian definition of momentum, we may write

$$\begin{aligned}\sum_\alpha {}_\alpha\dot{\mathbf{p}}\cdot\delta\ {}_\alpha\mathbf{r} &= \sum_\alpha\sum_a m_\alpha\ {}_\alpha\ddot{\mathbf{r}}\cdot\frac{\partial\ {}_\alpha\mathbf{r}}{\partial q_a}\,\delta q_a \\ &= \sum_{\alpha,a}\left[\frac{d}{dt}\left(m_\alpha\ {}_\alpha\dot{\mathbf{r}}\cdot\frac{\partial\ {}_\alpha\mathbf{r}}{\partial q_a}\right) - m_\alpha\ {}_\alpha\dot{\mathbf{r}}\cdot\frac{d}{dt}\left(\frac{\partial\ {}_\alpha\mathbf{r}}{\partial q_a}\right)\right]\delta q_a\end{aligned} \tag{B.10}$$

But

$$\frac{d}{dt}\frac{\partial\ {}_\alpha\mathbf{r}}{\partial q_a} = \sum_b \frac{\partial^2\ {}_\alpha\mathbf{r}}{\partial q_a\,\partial q_b}\dot{q}_b + \frac{\partial^2\ {}_\alpha\mathbf{r}}{\partial q_a\,\partial t} = \frac{\partial\ {}_\alpha\mathbf{v}}{\partial q_a} \tag{B.11}$$

and

$$\frac{\partial\, {}_\alpha\mathbf{v}}{\partial \dot{q}_a} = \frac{\partial\, {}_\alpha\mathbf{r}}{\partial q_a} \tag{B.12}$$

Inserting (B.11) and (B.12) into (B.10), we obtain

$$\sum_\alpha {}_\alpha\dot{\mathbf{p}} \cdot \delta\, {}_\alpha\mathbf{r} = \sum_{\alpha,a} \left[\frac{d}{dt} \left(m_\alpha\, {}_\alpha\mathbf{v} \cdot \frac{\partial\, {}_\alpha\mathbf{v}}{\partial \dot{q}_a} \right) - m_\alpha\, {}_\alpha\mathbf{v} \cdot \frac{\partial\, {}_\alpha\mathbf{v}}{\partial q_a} \right] \delta q_a \tag{B.13}$$

and (B.5) may be written

$$\sum_a \left[\frac{d}{dt} \left(\frac{\partial}{\partial \dot{q}_a} \left(\sum_\alpha \tfrac{1}{2} m_\alpha\, {}_\alpha v^2 \right) \right) - \frac{\partial}{\partial q_a} \left(\sum_\alpha \tfrac{1}{2} m_\alpha\, {}_\alpha v^2 \right) - Q_a \right] \delta q_a \tag{B.14}$$

But

$$\sum_\alpha \tfrac{1}{2} m_\alpha\, {}_\alpha v^2 = T = \text{kinetic energy}$$

and (B.14) becomes

$$\sum_a \left[\frac{d}{dt} \frac{\partial T}{\partial \dot{q}_a} - \frac{\partial T}{\partial q_a} - Q_a \right] \delta q_a = 0 \tag{B.15}$$

Since the δq_a are chosen to be independent of each other, (B. 15) can be satisfied for arbitrary δq_a only if the coefficients are separately zero, i.e.,

$$\frac{d}{dt} \frac{\partial T}{\partial \dot{q}_a} - \frac{\partial T}{\partial q_a} = Q_a, \qquad a = 1, 2, \ldots, N \tag{B.16}$$

These equations are the Lagrange equations of motion. If the system is conservative, i.e., if the forces may be derived from a scalar function of the coordinates (the potential energy),

$${}_\alpha\mathbf{F} = -\nabla_\alpha V \tag{B.17}$$

the generalized forces may be written as

$$Q_a = \sum_\alpha {}_\alpha\mathbf{F} \cdot \frac{\partial\, {}_\alpha\mathbf{r}}{\partial q_a} = -\sum_\alpha (\nabla_\alpha V) \cdot \frac{\partial\, {}_\alpha\mathbf{r}}{\partial q_a} = -\frac{\partial V}{\partial q_a} \tag{B.18}$$

where

$$\nabla_\alpha = \frac{\partial}{\partial\, {}_\alpha x} + \frac{\partial}{\partial\, {}_\alpha y} + \frac{\partial}{\partial\, {}_\alpha z} \tag{B.19}$$

If the potential energy is not a function of the velocities, (B. 16) may be written

$$\frac{d}{dt}\frac{\partial L}{\partial \dot{q}_a} - \frac{\partial L}{\partial q_a} = 0 \qquad \text{(B.20)}$$

where L, the Lagrange function or Lagrangian, is defined as

$$L = T - V \qquad \text{(B.21)}$$

The system of equations (B. 20) is the most common form of the Lagrange equations of motion. If some of the forces are derivable from a scalar function of the coordinates only and some are not, the equations of motion may be written quite generally as

$$\frac{d}{dt}\frac{\partial L}{\partial \dot{q}_a} - \frac{\partial L}{\partial q_a} = Q_a \qquad \text{(B.22)}$$

It is also quite simple to derive Hamilton's principle from Newton's equations. Using (B.17), (B.5) may be written

$$\sum_\alpha {}_\alpha\mathbf{F} \cdot \delta\, {}_\alpha\mathbf{r} = -\,\delta V = \sum_\alpha m_\alpha\, {}_\alpha\ddot{\mathbf{r}} \cdot \delta\, {}_\alpha\mathbf{r} \qquad \text{(B.23)}$$

But

$$\frac{d}{dt}({}_\alpha\dot{\mathbf{r}} \cdot \delta\, {}_\alpha\mathbf{r}) = {}_\alpha\ddot{\mathbf{r}} \cdot \delta\, {}_\alpha\mathbf{r} + \delta(\tfrac{1}{2}\, {}_\alpha\dot{\mathbf{r}} \cdot {}_\alpha\dot{\mathbf{r}}) \qquad \text{(B.24)}$$

Thus (B.23) becomes

$$-\,\delta V = \frac{d}{dt}(\sum_\alpha m_\alpha\, {}_\alpha\dot{\mathbf{r}} \cdot \delta\, {}_\alpha\mathbf{r}) - \delta T \qquad \text{(B.25)}$$

or

$$\int_{t_1}^{t_2} \delta(T - V)\, dt = (\sum_\alpha m_\alpha\, {}_\alpha\dot{\mathbf{r}} \cdot \delta\, {}_\alpha\mathbf{r})\Big|_{t_1}^{t_2} \qquad \text{(B.26)}$$

But if we require

$$\delta_\alpha\mathbf{r}(t_2) = \delta\, {}_\alpha\mathbf{r}(t_1) = 0$$

(B.26) becomes

$$\delta \int_{t_1}^{t_2} L\, dt = 0 \qquad \text{Hamilton's principle} \qquad \text{(B.27)}$$

The above formalism is readily extended to systems having constraints and forces derivable from velocity dependent potentials. It is not difficult to show [2] that any system of differential equations is derivable from the variation of some functional.

REFERENCES

1. H. Goldstein, "Classical Mechanics," Addison-Wesley, Reading, Massachusetts, 1965.
2. M. Becker, "The Principles and Applications of Variational Methods," MIT Press, Cambridge, Massachusetts, 1964.

Appendix C

LEUTWYLER'S PROOF OF THE NO-INTERACTION THEOREM

C.1. INTRODUCTION

In this Appendix we present Leutwyler's [1] elegant group theoretic proof of the No-Interaction Theorem. The theorem issues from an effort to construct a classical Hamiltonian particle dynamics that is compatible with either Einstein special relativity or Newton–Galilean relativity. Here we summarize only briefly the underlying assumptions; anyone who wishes to understand the connection between relativistic symmetry, Lie group formalism, and the mathematical representation of both classical and quantum dynamical systems is encouraged to consult the truly remarkable review article on the subject by Currie *et al.* [2] and at least the introduction of a paper by Foldy [3].

The theory must be symmetric under the transformations of the relativity group. This will be attained by assuming that the theory has the ten generators of the relativity group H, $\mathbf{P}$, $\mathbf{J}$, and $\mathbf{K}$ which satisfy the Lie bracket equations.

$$[P_i, P_j] = 0, \qquad [P_i, H] = 0, \qquad [J_i, H] = 0 \tag{C.1}$$

$$[J_i, J_j] = \varepsilon_{ijk} J_k, \qquad [P_i, J_j] = \varepsilon_{ijk} P_k \tag{C.2}$$

$$[J_i, K_j] = \varepsilon_{ijk} K_k, \qquad [K_j, H] = P_j \tag{C.3}$$

$$[K_i, K_j] = -\varepsilon_{ijk} J_k, \qquad [K_i, P_j] = \delta_{ij} H \tag{C.4a}$$

$$[K_i, K_j] = 0, \qquad [K_i, P_j] = \delta_{ij} M \tag{C.4b}$$

where (C.1)–(C.3) are valid for both the Lorentz and Galilean groups,

(C.4a) is valid for the Lorentz group only, and (C.4b) is valid only for the Galilean group. H, **P**, **J**, and **K** are the generators of time translations, space translations, spatial rotations, and pure Lorentz (i.e., velocity) transformations, respectively. M is a neutral element which is the "nonrelativistic" limit of H. The meaning of neutral elements, generators, Lie brackets, and so on, was presented in Appendix A.

In the canonical formalism the dynamical variables are the coordinates and momenta of the particles comprising the system. In Chapter 4 it was shown that the Hamiltonian is the generator of the motion of the system in time, the momentum is the generator of a spatial displacement of the system, and so on. Thus we establish contact between the abstract group (algebra) formalism and the physical system by identifying the elements of the Lie algebra H, **P**, and **J** with the Hamiltonian, the total linear momentum, and the total angular momentum. Furthermore, we assume that the Lie brackets are just the Poisson brackets of the canonical formalism. If we were to make no additional assumptions it could be shown that there exist nontrivial theories which permit a wide variety of interactions between particles [3]. However, an additional requirement seems necessary, viz., that the particle positions transform in the usual way under the group transformations. It is not difficult [2] to show that this is equivalent to requiring that the coordinates satisfy the bracket relations

$$[{}_\alpha q_i, P_j] = \delta_{ij}, \qquad [{}_\alpha q_i, J_j] = -\varepsilon_{ijk}\, {}_\alpha q_k \tag{C.5}$$

$$[{}_\alpha q_i, K_j] = {}_\alpha q_j [{}_\alpha q_i, H] \tag{C.6a}$$

$$[{}_\alpha q_i, K_j] = 0 \tag{C.6b}$$

where (C. 5) is valid for both the Lorentz and Galilean groups, (C. 6a) is valid only for the Lorentz group, and (C. 6a) is valid only for the Galilean group. No assumptions are made about the transformation properties of the momenta.

It will now be shown that it follows from the previous assumptions that when (C. 6a) is valid the acceleration of each particle is zero. The proof is accomplished in two steps. First it is demonstrated that the generators **P** and **J** may be brought to their free-particle form

$$P_i^0 = \sum_\alpha p_i, \qquad J_i^0 = \sum_\alpha \varepsilon_{ijk}\, {}_\alpha q_j\, {}_\alpha p_k \tag{C.7}$$

by a canonical transformation. The second step is to show that the generators

H and **K** can also be brought to their free-particle form

$$H^0 = \sum_\alpha ({}_\alpha p^2 + m_\alpha{}^2)^{1/2}, \qquad K_i^0 = \sum_\alpha ({}_\alpha p^2 + m_\alpha{}^2)^{1/2}\ {}_\alpha q_i \tag{C.8}$$

by a canonical transformation that leaves the P_i and J_i unchanged. Thus the proof consists in showing that the dynamics can always be reduced to free-particle motion by appropriate canonical transformations.

C.2. REDUCTION OF P AND J TO FREE PARTICLE FORM

Writing explicitly the Poisson bracket of (C.5), we find

$$\sum_\beta \frac{\partial\ {}_\alpha q_i}{\partial\ {}_\beta q_k} \frac{\partial P_j}{\partial\ {}_\beta p_k} = \delta_{ik}\,\delta_{\alpha\beta} \frac{\partial P_j}{\partial\ {}_\beta p_k} = \frac{\partial P_j}{\partial\ {}_\alpha p_i} = \delta_{ij} \tag{C.9}$$

This equation is satisfied by

$$P_j = P_j{}^0 + w_j(q) \tag{C.10}$$

where $P_j{}^0$ is defined in (C. 7) and W_j is an arbitrary function of the coordinates.

Proceeding in the same fashion for the **J** appearing in (C.5), we find

$$J_j = J_j^0 + F_j(q) \tag{C.11}$$

where $J_j{}^0$ is defined in (C. 7) and F_j is an arbitrary function of the coordinates. It follows [4] from (C.2), (C.10), and (C.11) that

$$F_i = [J_i, F], \qquad W_i = [P_i, F] \tag{C.12}$$

where $F = F(q)$.

The canonical transformation

$${}_\alpha q_i' = {}_\alpha q_i, \qquad {}_\alpha p_i' = {}_\alpha p_i + [{}_\alpha p_i, F] \tag{C.13}$$

reduces the generators to their free particle form, i.e.,

$$P_i \to \sum_\alpha ({}_\alpha p_i' - [{}_\alpha p_i, F]) + W_i' = P_i^{0\prime} - [P_i^0, F] + [P_i, F] = P_i^{0\prime} = P_i' \tag{C.14}$$

$$J_i \to J_i^{0\prime} - [J_i^0, F] + [J_i, F] = J_i^{0\prime} = J_i' \tag{C.15}$$

C.3. REDUCTION OF K AND H TO FREE-PARTICLE FORM

Equation (C.6a) implies that

$$({}_\alpha q_j - {}_\beta q_j)\frac{\partial^2 H}{\partial\, {}_\alpha p_l\, \partial\, {}_\beta p_i} = 0 \tag{C.16}$$

and therefore

$$H = \sum_\alpha {}_\alpha h({}_\alpha p, q) \tag{C.17}$$

where ${}_\alpha h$ is a function of the momentum of the αth particle only. From (C.1) it follows that H and therefore each ${}_\alpha h$ is invariant under translation and rotation. Inserting (C.17) into (C.6a), we find that

$$K_i = \sum_\alpha {}_\alpha h\, {}_\alpha q_i + k_i \tag{C.18}$$

where k_i is a translation invariant vector function of the coordinates only.

Evaluating the bracket containing H in (C.3) yields

$$\frac{1}{2}\sum_\alpha \frac{\partial}{\partial\, {}_\alpha p_i}({}_\alpha h^2 - {}_\alpha p^2) + \sum_\alpha \frac{\partial k_i}{\partial\, {}_\alpha q_j}\frac{\partial\, {}_\alpha h}{\partial\, {}_\alpha p_j} + \sum_\alpha \frac{\partial\, {}_\beta h}{\partial\, {}_\alpha q_j}\frac{\partial\, {}_\alpha h}{\partial\, {}_\alpha p_j}({}_\beta q_i - {}_\alpha q_i) \tag{C.19}$$

Differentiating (C.19) with respect to ${}_\gamma p_l$ and ${}_\delta p_m$, we obtain a contribution from the last term only. If $\gamma \neq \delta$

$$\frac{\partial^2\, {}_\gamma h}{\partial\, {}_\delta q_j\, \partial\, {}_\gamma p_l}\frac{\partial^2\, {}_\delta h}{\partial\, {}_\delta p_j\, \partial\, {}_\delta p_m} - \frac{\partial^2\, {}_\delta h}{\partial\, {}_\gamma q_j\, \partial\, {}_\delta p_m}\frac{\partial^2\, {}_\gamma h}{\partial\, {}_\gamma p_j\, \partial\, {}_\gamma p_l} = 0 \tag{C.20}$$

The canonical equations

$${}_\alpha \dot{q}_i = \frac{\partial H}{\partial\, {}_\alpha p_i} = \frac{\partial\, {}_\alpha h}{\partial\, {}_\alpha p_i} \tag{C.21}$$

provide a system of equations relating the velocities and the moments. In order to be able to solve (C. 21) for the momenta in terms of the velocities it is necessary that the inverse of the equations (C. 21) exists, i.e., that the Jacobian not equal zero:

$$\det \frac{\partial^2\, {}_\alpha h}{\partial\, {}_\alpha p_i\, \partial\, {}_\alpha p_j} \neq 0 \tag{C.22}$$

Designating the inverse of the matrix $\partial^2\, {}_\alpha b/\partial\, {}_\alpha p_i\, \partial\, {}_\alpha p_j$ by ${}_\alpha h_{ij}^{-1}$, we may write (C.20) as

$$\frac{\partial^2\, {}_\gamma h}{\partial\, {}_\delta q_i\, \partial\, {}_\gamma p_l}\, {}_\alpha h_{kl}^{-1} - \frac{\partial^2\, {}_\delta h}{\partial\, {}_\gamma q_k\, \partial\, {}_\delta p_m}\, {}_\delta h_{im}^{-1} = 0 \tag{C.23}$$

Since the first term must be independent of ${}_\delta p_j$, the second term must also be independent of ${}_\delta p_j$. We conclude that both the first and second terms can be functions of the coordinates only, say ${}_{\delta\gamma}\lambda_{ik}$ which from (C.23) must satisfy

$$ {}_{\delta\gamma}\lambda_{ik} = {}_{\gamma\delta}\lambda_{ki} \tag{C.24} $$

and

$$ \frac{\partial^2\, {}_\delta h}{\partial\, {}_\gamma q_k\, \partial\, {}_\delta p_j} = {}_{\gamma\delta}\lambda_{ki}\, {}_\delta h_{ji} = {}_{\gamma\delta}\lambda_{ki} \frac{\partial^2\, {}_\delta h}{\partial\, {}_\delta p_j\, \partial\, {}_\delta p_i} \tag{C.25} $$

This equation may be integrated to yield

$$ \frac{\partial\, {}_\delta h}{\partial\, {}_\gamma q_k} = {}_{\gamma\delta}\lambda_{ki} \frac{\partial\, {}_\delta h}{\partial\, {}_\delta p_i} + {}_{\gamma\delta}\mu_k \tag{C.26} $$

where ${}_{\gamma\delta}\mu_k$ is a function of the coordinates only. Actually it can be shown [1] that (C.26) is also valid if $\gamma = \delta$.

The integrability condition for (C.26) is

$$ \left(\frac{\partial\, {}_{\gamma\delta}\lambda_{ki}}{\partial\, {}_\alpha q_j} - \frac{\partial\, {}_{\alpha\delta}\lambda_{ji}}{\partial\, {}_\gamma q_k} \right) \frac{\partial\, {}_\delta h}{\partial\, {}_\delta p_i} + \frac{\partial\, {}_{\gamma\delta}\mu_k}{\partial\, {}_\alpha q_j} - \frac{\partial\, {}_{\alpha\delta}\mu_j}{\partial\, {}_\gamma q_k} = 0 \tag{C.27} $$

Differentiating (C.27) with respect to ${}_\delta p_l$ and using the inverse of ${}_\delta h_{il}$ we obtain

$$ \frac{\partial\, {}_{\gamma\delta}\lambda_{ki}}{\partial\, {}_\alpha q_j} = \frac{\partial\, {}_{\alpha\delta}\lambda_{ji}}{\partial\, {}_\gamma q_k}, \qquad \frac{\partial\, {}_{\gamma\delta}\mu_k}{\partial\, {}_\alpha q_j} = \frac{\partial\, {}_{\alpha\delta}\mu_j}{\partial\, {}_\gamma q_k} \tag{C.28} $$

It follows from (C.28) and (C.24) that

$$ {}_{\gamma\delta}\lambda_{ki} = \frac{\partial^2 L}{\partial\, {}_\gamma q_k\, \partial\, {}_\delta q_i}, \qquad {}_{\gamma\delta}\mu_k = \frac{\partial\, {}_\delta M}{\partial\, {}_\gamma q_k} \tag{C.29} $$

where L and ${}_\delta M$ are functions of the coordinates only and from (C. 25), (C. 26), and the properties of ${}_\alpha h$ it follows that they must be invariant under translations and rotations. Thus we conclude that (C. 26) may be written

$$ \frac{\partial\, {}_\delta h}{\partial\, {}_\gamma q_k} = \frac{\partial^2 L}{\partial\, {}_\gamma q_k\, \partial\, {}_\delta q_i} \frac{\partial\, {}_\delta h}{\partial\, {}_\delta p_i} + \frac{\partial\, {}_\delta M}{\partial\, {}_\gamma q_k} \tag{C.30} $$

A simple way of solving (C.30) is to perform the canonical transformation

$$ {}_\alpha q_i' = {}_\alpha q_i, \qquad {}_\alpha p_i' = {}_\alpha p_i - \frac{\partial L}{\partial\, {}_\alpha q_i} \tag{C.31} $$

Inserting (C.31) into (C.30) we find

$$\frac{\partial\ {}_\delta h'}{\partial\ {}_\gamma q_k'} = \frac{\partial\ {}_\delta M'}{\partial\ {}_\gamma q_k'} \tag{C.32}$$

where ${}_\delta h'(q', p') = {}_\delta h(p', q')$ and ${}_\delta M'(q') = {}_\delta M(q)$. Integrating (6.32), we obtain

$${}_\delta h'(q', p') = {}_\delta M'(q') + {}_\delta N(p') \tag{C.33}$$

i.e., in terms of the primed variables ${}_\delta h$ separates into a function of coordinates only and a function of momenta only. It is not difficult to show that the canonical transformation (C.31) leaves the free particle forms of **J** and **P** invariant.

Since (C. 31) is canonical, (C. 3) is invariant under the transformation; therefore, (C. 19) has the same form in terms of the primed variables as the unprimed variables. With (C.33) it may be written

$$\frac{1}{2}\sum_\alpha \frac{\partial}{\partial\ {}_\alpha p_i'}\left({}_\alpha N^2 - {}_\alpha p'^2\right) + \sum_\alpha \frac{\partial\ {}_\alpha N}{\partial\ {}_\alpha p_j'}\ {}_\alpha C_{ij} = 0 \tag{C.34}$$

where

$${}_\alpha C_{ij} = \frac{\partial}{\partial\ {}_\alpha q_j'}\left(\sum_\beta {}_\beta M'\ {}_\beta q_i' + k_i\right) = {}_\alpha q_i' \frac{\partial}{\partial\ {}_\alpha q_k'} \sum_\beta {}_\beta M' \tag{C.35}$$

By taking the second derivative of (C.34) with respect to ${}_\gamma p_l, {}_\delta q_m$, we find that ${}_\alpha C_{ij}$ is a constant and since M' is rotation invariant ${}_\alpha C_{ij}$ transforms like a second rank tensor. But the only numerically invariant second rank tensor is given by the Kronecker delta, therefore,

$${}_\alpha C_{ij} = {}_\alpha C\ \delta_{ij} \tag{C.36}$$

and (C.34) becomes

$$\sum_\alpha \frac{\partial}{\partial\ {}_\alpha p_i'}\left[({}_\alpha N + {}_\alpha C)^2 - {}_\alpha p'^2\right] = 0 \tag{C.37}$$

From (C.37) it follows that

$$\frac{\partial^2}{\partial\ {}_\alpha p_i'\ \partial\ {}_\alpha p_j'}\left[({}_\alpha N + {}_\alpha C)^2 - {}_\alpha p'^2\right] = 0 \tag{C.38}$$

and therefore

$$({}_\alpha N + {}_\alpha C)^2 - {}_\alpha p'^2 = {}_\alpha d_i\ {}_\alpha p_i' + {}_\alpha d \tag{C.39}$$

Rotational invariance of ${}_\alpha M'$ and ${}_\alpha h'$ requires the rotational invariance of

${}_\alpha N$ and therefore from (C.39) ${}_\alpha d_i$ must be a numerically invariant vector, but no such object exists and we must then require that ${}_\alpha d_i$ be zero. Thus

$$ {}_\alpha N + {}_\alpha C = ({}_\alpha p'^2 + {}_\alpha d)^{1/2} \tag{C.40} $$

Using (C.36), we may write (C.35) as

$$ {}_\alpha C \, \delta_{ij} = \frac{\partial f_i}{\partial \, {}_\alpha q_j'} - {}_\alpha q_i' \frac{\partial f}{\partial \, {}_\alpha q_j'} \tag{C.41} $$

where

$$ f = \sum_\alpha {}_\alpha M', \qquad f_i = \sum_\alpha {}_\alpha M \, {}_\alpha q_i + k_i \tag{C.42} $$

The integrability conditions for the system of equations (C.41) are

$$ \delta_{\alpha\beta}\left(\delta_{ij} \frac{\partial f}{\partial \, {}_\alpha q_k'} - \delta_{jk} \frac{\partial f}{\partial \, {}_\alpha q_i'} \right) + ({}_\alpha q_j' - {}_\beta q_j') \frac{\partial^2 f}{\partial \, {}_\alpha q_k' \, \partial \, {}_\beta q_j'} = 0 \tag{C.43} $$

When $\alpha = \beta$ and $i = j \neq k$, we find

$$ \frac{\partial f}{\partial \, {}_\alpha q_k'} = 0 \qquad \text{or} \quad f = \text{const} \tag{C.44} $$

and

$$ f_i = \sum_\alpha {}_\alpha C \, {}_\alpha q_i' + b_i \tag{C.45} $$

For f_i to transform as a vector under rotations, b_i must vanish. Therefore,

$$ k_i = \sum_\alpha ({}_\alpha C - {}_\alpha M') \, {}_\alpha q_i' \tag{C.46} $$

Since k_i and ${}_\alpha M'$ are translation invariant it is necessary that

$$ \sum_\alpha ({}_\alpha C - {}_\alpha M') = 0 \tag{C.47} $$

Therefore,

$$ H = \sum_\alpha ({}_\alpha N + {}_\alpha M') = \sum_\alpha ({}_\alpha p'^2 + {}_\alpha d)^{1/2} \tag{C.48} $$

and

$$ K_i = \sum_\alpha {}_\alpha h' \, {}_\alpha q_i' + k_i = \sum_\alpha ({}_\alpha p'^2 + {}_\alpha d)^{1/2} \, {}_\alpha q_i' \tag{C.49} $$

If we require that the particle velocities

$$ {}_\alpha \dot{q}_i' = \frac{\partial H}{\partial \, {}_\alpha p_i'} = {}_\alpha p_i' ({}_\alpha p'^2 + {}_\alpha d)^{-1/2} \tag{C.50} $$

be smaller than the velocity of light implies

$$ {}_{\alpha}d > 0 \tag{C.51} $$

and we may therefore write

$$ {}_{\alpha}d = {}_{\alpha}m^2 \tag{C.52} $$

Thus we have brought H, $\mathbf{K}$, $\mathbf{J}$, and $\mathbf{P}$ to their free-particle form by a canonical transformation and we conclude that *there can be no interaction in a canonical formalism based on the assumptions listed at the beginning of this Appendix.*

The no interaction result does not hold for Galilean invariant systems since the proof in the Lorentz invariant case proceeds from (C.6a) but this equation must be replaced by (C.6b) in the Galilean case.

REFERENCES

1. H. Leutwyler, *Nuovo Cimento* **37**, 556 (1965).
2. D. G. Currie, T. F. Jordan, and E. C. G. Sudarshan, *Rev. Mod. Phys.* **35**, 350 (1963).
3. L. L. Foldy, *Phys. Rev.* **122**, 275 (1961).
4. Leutwyler, *Nuovo Cimento* **37**, 543 (1965).

Appendix D

EULER ANGLES

In describing the motion of a rigid body, it is often advantageous to locate the Cartesian coordinate axes attached to the rigid body relative to another Cartesian set of axes which maintain their orientation in space. To do this requires three independent parameters, and it has become conventional to use the *Euler angles* (ϕ, θ, ψ) defined in Fig. D.1. The Cartesian axes XYZ are located relative to the axes $x^{(3)}$, $y^{(3)}$, $z^{(3)}$ by the angles ϕ, θ, ψ. Figure D.1a shows the definition of ϕ, viz., it is a counterclockwise rotation of the x, y, z axes, which are originally coincident with XYZ, through an an-

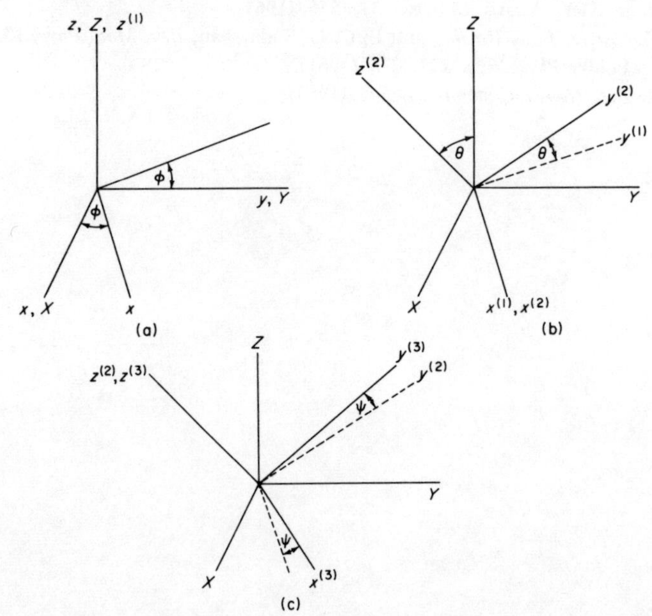

Fig. D.1. The Euler angles. (a) Angle ϕ defined as a counterclockwise rotation about Z. (b) Angle θ defined as a counterclockwise rotation about $x^{(1)}$. (c) Angle ψ defined as a counterclockwise rotation about $z^{(2)}$.

gle ϕ about the Z axis. This rotation may be represented by the operator R_ϕ.

$$R_\phi = \begin{bmatrix} \cos\phi & \sin\phi & 0 \\ -\sin\phi & \cos\phi & 0 \\ 0 & 0 & 1 \end{bmatrix}, \qquad \mathfrak{X}^{(1)} = R_\phi \mathfrak{X} \tag{D.1}$$

where

$$\mathfrak{X}^{(1)} = \begin{bmatrix} x^{(1)} \\ y^{(1)} \\ z^{(1)} \end{bmatrix}, \qquad \mathfrak{X} = \begin{bmatrix} x \\ y \\ z \end{bmatrix}$$

Figure D. 16 defines θ as the counterclockwise rotation of the $x^{(1)}$, $y^{(1)}$, $z^{(1)}$ Cartesian system about the $x^{(1)}$ axis through an angle θ. This rotation may be represented by the operator R_θ.

$$R_\theta = \begin{bmatrix} 1 & 0 & 0 \\ 0 & \cos\theta & \sin\theta \\ 0 & -\sin\theta & \cos\theta \end{bmatrix}, \qquad \mathfrak{X}^{(2)} = R_\theta \mathfrak{X}^{(1)} \tag{D.2}$$

Figure D.1c defines ψ as a counterclockwise rotation of the $x^{(2)}$, $y^{(2)}$, $z^{(2)}$ Cartesian system about the $z^{(2)}$axis through an angle ψ. This rotation may be represented by the operator R_ψ.

$$R_\psi = \begin{bmatrix} \cos\psi & \sin\psi & 0 \\ -\sin\psi & \cos\psi & 0 \\ 0 & 0 & 1 \end{bmatrix}, \qquad \mathfrak{X}^{(3)} = R_\psi \mathfrak{X}^{(2)} \tag{D.3}$$

By combining (D.1), (D.2), and (D.3), we may write

$$\mathfrak{X}^{(3)} = R_\psi R_\theta R_\phi \mathfrak{X} = R_E \mathfrak{X} \tag{D.4}$$

$$R_E = \begin{bmatrix} \cos\psi\cos\phi - \sin\psi\sin\phi\cos\theta & \cos\psi\sin\phi + \sin\psi\cos\phi\cos\theta & \sin\psi\sin\theta \\ -\sin\psi\cos\phi - \cos\psi\sin\phi\cos\theta & -\sin\psi\sin\phi + \cos\psi\cos\phi\cos\theta & \cos\psi\sin\theta \\ \sin\theta\sin\phi & -\sin\theta\cos\phi & \cos\theta \end{bmatrix} \tag{D.5}$$

$$R_E^{-1} = \begin{bmatrix} \cos\psi\cos\varphi - \cos\theta\sin\varphi\sin\psi & -\sin\psi\cos\varphi - \cos\theta\sin\varphi\cos\psi & \sin\theta\sin\varphi \\ \cos\psi\sin\varphi + \cos\theta\cos\varphi\sin\psi & -\sin\psi\sin\varphi + \cos\theta\cos\varphi\cos\psi & -\sin\theta\cos\varphi \\ \sin\theta\sin\psi & \sin\theta\cos\psi & \cos\theta \end{bmatrix} \tag{D.6}$$

Appendix E

SOME GROUP THEORY

This Appendix presents some of the essential elements of group theory necessary for an understanding of Chapter 6. Group theory is an extensive subject, and only a fragmentary account is given here. The reader is encouraged to consult the references at the end of Chapter 6 (expecially [1–6] and [16–19]) for a comprehensive presentation of the proofs and implications of the material contained in this Appendix and for the vast amount of material not contained in this Appendix.

E.1. FUNDAMENTALS

A set of elements $G = \{G_1, G_2, \ldots, G_\alpha, \ldots\}$ is a *group* if there is a binary law of composition (usually called a *product* and indicated by $G_\alpha G_\beta$) defined between the elements with the following properties:

(1) *Closure*, i.e., if $G_\alpha \in G$ and $G_\beta \in G$, then $G_\alpha G_\beta = G_\gamma \in G$.
(2) *Associative Law Holds*, i.e., $G_\alpha(G_\beta G_\gamma) = (G_\alpha G_\beta)G_\gamma$.
(3) A *unit element* G_1 exists such that for any $G_\alpha \in G$

$$G_1 G_\alpha = G_\alpha G_1 = G_\alpha$$

(4) An *inverse* G_α^{-1} exists for each element G_α such that

$$G_\alpha^{-1} G_\alpha = G_1$$

A group is *Abelian* if the product is commutative for every pair of elements in the group, i.e., $G_\alpha G_\beta = G_\beta G_\alpha$.

The *order* g of a group is the number of elements in the group. A group is either *finite* or *infinite* according to whether its order is finite or infinite.

A subset H of G is a *subgroup* of G if its elements form a group with respect to the same law of composition defined for G.

The law of composition of a group is conveniently presented in the *group table* (also called the multiplication table). Thus, for a group of order six, we have Table E.1. In this table the element at the intersection of the ith row and jth column is equal to the product G_iG_j. Thus $G_2G_5 = G_6$. Notice that no element appears more than once in any one column or row.

TABLE E.1

	G_1	G_2	G_3	G_4	G_5	G_6
G_1	G_1	G_2	G_3	G_4	G_5	G_6
G_2	G_2	G_3	G_1	G_5	G_6	G_4
G_3	G_3	G_1	G_2	G_6	G_4	G_5
G_4	G_4	G_6	G_5	G_1	G_3	G_2
G_5	G_5	G_4	G_6	G_2	G_1	G_3
G_6	G_6	G_5	G_4	G_3	G_2	G_1

Let us consider two groups $G = \{G_1, G_2, \ldots, G_g\}$ and $H = \{H_1, H_2, \ldots, H_h\}$. If one can establish a *correspondence* (*mapping*) of G onto H such that to any element of G there corresponds one and only one element of H

$$G_1 \rightarrow H_1, \qquad G_\alpha \rightarrow H_\alpha, \qquad \text{any } \alpha$$

such that

$$\text{if } G_\alpha G_\beta = G_\gamma, \qquad \text{then } H_\alpha H_\beta = H_\gamma$$

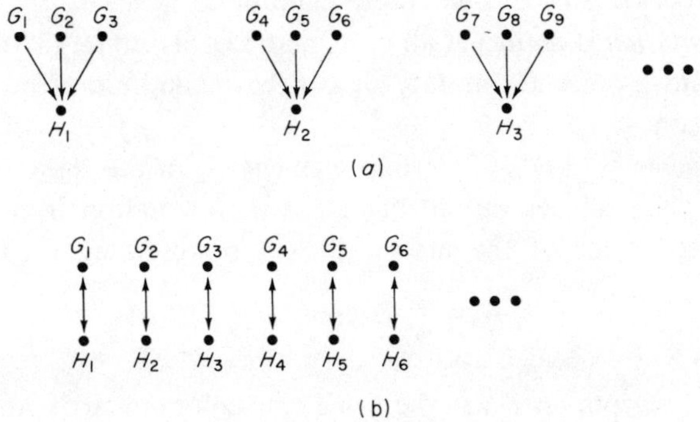

Fig. E.1. (a) The homomorphic mapping of G onto H. (b) The isomorphic mapping of G and H.

then H is said to be *homomorphic* to G. If the mapping is *one to one*, i.e., to any element of G there corresponds one and only one element of H and vice versa, then the mapping is an *isomorphism* (see Fig. E.1).

E.2. REPRESENTATION

Suppose that a set of $n \times n$ matrices $D(\alpha)$ forms a group $\mathfrak{D}$, where α distinguishes one matrix from another, i.e., $\alpha = \{\alpha_1, \alpha_2, \ldots, \alpha_g\}$, where g is the order of the group. If we can establish a homomorphism between an abstract group G and the group of matrices $D(\alpha)$, then the group $\mathfrak{D}$ is said to be an *n-dimensional representation* of the group G. The space R_n spanned by n linearly independent base vectors on which the $n \times n$ matrices operate, is called the *representation space* of the group. If the mapping of G onto $\mathfrak{D}$ is an isomorphism the representation is said to be faithful.

A group may, in general, have a number of different representations. These different representations may have the same or different dimensionalities. Two representations $\mathfrak{D}'$ and $\mathfrak{D}$ of the same dimensionality n are *equivalent* if there exists a nonsingular $n \times n$ matrix S such that for every α

$$D'(\alpha) = S^{-1}D(\alpha)S \tag{E.1}$$

The transformation from D to D' defined in (E. 1) is a *similarity* transformation. There are always an infinite number of equivalent representations, but it can be shown that they differ from each other only in that they have different bases in the n-dimensional representation space. Thus, we do not distinguish between equivalent representations.

A *representation* is *unitary* if all of its matrices are unitary. All representations of finite groups are unitary or can be made unitary by a similarity transformation.

The *character* $\chi^{(a)}(\alpha)$ of a group element G_α in the representation $\mathfrak{D}^{(a)}$ [where (a) distinguishes one inequivalent representation from another] is defined as the trace of the matrix $D^{(a)}(\alpha)$ representing G_α in $\mathfrak{D}^{(a)}$; i.e.,

$$\chi^{(a)}(\alpha) \equiv \mathrm{T}_r\, D^{(a)}(\alpha) = \sum_i D_{ii}^{(a)}(\alpha) \tag{E.2}$$

where the subscripts on D are the usual row-column matrix notations.

A representation $\mathfrak{D}^{(a)}$ is *reducible* if all the matrices of the representation can be brought into block diagonal form by a single similarity transformation. Block diagonal form occurs when the elements of the matrix appear in submatrices along the diagonals with zeros everywhere else, i.e.,

$$\begin{bmatrix} D^{(1)}(\alpha) & & & & 0 \\ & D^{(2)}(\alpha) & & & \\ & & D^{3}(\alpha) & & \\ & 0 & & \ddots & \\ & & & & D^{(r)}(\alpha) \end{bmatrix} \tag{E.3}$$

The reduction may be expressed as the *direct sum* of representations of lower dimensionality:

$$\mathfrak{D} = \mathfrak{D}^{(1)} \oplus \mathfrak{D}^{(2)} \oplus \mathfrak{D}^{(3)} \oplus \cdots \oplus \mathfrak{D}^{(r)} \tag{E.4}$$

When the matrices of the representation have been transformed so as to make the submatrices as small as possible, the representation is said to be *completely reduced* and the submatrices along the diagonal form *irreducible representations.*

If a representation $\mathfrak{D}$ is reducible, there exists in the representation space R_n a subspace $R_m (m < n$, m and n designating the dimension of the space) whose vectors are always transformed among themselves by the representation matrices, i.e., for a reducible representation there exists an invariant subspace R_m in R_{D}. When a representation is completely reduced according to (E. 4), the representation space R_n is composed of invariant subspaces which serve as the invariant representation spaces of the irreducible representations.

It is clear that the sum of the dimensions of the irreducible representations appearing in (E. 4) must equal the dimension of the reducible representation. The decomposition of a reducible representation into its irreducible parts is *unique* up to equivalence. We may write (E.4) as

$$\mathfrak{D} = \sum_{l} n_l \, \mathfrak{D}^{(l)} \tag{E.5}$$

where the n_l are nonnegative integers, i.e., a given irreducible representation may appear more than once in a decomposition. The character of an element of the reducible representation is related to the characters of the elements of the irreducible representations according to

$$\chi(\alpha) = \sum_{l} n_l \chi^{(l)}(\alpha) \tag{E.6}$$

Two completely reducible representations are equivalent if, and only if, they have the same set of characters.

Let $G_\alpha \in G$ and $G_\beta \in G$. The element $G_\gamma = G_\beta^{-1} G_\alpha G_\beta$ is said to be *conjugate* to G_α with respect to G_β. The set of conjugate elements constructed by allowing G_β to be all the other elements of the group is called the *class* C_α of the element G_α, i.e.,

$$C_\alpha = \{G_1^{-1} G_\alpha G_1, G_2^{-1} G_\alpha G_2, \ldots, G_\beta^{-1} G_\alpha G_\beta, \ldots\} \tag{E.7}$$

The set of all elements that are mutually conjugate to each other form a class. The group elements may be uniquely classified into mutually exclusive classes. The identity is always in a class by itself. In an *Abelian* group each element is in a class by itself. All the elements of a class have the same character. *The number of irreducible representations is equal to the number of classes*. The dimensionality d_a of any irreducible representation $\mathfrak{D}^{(a)}$ is a divisor of the order of the group, i.e.,

$$g/d_a = \text{integer} \tag{E.8}$$

also

$$g = \sum_{a=1}^{g_c} d_a^{\,2} \tag{E.9}$$

where g_c is the number of irreducible representations (i.e., number of classes) of the group.

Classes may be multiplied. Thus for two classes $C_\nu = \{G_{\nu 1}, G_{\nu 2}, \ldots, G_{\nu N_\nu}\}$ and $C_\mu = \{G_{\mu 1}, G_{\mu 2}, \ldots, G_{\mu N_\mu}\}$, we define the multiplication of the classes as

$$C_\nu C_\mu = \{G_{\nu 1} G_{\mu 1}, G_{\nu 1} G_{\mu 2}, \ldots, G_{\nu 1} G_{\mu N_\mu}, G_{\nu 2} G_{\mu 1}, \ldots, G_{\nu N_\nu} G_{\mu N_\mu}\} \tag{E.10}$$

The elements contained within the set on the right side of (E. 10) are just the group elements which may be arranged according to their class membership in such a way that we may write

$$\mathcal{C}_\nu \mathcal{C}_\mu = \sum_{\varrho=1}^{g_c} c_{\nu\mu\varrho} \mathcal{C}_\varrho \tag{E.11}$$

where the coefficients $c_{\nu\mu\varrho}$ give the number of times the class C_ϱ appears in the set on the right side of (E.10). To illustrate, consider the point group D_3 which has the multiplication table shown in Table E.2.

TABLE E.2

	G_1	G_2	G_3	G_4	G_5	G_6
G_1	G_1	G_2	G_3	G_4	G_5	G_6
G_2	G_2	G_3	G_1	G_5	G_6	G_4
G_3	G_3	G_1	G_2	G_6	G_4	G_5
G_4	G_4	G_6	G_5	G_1	G_3	G_2
G_5	G_5	G_4	G_6	G_2	G_1	G_3
G_6	G_6	G_5	G_4	G_3	G_2	G_1

There are three classes:

$$C_1 = \{G_1\}, \qquad C_2 = \{G_2, G_3\}, \qquad C_3 = \{G_4, G_5, G_6\}$$

Now

$$C_1C_2 = \{G_1G_2, G_1G_3\} = \{G_2, G_3\} = C_2 = c_{121}C_1 + c_{122}C_2 + c_{123}C_3$$

Therefore,

$$c_{121} = c_{123} = 0, \qquad c_{122} = 1$$

Similarly,

$$C_2C_2 = \{G_2G_2, G_2G_3, G_3G_2, G_3G_3\} = \{G_3, G_1, G_1, G_2\} = 2\{G_1\} + \{G_2, G_3\}$$

$$C_2C_2 = 2C_1 + C_2 = c_{221}C_1 + c_{222}C_2 + c_{223}C_3$$

Therefore,

$$C_{221} = 2, \qquad C_{221} = 1, \qquad C_{223} = 0 \tag{E.12}$$

One finds for the point group D_3

$$\begin{aligned} &C_1C_1 = C_1, && C_1C_2 = C_2C_1 = C_2, && C_1C_3 = C_3C_1 = C_3 \\ &C_2C_2 = 2C_1 + C_2, && C_2C_3 = C_3C_2 = 2C_3, && C_3C_3 = 3C_1 + 3C_2 \end{aligned} \tag{E.13}$$

If $\mathfrak{D}^{(a)} = \{\ldots, D^{(a)}(\alpha), \ldots\}$ and $\mathfrak{D}^{(b)} = \{\ldots, D^{(b)}(\alpha), \ldots\}$ are two irreducible representations of the group G, then they satisfy the *orthogonality relation*

$$\sum_{\alpha=1}^{g} D_{ik}^{(a)}(\alpha)\, D_{jl}^{(b)-1}(\alpha) = (g/d_a)\, \delta_{ab}\, \delta_{il}\, \delta_{kj} \tag{E.14}$$

where g is the order of the group and d_a is the dimensionality of the $\mathfrak{D}^{(a)}$ representation. Since the representation matrices of any finite group may be brought to unitary form by a similarity transformation, (E.14) may be written

$$\sum_{\alpha=1}^{g} D_{ik}^{(a)}(\alpha)\, (D_{jl}^{(b)}(\alpha))^{\dagger} = (g/d_a)\, \delta_{ab}\, \delta_{il}\, \delta_{ki} \tag{E.15}$$

where the "dagger" (†) signifies the combined operations of transpose and complex conjugation. The characters of irreducible representations also satisfy orthogonality relations. For unitary representations

$$\sum_{\alpha=1}^{g} \chi^{(a)}(\alpha)\bar{\chi}^{(b)}(\alpha) = g\, \delta_{ab} \tag{E.16}$$

where the overbar indicates complex conjugation. Since the character is the same for all members of a given class, i.e., the character is a class function, it follows that (E.16) may be written

$$\sum_{\nu=1}^{g_c} [(N_\nu/g)^{1/2}\chi^{(a)}(\alpha_\nu)][(N_\nu/g)^{1/2}\bar{\chi}^{(b)}(\alpha_\nu)] = \delta_{ab} \tag{E.17}$$

where N_ν is the number of members in the class C_ν, q_c is the number of classes, and α_ν is an arbitrary member of the class C_ν.

A second orthogonality relation satisfied by the characters is

$$\sum_{a=1}^{g_c} \bar{\chi}^{(a)}(\mathcal{C}_\nu)\chi^{(a)}(\mathcal{C}_\varrho) = (g/N_\nu)\, \delta_{\nu\varrho} \tag{E.18}$$

where g_c is the number of irreducible representations.

From (E.6) and (E.16) it follows that

$$n_l = (1/g) \sum_{\alpha=1}^{g} \chi(\alpha)\bar{\chi}^{(l)}(\alpha) \tag{E.19}$$

Equation (E. 19) is an extremely important equation which provides us with a method of uniquely determining the number of times n_l that the irreducible representation $\mathfrak{D}^{(l)}$ appears in a decomposition of a reducible representation. Similarly, it follows that

$$\sum_{\alpha=1}^{g} | \chi(\alpha) |^2 = g \sum_{l} n_l^2 \tag{E.20}$$

From (E.20) there follows a most useful *criterion of irreducibility*;

$$(1/g) \sum_{\alpha=1}^{g} | \chi(\alpha) |^2 \begin{cases} > 1 & \text{if the representation is reducible} \\ = 1 & \text{if the representation is irreducible} \end{cases} \tag{E.21}$$

Another important *criterion of irreducibility* is *Schur's lemma* which states that a representation is irreducible if and only if the only matrix which commutes with all of the matrices $D(\alpha)$ of the representation is a scalar multiple of the unit matrix. It follows from this that all of the irreducible representations of an Abelian group are one dimensional.

From (E.14) it follows that

$$\langle \psi_i^{(a)} / \psi_j^{(b)} \rangle = h^{(a)} \delta_{ab} \delta_{ij} \tag{E.22}$$

where the vectors $\psi_j^{(b)}$ $(j = 1, 2, \ldots, d_b)$ and $\psi_i^{(a)}$ $(i = 1, 2, \ldots, d_a)$ span the representation spaces of the $\mathfrak{D}^{(b)}$ and $\mathfrak{D}^{(a)}$ irreducible representations, respectively, and $h^{(a)}$ is a constant independent of i and j. $\langle \psi_i^{(a)} / \psi_j^{(b)} \rangle$ is the scalar product of the vectors and thus (E.22) is an orthogonality relation for the base vectors of the irreducible representation spaces.

E.3. CALCULATION OF CHARACTER TABLES

In most applications of symmetry groups to physical problems, the character table of the group must be known. The *character table* is a square table as shown in Table E.3 where $\mathfrak{D}^{(1)}$, $\mathfrak{D}^{(2)}$, $\mathfrak{D}^{(3)}$ are the irreducible representations of the group, and C_ν is the νth class.

TABLE E.3

	$\mathcal{C}_1 = \mathcal{C}_E$	$N_2\, \mathcal{C}_2$	$N_3\, \mathcal{C}_3$
$\mathfrak{D}^{(1)}$	$\chi^{(1)}(\mathcal{C}_1)$	$\chi^{(1)}(\mathcal{C}_2)$	$\chi^{(1)}(\mathcal{C}_3)$
$\mathfrak{D}^{(2)}$	$\chi^{(2)}(\mathcal{C}_1)$	$\chi^{(2)}(\mathcal{C}_2)$	$\chi^{(2)}(\mathcal{C}_3)$
$\mathfrak{D}^{(3)}$	$\chi^{(3)}(\mathcal{C}_1)$	$\chi^{(3)}(\mathcal{C}_3)$	$\chi^{(3)}(\mathcal{C}_3)$

The table is square since the number of irreducible representations is equal to the number of classes of the group. To find the characters within the table, we follow these rules:

(1) Every finite group has an *identity irreducible representation* in which each group element is represented by the scalar 1. The identity representation is usually placed in the first row; i.e.,

$$\chi^{(1)}(\mathcal{C}_1) = \chi^{(1)}(\mathcal{C}_2) = \chi^{(1)}(\mathcal{C}_3) = \cdots = \chi^{(1)}(\mathcal{C}_c) = 1 \tag{E.23}$$

(2) Since the first column contains the character of the representative of the unit element, which is just the identity matrix, we conclude that

$$\chi^{(a)}(\mathcal{C}_1) = d_a \qquad \text{where} \quad d_a \text{ is the dimension of } \mathfrak{D}^{(a)} \tag{E.24}$$

To this point Table E. 3 has been brought to the form shown in Table E.4.

TABLE E.4

	$\mathcal{C}_1$	$N_2\,\mathcal{C}_2$	$N_3\,\mathcal{C}_3$
$\mathfrak{D}^{(1)}$	1	1	1
$\mathfrak{D}^{(2)}$	d_2	$\chi^{(2)}(\mathcal{C}_2)$	$\chi^{(2)}(\mathcal{C}_3)$
$\mathfrak{D}^{(3)}$	d_3	$\chi^{(3)}(\mathcal{C}_2)$	$\chi^{(3)}(\mathcal{C}_3)$

(3) Often the application of (E.9) is sufficient to determine the characters associated with the dimension of the representation; i.e.,

$$\sum_{a=1}^{g_c} d_a^{\,2} = g \tag{E.25}$$

(4) Next the first orthogonality condition, (E.17) may be employed.
(5) Next the second orthogonality relation, (E.18), may be applied.
(6) Finally, it can be shown that

$$N_\mu \chi^{(a)}(\mathcal{C}_\mu) N_\nu \chi^{(a)}(\mathcal{C}_\nu) = d_a \sum_{\varrho=1}^{g_c} c_{\mu\nu\varrho} N_\varrho \chi^{(a)}(\mathcal{C}_\varrho) \tag{E.26}$$

where the $c_{\mu\nu\varrho}$ are the coefficients of class multiplication defined in (E.11).

We illustrate the method by determining the character table of the point group D_3 whose multiplication table and class structure are presented in Table E.2 and (E.13). From Rules 1 and 2 we may write Table E.5.

TABLE E.5

D_3	$\mathcal{C}_1$	$2\mathcal{C}_2$	$3\mathcal{C}_3$
$\mathfrak{D}^{(1)}$	1	1	1
$\mathfrak{D}^{(2)}$	d_2	$\chi^{(2)}(\mathcal{C}_2)$	$\chi^{(2)}(\mathcal{C}_3)$
$\mathfrak{D}^{(3)}$	d_3	$\chi^{(3)}(\mathcal{C}_2)$	$\chi^{(3)}(\mathcal{C}_3)$

Now, Rule 3 yields

$$1 + d_2^2 + d_3^2 = 6 \tag{E.27}$$

Since d_2 and d_3 are integers, this is satisfied if and only if $d_2 = 1$, $d_3 = 2$. Thus Table E.5 becomes Table E.6.

TABLE E.6

	$\mathcal{C}_1$	$2\mathcal{C}_2$	$3\mathcal{C}_3$
$\mathfrak{D}^{(1)}$	1	1	1
$\mathfrak{D}^{(2)}$	1	$\chi^{(2)}(\mathcal{C}_2)$	$\chi^{(2)}(\mathcal{C}_3)$
$\mathfrak{D}^{(3)}$	2	$\chi^{(3)}(\mathcal{C}_2)$	$\chi^{(3)}(\mathcal{C}_3)$

Application of Rule 4 yields

$$1 + 2\,|\,\chi^{(2)}(\mathcal{C}_2)\,|^2 + 3\,|\,\chi^{(2)}(\mathcal{C}_3)\,|^2 = 6 \qquad \text{for} \quad a = b = 2 \tag{E.28}$$

$$4 + 2\,|\,\chi^{(3)}(\mathcal{C}_2)\,|^2 + 3\,|\,\chi^{(3)}(\mathcal{C}_3)\,|^2 = 6 \qquad \text{for} \quad a = b = 3 \tag{E.29}$$

$$1 + 2\chi^{(2)}(\mathcal{C}_2) + 3\chi^{(2)}(\mathcal{C}_3) = 0 \qquad \text{for} \quad a = 1, b = 2 \tag{E.30}$$

$$2 + 2\chi^{(3)}(\mathcal{C}_2) + 3\chi^{(3)}(\mathcal{C}_3) = 0 \qquad \text{for} \quad a = 1, b = 3 \tag{E.31}$$

$$2 + 2\chi^{(2)}(\mathcal{C}_2)\bar{\chi}^{(3)}(\mathcal{C}_2) + 3\chi^{(2)}(\mathcal{C}_3)\bar{\chi}^{(3)}(\mathcal{C}_3) = 0 \qquad \text{for} \quad a = 2, b = 3 \tag{E.32}$$

Equations (E.28) thru (E.32) are not sufficient to determine the unknown characters; however, applying Rule 6 we find from (E.25) and (E.13)

$$6\chi^{(2)}(\mathcal{C}_2)\chi^{(2)}(\mathcal{C}_3) = 6\chi^{(2)}(\mathcal{C}_3) \qquad \text{for} \quad a = 2,\ \mu = 2,\ \nu = 3 \tag{E.33}$$

$$6\chi^{(3)}(\mathcal{C}_2)\chi^{(3)}(\mathcal{C}_3) = 12\chi^{(3)}(\mathcal{C}_3) \qquad \text{for} \quad a = 3,\ \mu = 2,\ \nu = 3 \tag{E.34}$$

Now from (E.33) it follows that either $\chi^{(2)}(\mathcal{C}_2) = 1$ or $\chi^{(2)}(\mathcal{C}_3) = 0$. Substituting these possibilities into (E.30) and (E.28) we conclude that $\chi^{(2)}(\mathcal{C}_2)$

$= 1$ and $\chi^{(2)}(\mathcal{C}_3) = -1$ [i.e., $\chi^{(2)}(\mathcal{C}_3) = 0$ solution is incompatible with (E.30) and (E.28)]. Thus we have determined the character table of D_3 to be Table E.7.

TABLE E.7

D_3	$\mathcal{C}_1$	$2\mathcal{C}_2$	$3\mathcal{C}_3$
$\mathfrak{D}^{(1)}$	1	1	1
$\mathfrak{D}^{(2)}$	1	1	-1
$\mathfrak{D}^{(3)}$	2	-1	0

E.4. CONTINUOUS GROUPS

A continuous group has an infinite set of elements, each of which may be labeled by a set of r continuously varying real parameters, $\{\alpha_1, \alpha_2, \ldots, \alpha_r\}$. The group is then said to be an *r-parameter continuous group*. The set of parameters may be considered as the components of a vector in an r-dimensional space. Thus, an element of the group may be designated by

$$G(\alpha_1, \alpha_2, \ldots, \alpha_r) \equiv G(\boldsymbol{\alpha}) \tag{E.35}$$

A one parameter continuous group is Abelian.

A continuous group satisfies the four defining group properties and the group composition law

$$G(\boldsymbol{\alpha})G(\boldsymbol{\beta}) = G(\boldsymbol{\gamma}) \tag{E.36}$$

requires that the parameter be related by

$$\boldsymbol{\gamma} = \boldsymbol{\varphi}(\boldsymbol{\alpha}, \boldsymbol{\beta}) \tag{E.37}$$

where $\boldsymbol{\varphi}$ is a real function of the real parameters $\boldsymbol{\alpha}, \boldsymbol{\beta}$.

The unit element of the group is often (but not necessarily) characterized by the set of parameters $\alpha_1 = \alpha_2 = \cdots = \alpha_r = 0$; i.e., the origin of the r-dimensional parameter space. Two elements, all of whose corresponding parameters differ only by an infinitesimal amount (e.g., $\beta_i = \alpha_i + \varepsilon_i$, where $|\varepsilon_i| \ll 1$), are said to be infinitesimally close to one another.

If the functions $\boldsymbol{\varphi}$ appearing in (E.37) are analytic functions, the group is a *Lie Group*. For Lie groups the theorems previously presented for finite groups (e.g., the orthogonality relations) are valid provided that all sums over groups elements are replaced with integrals over the "differential" of the group element. Thus, if $F(G_\alpha)$ is a function of the group element G_α, then the sums

$$\sum_{\alpha=1}^{g} F(G_\alpha) = \sum_{\alpha=1}^{g} F(G_\beta G_\alpha) \tag{E.38}$$

become

$$\int_G F(G(\alpha))\, dG(\alpha) = \int_G F(G(\alpha)G(\beta))\, dG(\alpha) \tag{E.39}$$

The "differential" of the group element is defined as

$$dG(\alpha) = J(\boldsymbol{\alpha})\, d\alpha_1\, d\alpha_2 \cdots d\alpha_r \tag{E.40}$$

where $J(\boldsymbol{\alpha})$ is the Jacobian of the transformation appearing in (E.37); i.e.,

$$J(\boldsymbol{\alpha}) = \left[\frac{\partial(\varphi_1(\boldsymbol{\alpha}, \boldsymbol{\beta}), \varphi_2(\boldsymbol{\alpha}, \boldsymbol{\beta}), \ldots, \varphi_r(\boldsymbol{\alpha}, \boldsymbol{\beta}))}{\partial(\alpha_1, \alpha_2, \alpha_3, \ldots, \alpha_r)}\right]_{\boldsymbol{\beta}=(\boldsymbol{\alpha})^{-1}}$$

$$\equiv \det \begin{bmatrix} \dfrac{\partial\varphi_1}{\partial\alpha_1} & \dfrac{\partial\varphi_1}{\partial\alpha_2} & \cdots & \dfrac{\partial\varphi_1}{\partial\alpha_r} \\ \dfrac{\partial\varphi_2}{\partial\alpha_1} & \dfrac{\partial\varphi_2}{\partial\alpha_2} & \cdots & \dfrac{\partial\varphi_2}{\partial\alpha_r} \\ \vdots & & & \\ \dfrac{\partial\varphi_r}{\partial\alpha_1} & \cdot & \cdots & \dfrac{\partial\varphi_r}{\partial\alpha_r} \end{bmatrix}_{\boldsymbol{\beta}=(\boldsymbol{\alpha})^{-1}} \tag{E.41}$$

where $\boldsymbol{\beta} = (\boldsymbol{\alpha})^{-1}$ means that the parameters $\{\beta_1, \beta_2, \ldots, \beta_r\}$ are to be evaluated after differentiation at $\{\beta_1 = \alpha_1^{-1}, \beta_2 = \alpha_2^{-1}, \ldots, \beta_r = \alpha_r^{-1}\}$. The *volume of the group* is defined as

$$V \equiv \int dG(\alpha) \equiv \int J(\boldsymbol{\alpha})\, d\alpha_1 \cdots d\alpha_r \tag{E.42}$$

A Lie group is *compact* if every infinite sequence of its elements has a limit element in the group; if one or more of the parameters of a group is not bounded, the group will not be compact. Compact groups have these delightful properties: (a) all representations are discreet; (b) all representations are unitary (up to equivalence); (c) all irreducible representations are finite dimensional; (d) the density function $J(\boldsymbol{\alpha})$ exists and is unique.

The group elements $G(\boldsymbol{\alpha})$ are functions defined on the r-dimensional parameter space. Curves may be defined in this space that depend only on a single parameter ε [i.e., in general $\boldsymbol{\alpha} = \boldsymbol{\alpha}(\varepsilon)$]; the subset of group elements defining this curve will be designated as $G(\varepsilon)$. If the set of elements $G(\varepsilon)$ also form a group, they are said to constitute a *one parameter subgroup* of $G(\boldsymbol{\alpha})$. Thus the set of rotations about a given axis in a three-dimensional space constitute a one-parameter subgroup of the three-parameter rotation group. There are many choices of parameter that one might make to describe the rotation, the most obvious being $\varepsilon = \theta$ or perhaps $\varepsilon = \sin\theta$. The form of the composition law, (E.37), depends on the choice of parameter. It can be shown that it is always possible to parametrize the subgroup such that the group composition takes the standard form

$$G(\varepsilon_1)G(\varepsilon_2) = G(\varepsilon_1 + \varepsilon_2) \tag{E.43}$$

Now if $G(0)$ is the identity element, it follows that

$$G^{-1}(\varepsilon) = G(-\varepsilon) \tag{E.44}$$

and

$$G(\varepsilon) = e^{\varepsilon I} \tag{E.45}$$

Where I is the *infinitesimal generator* of the one parameter group. Equation (E.45) is so widely used that we present a proof. Let us differentiate $G(\varepsilon)G^{-1}(\varepsilon) = E$ with respect to the parameter ε. Then

$$\frac{dG(\varepsilon)}{d\varepsilon}G^{-1}(\varepsilon) + G(\varepsilon)\frac{dG^{-1}(\varepsilon)}{d\varepsilon} = 0$$

Multiplying from the right with $G(\varepsilon)$, we obtain

$$\frac{dG(\varepsilon)}{d\varepsilon} + G(\varepsilon)\frac{gG^{-1}(\varepsilon)}{d\varepsilon}G(\varepsilon) = 0 \tag{E.46}$$

Using (E.44), we may write

$$\begin{aligned} G(\varepsilon)\frac{dG^{-1}(\varepsilon)}{d\varepsilon} &= G(\varepsilon)\lim_{\Delta\varepsilon\to 0}\left[\frac{G(-\varepsilon-\Delta\varepsilon) - G(-\varepsilon)}{\Delta\varepsilon}\right] \\ &= \lim_{\Delta\varepsilon\to 0}\left[\frac{G(\varepsilon)G(-\varepsilon-\Delta\varepsilon) - G(\varepsilon)G(-\varepsilon)}{\Delta\varepsilon}\right] \end{aligned} \tag{E.47}$$

But from (E.43) it follows that (E.47) becomes

$$G(\varepsilon)\frac{dG^{-1}(\varepsilon)}{d\varepsilon} = \lim_{\Delta\varepsilon\to 0}\left[\frac{G(0-\Delta\varepsilon) - G(0)}{\Delta\varepsilon}\right] = -\left.\frac{dG}{d\varepsilon}\right|_{\varepsilon=0} \equiv -I \tag{E.48}$$

Substituting (E.48) into (E.46), we find

$$\frac{dG(\varepsilon)}{d\varepsilon} = IG(\varepsilon) \tag{E.49}$$

the solution of which is $G(\varepsilon) = e^{\varepsilon I}$ where the infinitesimal generator is defined as

$$I = \left.\frac{dG}{d\varepsilon}\right|_{\varepsilon=0} \tag{E.50}$$

Since in realizations of the group the elements are operators, it follows that the generators also are operators and the exponential in (E.45) is defined by its power series expansion:

$$G(\varepsilon) = E + \varepsilon I + \tfrac{1}{2}\varepsilon^2 I^2 + \cdots \tag{E.51}$$

For infinitesimal transformations (i.e., $|\varepsilon| \ll 1$) (E.51) becomes

$$G(\varepsilon) = E + \varepsilon I \tag{E.52}$$

Now although the previous results are for a one parameter subgroup, it can be shown that in the vicinity of the identity any element of the p-parameter group can be written in the form

$$G(\boldsymbol{\epsilon}) = E + \sum_{t=1}^{r} \varepsilon_t I_t \tag{E.53}$$

where the parameters are infinitesimal real numbers, and the I_t are the r *infinitesimal generators* of the group which are obtained from the group element by

$$I_t = \left(\frac{\partial G(\boldsymbol{\epsilon})}{\partial \varepsilon_t}\right)_{\varepsilon_1=\varepsilon_2=\cdots=\varepsilon_r=0} \tag{E.54}$$

A *Lie algebra* is a linear vector space for which there is defined a *Lie product* $[X_i, X_j]$ satisfying the properties

$$\begin{aligned}
[X_i, X_j] &= X_i X_j - X_j X_i = c_{ij}^k X_j \\
[\alpha X_i + \beta X_j, X_k] &= \alpha[X_i, X_k] + \beta[X_j, X_k] \\
[X_i, \alpha X_j + \beta X_k] &= \alpha[X_i, X_j] + \beta[X_i, X_k] \\
[X_i, X_j] &= -[X_j, X_i]
\end{aligned} \tag{E.55}$$

$$[X_i, [X_j, X_k]] + [X_k, [X_i, X_j]] + [X_j, [X_k, X_i]] = 0 \tag{E.56}$$

where the c_{ij}^k are the structure constants of the algebra, the X_i are the elements of the algebra, and α and β belong to the field of complex numbers. The structure constants are independent of the representation of the elements of the algebra, and, therefore, they characterize the properties of the algebra.

The infinitesimal generators of the r-parameter Lie group form a Lie algebra. Thus

$$[I_a, I_b] = C_{ab}^c I_c, \qquad a, b, c = 1, 2, \ldots, r \tag{E.57}$$

The elements of the algebra are defined in terms of the group elements. On the other hand, if we are given the Lie algebra, it is only possible to construct from this algebra a restricted number of Lie groups which have the property that they are isomorphic in the vicinity of the identity (i.e., *locally*) although their *global* properties (i.e., properties for elements far removed from the identity) may diverge.

Appendix F

POINT GROUPS AND THEIR CHARACTER TABLES

F.1. DEFINITIONS

The point groups are normally defined in terms of their operations on an ordinary three-dimensional representation space of the coordinates. A point group is a group whose elements are symmetry transformations which leave at least one point in the space unchanged. All point group symmetry operations are *orthogonal transformations*, i.e., the determinant of the representation matrix is ± 1. Thus the point group operations are rotations and reflections in a three-dimensional space.

The point groups are classified as follows.

A. The $\mathcal{C}_n$ Groups

These groups consist of proper rotations (i.e., determinant of the representation matrix is $+1$) about an axis through an angle of $360°/n$. There are n elements in the group, and each element is in a class by itself. For example, the group $\mathcal{C}_4$ corresponds to rotations through an angle of $360°/4 = 90°$. The four elements of the group are: $\mathcal{C}_4(90°)$, $\mathcal{C}_4^2(180°)$, $\mathcal{C}_4^3(270°)$, $\mathcal{C}_4^4(360°) \equiv E$ (identity). The axis of rotation is called an *n-fold* axis.

B. The $\mathcal{C}_{nh}$ Groups

These groups are obtained by adding to the $\mathcal{C}_n$ group the n operations obtained by multiplying each proper rotation operator by a reflection about a horizontal plane, designated by the symbol σ_h. The intersection of the axis of rotation and the horizontal plane is the invariant point. The $\mathcal{C}_{nh}$ groups contain $2n$ elements. Thus for $\mathcal{C}_{4h}$ we have

$$\{\mathcal{C}_4, \mathcal{C}_4^2, \mathcal{C}_4^3, \mathcal{C}_4^4 \equiv E, \sigma_h\mathcal{C}_4, \sigma_h\mathcal{C}_4^2, \sigma_h\mathcal{C}_4^3, \sigma_h\mathcal{C}_4^4 \equiv \sigma_h\}$$

C. The $\mathcal{C}_{nv}$ Groups

These groups are characterized by adding to the proper rotations reflections about vertical planes. There are n vertical reflection planes whose intersection is the axis of rotation of $\mathcal{C}_n$. If n is odd, all n planes belong to the same class; but, if n is even, there are two distinct classes of planes, each having $n/2$ operators. Because of the vertical reflection planes, the operators $\mathcal{C}_n{}^m$ and $\mathcal{C}_n^{0-m}$ belong to the same class (m is an integer between zero and n). The reflection about a vertical plane is designated by σ_v. Thus the elements of the $\mathcal{C}_{nv}$ group are

$$E \equiv \mathcal{C}_n{}^n, 2\mathcal{C}_n, 2\mathcal{C}_n{}^2, \ldots, 2\mathcal{C}_n^{(n-1)/2}, n\sigma_v \qquad \text{for} \quad n = \text{odd integer}$$

$$E \equiv \mathcal{C}_n{}^n, 2\mathcal{C}_n{}^2, \ldots, \mathcal{C}_n^{n/2} \equiv \mathcal{C}_2, \tfrac{1}{2}n\sigma_v, \tfrac{1}{2}n\sigma_v' \qquad \text{for} \quad n = \text{even integer}$$

The angle between the vertical reflection planes in a given class is $(n - m)$ $(360°/n)$.

D. The D_n Groups

In addition to the $\mathcal{C}_n$ operations, this group has twofold rotation axes $\mathcal{C}_2'$ at right angles to the $\mathcal{C}_n$ axis of rotation. The $\mathcal{C}_n$ axis is called the *principal axis*. When n is odd, there are $n\mathcal{C}_2'$ operators of the same class whose axes are separeted by an angle of $(360°/n)$. When n is even, there are two distinct classes of twofold rotations, each containing $n/2$ operators.

Because of the twofold axes $\mathcal{C}_n{}^m$ and $\mathcal{C}_n^{n-m}$ are in the same class. When n is even $\mathcal{C}_n\mathcal{C}_2'$ generates a member $\mathcal{C}_2''$ of the class of twofold operations belonging to a class distinct from the class to which $\mathcal{C}_2'$ belongs. The elements of D_n are

$$E, 2\mathcal{C}_n, 2\mathcal{C}_n{}^2, \ldots, 2\mathcal{C}_n^{(n-1)/2}, n\mathcal{C}_2', \qquad n = \text{odd integer}$$

$$E, 2\mathcal{C}_n, 2\mathcal{C}_n{}^2, \ldots, \mathcal{C}_n^{n/2} \equiv \mathcal{C}_2, \tfrac{1}{2}n\mathcal{C}_2', \tfrac{1}{2}n\mathcal{C}_2'', \qquad n = \text{even integer}$$

E. The D_{nh} Groups

To the elements of D_n are added elements obtained by multiplying the D_n elements with σ_h. The elements are

$$E, 2\mathcal{C}_n, 2\mathcal{C}_n^{\,2}, \ldots, 2\mathcal{C}_n^{(n-1)/2}, n\mathcal{C}_2{}', \sigma_h, 2S_n, 2S_n{}^3, \ldots, 2S_n^{n-2}, n\sigma_v, \qquad n = \text{odd integer}$$

$$E, 2\mathcal{C}_n, 2\mathcal{C}_n^{\,2}, \ldots, \mathcal{C}_n^{n/2} \equiv \mathcal{C}_2, \tfrac{1}{2}n\mathcal{C}_2{}', \tfrac{1}{2}n\mathcal{C}_2'', \sigma_h, 2S_n, 2\mathcal{C}_n{}^2\sigma_h, \ldots, i, \tfrac{1}{2}n\sigma_v, \tfrac{1}{2}n\sigma_v{}', \qquad n = \text{even integer}$$

where (i) is the *inversion* operator which transforms **r** and −**r** and S_n is the *improper rotation* $\mathcal{C}_n\sigma_h$.

F. The D_{nd} Groups

To the elements of D_n are added the elements generated by the σ_d operation, which is defined as reflection about vertical planes which bisect the angle between two adjacent $\mathcal{C}_2{}'$ axes (n odd) or two adjacent $\mathcal{C}_2{}'$ and $\mathcal{C}_2''$ axes (n even). The elements are

$$E, 2\mathcal{C}_n, 2\mathcal{C}_n{}^2, \ldots, \mathcal{C}_n^{n/2} \equiv \mathcal{C}_2, n\mathcal{C}_2{}', n\sigma_d, 2S_{2n}, 2S_n{}^3, \ldots, 2S_n^{n-1}, \qquad n = \text{even integer}$$

$$E, 2\mathcal{C}_n, 2\mathcal{C}_n{}^2, \ldots, 2\mathcal{C}_n^{(n-1)/2}, n\mathcal{C}_2{}^1, n\sigma_d, 2S_{2n}, 2S_{2n}^3, \ldots, S_{2n}^n \equiv i, \qquad n = \text{odd integer}$$

G. The S_n Groups

When n is odd, these groups are identical with $\mathcal{C}_{nh}$ groups. When n is even, the elements are

$$E, S_n, \mathcal{C}_n, S_n{}^3, \mathcal{C}_n{}^2, \ldots, \mathcal{C}_n^{(n-1)/2}, S_n^{n-1}, \qquad n \text{ even}, n/2 \text{ even}$$

$$E, S_n, \mathcal{C}_n, S_n{}^3, \ldots, S_n^{n/2} \equiv i, \ldots, \mathcal{C}_n^{(n-1)/2}, S_n^{n-1}, \qquad n \text{ even}, n/2 \text{ odd}$$

H. The S_{nh} Groups

These groups are identical with the $\mathcal{C}_{nh}$ groups.

I. The S_{nv} Groups

These groups are identical with the D_{nh} groups when n is odd. When n is even, S_{nv} is identical with $D_{(n/2)d}$ groups.

J. The Tetrahedral Group (T)

This group consists of rotations about four threefold axes, the angle between the axes being 109° 32′, and three twofold axes which are mutually perpendicular as shown in Fig. F. 1a. The group derives its name from the fact that it is a symmetry group of the tetrahedron; the relationship of the rotation axes to the tetrahedron is shown in Fig. F. 1b. The elements of this group are $E, 4\mathcal{C}_3, 4\mathcal{C}_3^2, 3\mathcal{C}_2$.

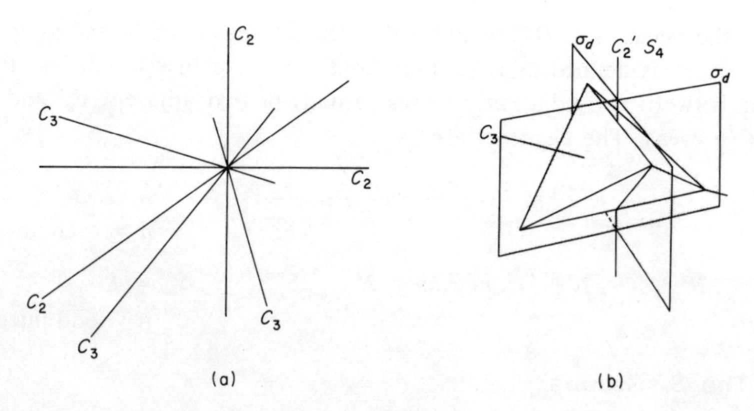

Fig. F.1. (a) The complete array of twofold and threefold axes of the tetrahedral group. (b) The relationship of the tetrahedral group rotation axes and symmetry planes to the tetrahedron.

K. The Group T_d

To the tetrahedral group are added reflections about the vertical planes shown in Fig. F. 1b and designated by the symbol σ_d. The elements of the group are $E, 8\mathcal{C}_3, 6\sigma_d, 3\mathcal{C}_2, 6S_4$.

L. The Group $T_i(T_h)$

To the tetrahedral group is added the inversion operation. The elements of the group are E, $8\mathcal{C}_3$, $3\mathcal{C}_2$, i, $8S_b$, $3\sigma_h$.

M. The Octahedral (Cubic) Group $\mathcal{O}$

This group consists of rotation operations which are symmetry operations on a cube (or octahedron). The rotation operations are defined in Table F.1 and Fig. F.2. The elements of the group are E, $6\mathcal{C}_4$, $8\mathcal{C}_3$, $3\mathcal{C}_2$, $6\mathcal{C}_2'$.

TABLE F.1

Elements of the octahedral group

Operator	Multiplicity	Definition	Typical Axis in Fig. F.2
E	1	The identity	
$\mathcal{C}_4$	6	$\pm 90°$ rotation about cube axes	A B
$\mathcal{C}_4^2$	3	180° rotation about cube axes	A B
$\mathcal{C}_2$	6	180° rotation about axes parallel to face diagonals	G H
$\mathcal{C}_3$	8	$\pm 120°$ rotation about body diagonals	I J

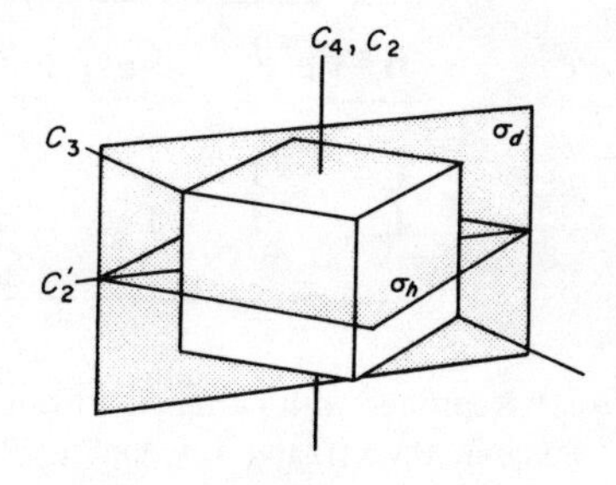

Fig. F.2. Cube illustrating the rotation axes of the octahedral group.

N. The Group $\mathcal{O}_h(\mathcal{O}_i)$

To the octahedral group is added the operation of inversion i yielding the elements E, $6\mathcal{C}_4$, $8\mathcal{C}_3$, $3\mathcal{C}_2$, $6\mathcal{C}_2'$, i, $6S_4$, $8S_6$, $3\sigma_h$, $6\sigma_v$.

O. The Icosahedral Group $I(P)$

The elements of this group are symmetry rotations of an icosahedron. There are six fivefold axes, ten threefold axes and fifteen twofold axes. The elements of the group are E, $12\mathcal{C}_5$, $12\mathcal{C}_5^2$, $20\mathcal{C}_3$, $15\mathcal{C}_2$.

P. The Group I_i

To the icosahedral group is added the inversion operation i. The elements of this group are E, $12\mathcal{C}_5$, $12\mathcal{C}_5^2$, $15\mathcal{C}_2$, i, $12S_{10}$, $12S_{10}^3$, $20S_6$, $15\sigma_h$.

F.2. CHARACTER TABLES OF THE POINT GROUPS*

C_1	E
A	1

$C_{1h} \equiv C_s$	E	σ_h		
A'	1	1	x, y, xy, x^2, y^2, z^2	R_z
A''	1	-1	z, xz, yz	R_x, R_y

$S_2 \equiv C_i$	E	$C_2\sigma_h = S_2 = i$		
A_g	1	1	$xy, xz, yz, x^2, y^2, z^2$	R_x, R_y, R_z
A_u	1	-1	x, y, z	

* Reprinted with permission from D. S. Urch, "Orbitals and Symmetry," pp. 211–236. Penguin Books Ltd., London, 1970. Copyright © D. S. Urch, 1970.

Twofold Axis

C_2	E	C_2		
A	1	1	z, xy, x^2, y^2, z^2	R_z
B	1	-1	x, y, xz, yz	R_x, R_y

C_{2v}	E	C_2	$\sigma_v(xz)$	$\sigma_v'(yz)$		
A_1	1	1	1	1	z, x^2, y^2, z^2	
A_2	1	1	-1	-1	xy	R_z
B_1	1	-1	1	-1	x, xz	R_y
B_2	1	-1	-1	1	y, yz	R_x

C_{2h}	E	C_2	σ_h	i		
A_g	1	1	1	1	xy, x^2, y^2, z^2	R_z
A_u	1	1	-1	-1	z	
B_g	1	-1	-1	1	xz, yz	R_x, R_y
B_u	1	-1	1	-1	x, y	

D_2	E	$C_2(z)$	$C_2(y)$	$C_2(x)$		
A_1	1	1	1	1	x^2, y^2, z^2	
B_1	1	1	-1	-1	z, xy	R_z
B_2	1	-1	1	-1	y, xz	R_y
B_3	1	-1	-1	1	x, yz	R_x

D_{2h}	E	$C_2(z)$	$C_2(y)$	$C_2(x)$	i	$\sigma_h(xy)$	$\sigma_h(xz)$	$\sigma_h(yz)$		
A_{1g}	1	1	1	1	1	1	1	1	x^2, y^2, z^2	
A_{1u}	1	1	1	1	−1	−1	−1	−1		
B_{1g}	1	1	−1	−1	1	1	−1	−1	xy	R_z
B_{1u}	1	1	−1	−1	−1	−1	1	1	z	
B_{2g}	1	−1	1	−1	1	−1	1	−1	xz	R_y
B_{2u}	1	−1	1	−1	−1	1	−1	1	y	
B_{3g}	1	−1	−1	1	1	−1	−1	1	yz	R_x
B_{3u}	1	−1	−1	1	−1	1	1	−1	x	

S_4	E	C_2	S_4	S_4^3		
A	1	1	1	1	z^2, x^2+y^2	R_z
B	1	1	−1	−1	z, x^2-y^2, xy	
E	1	−1	i	$-i$	$(x, y), (xz, yz)$	(R_x, R_y)
	1	−1	$-i$	i		

$D_{2d}(S_{4v})$	E	C_2	$2S_4$	$2C_2'$	$2\sigma_d$		
A_1	1	1	1	1	1	z^2, x^2+y^2	
A_2	1	1	1	−1	−1		R_z
B_1	1	1	−1	1	−1	x^2-y^2	
B_2	1	1	−1	−1	1	z, xy	
E	2	−2	0	0	0	$(x, y), (xz, yz)$	(R_x, R_y)

Threefold Axis

C_3	E	C_3	C_3^2		
A	1	1	1	z, z^2, x^2+y^2	R_z
E	$\left\{\begin{matrix}1\\1\end{matrix}\right.$	$\begin{matrix}\omega\\\omega^2\end{matrix}$	$\left.\begin{matrix}\omega^2\\\omega\end{matrix}\right\}$	$(x, y), (xz, yz)$ (x^2-y^2, xy)	(R_x, R_y)

$\omega = \exp\dfrac{2\pi i}{3}$

C_{3v}	E	$2C_3$	$3\sigma_v$		
A_1	1	1	1	z, z^2, x^2+y^2	
A_2	1	1	-1		R_z
E	2	-1	0	$\left\{\begin{matrix}(x, y), (xz, yz)\\(x^2-y^2, xy)\end{matrix}\right\}$	(R_x, R_y)

C_{3h}	E	C_3	C_3^2	σ_h	S_3	$C_3^2\sigma_h \equiv S_3^5$		
A'	1	1	1	1	1	1	z^2, x^2+y^2	R_z
A''	1	1	1	-1	-1	-1	z	
E'	$\left\{\begin{matrix}1\\1\end{matrix}\right.$	$\begin{matrix}\omega\\\omega^2\end{matrix}$	$\begin{matrix}\omega^2\\\omega\end{matrix}$	$\begin{matrix}1\\1\end{matrix}$	$\begin{matrix}\omega\\\omega^2\end{matrix}$	$\left.\begin{matrix}\omega^2\\\omega\end{matrix}\right\}$	$\left\{\begin{matrix}(x, y)\\(x^2-y^2, xy)\end{matrix}\right.$	
E''	$\left\{\begin{matrix}1\\1\end{matrix}\right.$	$\begin{matrix}\omega\\\omega^2\end{matrix}$	$\begin{matrix}\omega^2\\\omega\end{matrix}$	$\begin{matrix}-1\\-1\end{matrix}$	$\begin{matrix}-\omega\\-\omega^2\end{matrix}$	$\left.\begin{matrix}-\omega^2\\-\omega\end{matrix}\right\}$	(xz, yz)	R_x, R_y

$\omega = \exp\dfrac{2\pi i}{3}$

D_3	E	$2C_3$	$3C_2'$		
A_1	1	1	1	z^2, x^2+y^2	
A_2	1	1	-1	z	R_z
E	2	-1	0	$\left\{\begin{matrix}(xz, yz), (x, y)\\(x^2-y^2, xy)\end{matrix}\right\}$	(R_x, R_y)

D_{3h}	E	σ_h	$2C_3$	$2S_3$	$3C_2'$	$3\sigma_v$		
A_1'	1	1	1	1	1	1	z^2, x^2+y^2	
A_1''	1	−1	1	−1	1	−1		
A_2'	1	1	1	1	−1	−1		R_z
A_2''	1	−1	1	−1	−1	1	z	
E'	2	2	−1	−1	0	0	$(x, y), (xy, x^2-y^2)$	
E''	2	−2	−1	1	0	0	(xz, yz)	(R_x, R_y)

S_6	E	C_3	C_3^2	i	S_6	S_6^5		
A_g	1	1	1	1	1	1	z^2, x^2+y^2	R_z
A_u	1	1	1	−1	−1	−1	z	
E_g	1	ω	ω^2	1	ω^2	ω	$(xz, yz), (x^2-y^2, xy)$	(R_x, R_y)
	1	ω^2	ω	1	ω	ω^2		
E_u	1	ω	ω^2	−1	$-\omega^2$	$-\omega$	(x, y)	
	1	ω^2	ω	−1	$-\omega$	$-\omega^2$		

$$\omega = \exp\frac{2\pi i}{3}$$

$D_{3d}(S_{6v})$	E	$2C_3$	$3C_2'$	i	$2S_6$	σ_d		
A_{1g}	1	1	1	1	1	1	z^2, x^2+y^2	
A_{1u}	1	1	1	−1	−1	−1		
A_{2g}	1	1	−1	1	1	−1		R_z
A_{2u}	1	1	−1	−1	−1	1	z	
E_g	2	−1	0	2	−1	0	$(x^2-y^2, xy), (xz, yz)$	(R_x, R_y)
E_u	2	−1	0	−2	1	0	(x, y)	

Fourfold Axis

C_4	E	C_2	C_4	C_4^3		
A	1	1	1	1	z, z^2, x^2+y^2	R_z
B	1	1	-1	-1	xy, x^2-y^2	
E	$\left\{\begin{matrix}1\\1\end{matrix}\right.$	$\begin{matrix}-1\\-1\end{matrix}$	$\begin{matrix}i\\-i\end{matrix}$	$\left.\begin{matrix}-i\\i\end{matrix}\right\}$	$\left\{\begin{matrix}(xz, yz)\\(x, y)\end{matrix}\right\}$	(R_x, R_y)

C_{4v}	E	C_2	$2C_4$	$2\sigma_v$	$2\sigma'_v(2\sigma_d)$		
A_1	1	1	1	1	1	z, z^2, x^2+y^2	
A_2	1	1	1	-1	-1		R_z
B_1	1	1	-1	1	-1	x^2-y^2	
B_2	1	1	-1	-1	1	xy	
E	2	-2	0	0	0	$\left\{\begin{matrix}(x, y)\\(xz, yz)\end{matrix}\right\}$	(R_x, R_y)

C_{4h}	E	C_2	C_4	C_4^3	i	σ_h	S_4^3	S_4		
A_g	1	1	1	1	1	1	1	1	z^2, x^2+y^2	R_z
A_u	1	1	1	1	-1	-1	-1	-1	z	
B_g	1	1	-1	-1	1	1	-1	-1	xy, x^2-y^2	
B_u	1	1	-1	-1	-1	-1	1	1		
E_g	$\left\{\begin{matrix}1\\1\end{matrix}\right.$	$\begin{matrix}-1\\-1\end{matrix}$	$\begin{matrix}i\\-i\end{matrix}$	$\begin{matrix}-i\\i\end{matrix}$	$\begin{matrix}1\\1\end{matrix}$	$\begin{matrix}-1\\-1\end{matrix}$	$\begin{matrix}i\\-i\end{matrix}$	$\left.\begin{matrix}-i\\i\end{matrix}\right\}$	(xz, yz)	(R_x, R_y)
E_u	$\left\{\begin{matrix}1\\1\end{matrix}\right.$	$\begin{matrix}-1\\-1\end{matrix}$	$\begin{matrix}i\\-i\end{matrix}$	$\begin{matrix}-i\\i\end{matrix}$	$\begin{matrix}-1\\-1\end{matrix}$	$\begin{matrix}1\\1\end{matrix}$	$\begin{matrix}-i\\i\end{matrix}$	$\left.\begin{matrix}i\\-i\end{matrix}\right\}$	(x, y)	

D_4	E	C_2	$2C_4$	$2C'_2$	$2C''_2$		
A_1	1	1	1	1	1	z^2, x^2+y^2	
A_2	1	1	1	-1	-1	z	R_z
B_1	1	1	-1	1	-1	x^2-y^2	
B_2	1	1	-1	-1	1	xy	
E	2	-2	0	0	0	$(x, y), (xz, yz)$	(R_x, R_y)

D_{4h}	E	C_2	$2C_4$	$2C_2'$	$2C_2''$	i	σ_h	$2S_4$	$2\sigma_v = 2iC_2'$	$2\sigma_v' = 2iC_2''$		
A_{1g}	1	1	1	1	1	1	1	1	1	1	z^2, x^2+y^2	
A_{1u}	1	1	1	1	1	−1	−1	−1	−1	−1		
A_{2g}	1	1	1	−1	−1	1	1	1	−1	−1		R_z
A_{2u}	1	1	1	−1	−1	−1	−1	−1	1	1	z	
B_{1g}	1	1	−1	1	−1	1	1	−1	1	−1	x^2-y^2	
B_{1u}	1	1	−1	1	−1	−1	−1	1	−1	1		
B_{2g}	1	1	−1	−1	1	1	1	−1	−1	1	xy	
B_{2u}	1	1	−1	−1	1	−1	−1	1	1	−1		
E_g	2	−2	0	0	0	2	−2	0	0	0	(xz, yz)	(R_x, R_y)
E_u	2	−2	0	0	0	−2	2	0	0	0	(x, y)	

S_8	E	S_8	C_4	S_8^3	C_2	S_8^5	C_4^3	S_8^7		
A	1	1	1	1	1	1	1	1	z^2, x^2+y^2	R_z
B	1	-1	1	-1	1	-1	1	-1	z	
E_1	1	ω	ω^2	ω^3	-1	ω^5	ω^6	ω^7	(x, y)	(R_x, R_y)
	1	ω^7	ω^6	ω^5	-1	ω^3	ω^2	ω		
E_2	1	ω^2	-1	ω^6	1	ω^2	-1	ω^6	(x^2-y^2, xy)	
	1	ω^6	-1	ω^2	1	ω^6	-1	ω^2		
E_3	1	ω^3	ω^6	ω	-1	ω^7	ω^2	ω^5	(xz, yz)	
	1	ω^5	ω^2	ω^7	-1	ω	ω^6	ω^3		

$\omega = \exp \frac{2\pi i}{8}$

$D_{4d}(S_{8v})$	E	C_2	$2C_4$	$2S_8$	$2S_8^3$	$4C_2'$	$4\sigma_d$		
A_1	1	1	1	1	1	1	1	z^2, x^2+y^2	
A_2	1	1	1	1	1	-1	-1		R_z
B_1	1	1	1	-1	-1	1	-1		
B_2	1	1	1	-1	-1	-1	1	z	
E_1	2	-2	0	$\sqrt{2}$	$-\sqrt{2}$	0	0	(x, y)	
E_2	2	2	-2	0	0	0	0	(x^2-y^2, xy)	
E_3	2	-2	0	$-\sqrt{2}$	$\sqrt{2}$	0	0	(xz, yz)	(R_x, R_y)

Fivefold Axis

C_5	E	C_5	C_5^2	C_5^3	C_5^4		
A	1	1	1	1	1	z, x^2+y^2, z^2	R_z
E_1	1	ω	ω^2	ω^3	ω^4	$(x, y), (xz, yz)$	(R_x, R_y)
	1	ω^4	ω^3	ω^2	ω		
E_2	1	ω^2	ω^4	ω	ω^3	(x^2-y^2, xy)	
	1	ω^3	ω	ω^4	ω^2		

$\omega = \exp \frac{2\pi i}{5}$

C_{5v}	E	$2C_5$	$2C_5^2$	$5\sigma_v$		
A_1	1	1	1	1	z, z^2, x^2+y^2	
A_2	1	1	1	-1		R_z
E_1	2	$2\cos 72°$	$2\cos 144°$	0	$(x, y), (xz, yz)$	(R_x, R_y)
E_2	2	$2\cos 144°$	$2\cos 288°$	0	(x^2-y^2, xy)	

C_{5h}	E	C_5	C_5^2	C_5^3	C_5^4	σ_h	S_5	$\sigma_h C_5^2 \equiv S_5^7$	S_5^3	$\sigma_h C_5^4 \equiv S_5^9$		
A'	1	1	1	1	1	1	1	1	1	1	z^2, x^2+y^2	R_z
A''	1	1	1	1	1	-1	-1	-1	-1	-1	z	
E_1'	1	ω	ω^2	ω^3	ω^4	1	ω	ω^2	ω^3	ω^4	(x, y)	
	1	ω^4	ω^3	ω^2	ω	1	ω^4	ω^3	ω^2	ω		
E_1''	1	ω	ω^2	ω^3	ω^4	-1	$-\omega$	$-\omega^2$	$-\omega^3$	$-\omega^4$	(xz, yz)	(R_x, R_y)
	1	ω^4	ω^3	ω^2	ω	-1	$-\omega^4$	$-\omega^3$	$-\omega^2$	$-\omega$		
E_2'	1	ω^2	ω^4	ω	ω^3	1	ω^2	ω^4	ω	ω^3	(x^2, y^2, xy)	
	1	ω^3	ω	ω^4	ω^2	1	ω^3	ω	ω^4	ω^2		
E_2''	1	ω^2	ω^4	ω	ω^3	-1	$-\omega^2$	$-\omega^4$	$-\omega$	$-\omega^3$		
	1	ω^3	ω	ω^4	ω^2	-1	$-\omega^3$	$-\omega$	$-\omega^4$	$-\omega^2$		

$\omega = \exp\frac{2\pi i}{5}$

$\boldsymbol{D_5}$	E	$2C_5$	$2C_5^2$	$5C_2'$		
A_1	1	1	1	1	z^2, x^2+y^2	
A_2	1	1	1	−1	z	R_z
E_1	2	2 cos 72°	2 cos 144°	0	$(x, y), (xz, yz)$	(R_x, R_y)
E_2	2	2 cos 144°	2 cos 288°	0	(x^2-y^2, xy)	

$\boldsymbol{D_{5h}}$	E	$2C_5$	$2C_5^2$	$5C_2'$	σ_h	$2S_5$	$2S_5^3$	$5\sigma_v$		
A_1'	1	1	1	1	1	1	1	1	z^2, x^2+y^2	
A_1''	1	1	1	1	−1	−1	−1	−1		
A_2'	1	1	1	−1	1	1	1	−1		R_z
A_2''	1	1	1	−1	−1	−1	−1	1	z	
E_1'	2	2 cos 72°	2 cos 144°	0	2	2 cos 72°	2 cos 144°	0	(x, y)	
E_1''	2	2 cos 72°	2 cos 144°	0	−2	−2 cos 72°	−2 cos 144°	0	(xz, yz)	(R_x, R_y)
E_2'	2	2 cos 144°	2 cos 288°	0	2	2 cos 144°	2 cos 288°	0	(x^2-y^2, xy)	
E_2''	2	2 cos 144°	2 cos 288°	0	−2	−2 cos 144°	−2 cos 288°	0		

S_{10}	E	S_{10}	C_5	S_{10}^3	C_5^2	$i = (S_{10}^5)$	C_5^3	S_{10}^7	C_5^4	S_{10}^9		
A_g	1	1	1	1	1	1	1	1	1	1	z^2, x^2+y^2	R_z
A_u	1	1	1	1	1	-1	-1	-1	-1	-1	z	
E_{1g}	1	ω^3	ω	ω^4	ω^2	1	ω^3	ω	ω^4	ω^2	(xz, yz)	(R_x, R_y)
	1	ω^2	ω^4	ω	ω^3	1	ω^2	ω^4	ω	ω^3		
E_{1u}	1	$-\omega^3$	ω	$-\omega^4$	ω^2	-1	ω^3	$-\omega$	ω^4	$-\omega^2$	(x, y)	
	1	$-\omega^2$	ω^4	$-\omega$	ω^3	-1	ω^2	$-\omega^4$	ω	$-\omega^3$		
E_{2g}	1	ω	ω^2	ω^3	ω^4	1	ω	ω^2	ω^3	ω^4	(x^2-y^2, xy)	
	1	ω^4	ω^3	ω^2	ω	1	ω^4	ω^3	ω^2	ω		
E_{2u}	1	$-\omega$	ω^2	$-\omega^3$	ω^4	-1	ω	$-\omega^2$	ω^3	$-\omega^4$		
	1	$-\omega^4$	ω^3	$-\omega^2$	ω	-1	ω^4	$-\omega^3$	ω^2	$-\omega$		

$\omega = \exp\frac{2\pi i}{5}$

$D_{5d}(S_{10v})$	E	$2C_5$	$2C_5^2$	$5C_2$	i	$2S_{10}$	$2S_{10}^3$	$5\sigma_d$		
A_{1g}	1	1	1	1	1	1	1	1	z^2, x^2+y^2	
A_{1u}	1	1	1	1	−1	−1	−1	−1		
A_{2g}	1	1	1	−1	1	1	1	−1		R_z
A_{2u}	1	1	1	−1	−1	−1	−1	1	z	
E_{1g}	2	2 cos 72°	2 cos 144°	0	2	2 cos 144°	2 cos 72°	0	(xz, yz)	(R_x, R_y)
E_{1u}	2	2 cos 72°	2 cos 144°	0	−2	−2 cos 144°	−2 cos 72°	0	(x, y)	
E_{2g}	2	2 cos 144°	2 cos 288°	0	2	2 cos 288°	2 cos 144°	0	(x^2-y^2, xy)	
E_{2u}	2	2 cos 144°	2 cos 288°	0	−2	−2 cos 288°	−2 cos 144°	0		

Sixfold Axis

C_6	E	C_6	C_3	C_2	C_3^2	C_6^5		
A	1	1	1	1	1	1	z, z^2, x^2-y^2	R_z
B	1	-1	1	-1	1	-1		
E_1	$\begin{cases}1\\1\end{cases}$	$\begin{matrix}\omega\\\omega^5\end{matrix}$	$\begin{matrix}\omega^2\\\omega^4\end{matrix}$	$\begin{matrix}-1\\-1\end{matrix}$	$\begin{matrix}\omega^4\\\omega^2\end{matrix}$	$\left.\begin{matrix}\omega^5\\\omega\end{matrix}\right\}$	$\left\{\begin{matrix}(x, y)\\(xz, yz)\end{matrix}\right\}$	(R_x, R_y)
E_2	$\begin{cases}1\\1\end{cases}$	$\begin{matrix}\omega^2\\\omega^4\end{matrix}$	$\begin{matrix}\omega^4\\\omega^2\end{matrix}$	$\begin{matrix}1\\1\end{matrix}$	$\begin{matrix}\omega^2\\\omega^4\end{matrix}$	$\left.\begin{matrix}\omega^4\\\omega^2\end{matrix}\right\}$	(x^2-y^2, xy)	

$$\omega = \exp\frac{2\pi i}{6}$$

C_{6v}	E	C_2	$2C_3$	$2C_6$	$3\sigma_v$	$3\sigma'_v$		
A_1	1	1	1	1	1	1	z, z^2, x^2+y^2	
A_2	1	1	1	1	-1	-1		R_z
B_1	1	-1	1	-1	-1	1		
B_2	1	-1	1	-1	1	-1		
E_1	2	-2	-1	1	0	0	$(x, y), (xz, yz)$	(R_x, R_y)
E_2	2	2	-1	-1	0	0	(x^2-y^2, xy)	

S_{12}	E	S_{12}	C_6	S_4	C_3	S_{12}^5	C_2	S_{12}^7	C_3^2	S_4^3	C_6^5	S_{12}^{11}		
A	1	1	1	1	1	1	1	1	1	1	1	1	z^2, x^2+y^2	R_z
B	1	−1	1	−1	1	−1	1	−1	1	−1	1	−1	z	
E_1	1	ω	ω^2	ω^3	ω^4	ω^5	−1	ω^7	ω^8	ω^9	ω^{10}	ω^{11}	(x, y)	
	1	ω^{11}	ω^{10}	ω^9	ω^8	ω^7	−1	ω^5	ω^4	ω^3	ω^2	ω		
E_2	1	ω^2	ω^4	−1	ω^8	ω^{10}	1	ω^2	ω^4	−1	ω^8	ω^{10}	(x^2-y^2, xy)	
	1	ω^{10}	ω^8	−1	ω^4	ω^2	1	ω^{10}	ω^8	−1	ω^4	ω^2		
E_3	1	ω^3	−1	ω^9	1	ω^3	−1	ω^9	1	ω^3	−1	ω^9		
	1	ω^9	−1	ω^3	1	ω^9	−1	ω^3	1	ω^9	−1	ω^3		
E_4	1	ω^4	ω^8	1	ω^4	ω^8	1	ω^4	ω^8	1	ω^4	ω^8		
	1	ω^8	ω^4	1	ω^8	ω^4	1	ω^8	ω^4	1	ω^8	ω^4		
E_5	1	ω^5	ω^{10}	ω^3	ω^8	ω	−1	ω^{11}	ω^4	ω^9	ω^2	ω^7	(xz, yz)	(R_x, R_y)
	1	ω^7	ω^2	ω^9	ω^4	ω^{11}	−1	ω	ω^2	ω^3	ω^{10}	ω^5		

$$\omega = \exp\frac{2\pi i}{12}$$

$D_{6d}(S_{12v})$	E	$2S_{12}$	$2C_6$	$2S_4$	$2C_3$	$2S_{12}^5$	C_2	$6C_2'$	$6\sigma_d$		
A_1	1	1	1	1	1	1	1	1	1	z^2, x^2+y^2	
A_2	1	1	1	1	1	1	1	-1	-1		R_z
B_1	1	-1	1	-1	1	-1	1	1	-1		
B_2	1	-1	1	-1	1	-1	1	-1	1	z	
E_1	2	$\sqrt{3}$	1	0	-1	$-\sqrt{3}$	-2	0	0	(x, y)	
E_2	2	1	-1	-2	-1	1	2	0	0	(x^2-y^2, xy)	
E_3	2	0	-2	0	2	0	-2	0	0		
E_4	2	-1	-1	2	-1	-1	2	0	0		
E_5	2	$-\sqrt{3}$	1	0	-1	$\sqrt{3}$	-2	0	0	(xz, yz)	(R_x, R_y)

C_{6h}	E	C_6	C_3	C_2	C_3^2	C_6^5	σ_h	S_6	S_3	i	S_3^5	S_6^5		
A_g	1	1	1	1	1	1	1	1	1	1	1	1	z^2, x^2+y^2	R_z
A_u	1	1	1	1	1	1	−1	1	−1	1	−1	1	z	
B_g	1	−1	1	−1	1	−1	−1	−1	1	−1	1	−1		
B_u	1	−1	1	−1	1	−1	1	1	−1	−1	−1	1		
E_{1g}	1	ω	ω^2	−1	ω^4	ω^5	−1	ω^5	ω^4	1	ω^2	ω	(xz, yz)	(R_x, R_y)
	1	ω^5	ω^4	−1	ω^2	ω	−1	ω	ω^2	1	ω^4	ω^5		
E_{1u}	1	ω	ω^2	−1	ω^4	ω^5	1	$-\omega^5$	$-\omega^4$	−1	$-\omega^2$	$-\omega$	(x, y)	
	1	ω^5	ω^4	−1	ω^2	ω	1	$-\omega$	$-\omega^2$	−1	$-\omega^4$	$-\omega^5$		
E_{2g}	1	ω^2	ω^4	1	ω^2	ω^4	1	ω^4	ω^2	1	ω^4	ω^2	(x^2-y^2, xy)	
	1	ω^4	ω^2	1	ω^4	ω^2	1	ω^2	ω^4	1	ω^2	ω^4		
E_{2u}	1	ω^2	ω^4	1	ω^2	ω^4	−1	$-\omega^4$	$-\omega^2$	−1	$-\omega^4$	$-\omega^2$		
	1	ω^4	ω^2	1	ω^4	ω^2	−1	$-\omega^2$	$-\omega^4$	−1	$-\omega^2$	$-\omega^4$		

$$\omega = \exp\frac{2\pi i}{6}$$

$\mathbf{D_6}$	E	C_2	$2C_3$	$2C_6$	$3C_2'$	$3C_2''$		
A_1	1	1	1	1	1	1	z^2, x^2+y^2	
A_2	1	1	1	1	−1	−1	z	R_z
B_1	1	−1	1	−1	1	−1		
B_2	1	−1	1	−1	−1	1		
E_1	2	−2	−1	1	0	0	$(x, y), (xz, yz)$	(R_x, R_y)
E_2	2	2	−1	−1	0	0	(x^2-y^2, xy)	

$\mathbf{D_{6h}}$	E	C_2	$2C_3$	$2C_6$	$3C_2'$	$3C_2''$	i	σ_h	$2S_6$	$2S_3$	$3\sigma_v(=iC_2')$	$3\sigma_v'(=iC_2'')$		
A_{1g}	1	1	1	1	1	1	1	1	1	1	1	1	z^2, x^2+y^2	
A_{1u}	1	1	1	1	1	1	−1	−1	−1	−1	−1	−1		
A_{2g}	1	1	1	1	−1	−1	1	1	1	1	−1	−1		R_z
A_{2u}	1	1	1	1	−1	−1	−1	−1	−1	−1	1	1	z	
B_{1g}	1	−1	1	−1	1	−1	1	−1	1	−1	1	−1		
B_{1u}	1	−1	1	−1	1	−1	−1	1	−1	1	−1	1		
B_{2g}	1	−1	1	−1	−1	1	1	−1	1	−1	−1	1		
B_{2u}	1	−1	1	−1	−1	1	−1	1	−1	1	1	−1		
E_{1g}	2	−2	−1	1	0	0	2	−2	−1	1	0	0	(xz, yz)	R_x, R_y
E_{1u}	2	−2	−1	1	0	0	−2	2	1	−1	0	0	(x, y)	
E_{2g}	2	2	−1	−1	0	0	2	2	−1	−1	0	0	(x^2-y^2, xy)	
E_{2u}	2	2	−1	−1	0	0	−2	−2	1	1	0	0		

Tetrahedral Groups

T	E	$3C_2$	$4C_3$	$4C_3^2$		
A	1	1	1	1	$(x^2+y^2+z^2)$	
E	1	1	ω	ω^2	$(x^2-y^2), (2z^2-x^2+y^2)$	
	1	1	ω^2	ω		
T	3	-1	0	0	(x, y, z); (xy, xz, yz)	(R_x, R_y, R_z)

$$\omega = \exp\frac{2\pi i}{3}$$

$\mathbf{S_{10}}$	E	S_{10}	C_5	$(S_{10})^3$	$(C_5)^2$	$(S_{10})^5$	$(C_5)^3$	$(S_{10})^7$	$(C_5)^4$	$(S_{10})^9$		
A_g	1	1	1	1	1	1	1	1	1	1	z^2, x^2+y^2	R_z
A_u	1	1	1	1	1	-1	-1	-1	-1	-1	z	
E_{1g}	1	ω^3	ω	ω^4	ω^2	1	ω^3	ω	ω^4	ω^2	(xz, yz)	(R_x, R_y)
	1	ω^2	ω^4	ω	ω^3	1	ω^2	ω^4	ω	ω^3		
E_{1u}	1	$-\omega^3$	ω	$-\omega^4$	ω^2	-1	ω^3	$-\omega$	ω^4	$-\omega^2$	(x, y)	
	1	$-\omega^2$	ω^4	$-\omega$	ω^3	-1	ω^2	$-\omega^4$	ω	$-\omega^3$		
E_{2g}	1	ω	ω^2	ω^3	ω^4	1	ω	ω^2	ω^3	ω^4	(x^2-y^2, xy)	
	1	ω^4	ω^3	ω^2	ω	1	ω^4	ω^3	ω^2	ω		
E_{2u}	1	$-\omega$	ω^2	$-\omega^3$	ω^4	-1	ω	$-\omega^2$	ω^3	$-\omega^4$		
	1	$-\omega^4$	ω^3	$-\omega^2$	ω	-1	ω^4	$-\omega^3$	ω^2	$-\omega$		

D_5 (S_{10r})	E	$2C_5$	$2(C_5)^2$	$5C'_2$	i	$2S_{10}$	$2(S_{10})^3$	$5_{\sigma d}$		
A_{1g}	1	1	1	i	1	1	1	1	z^2, x^2+y^2	
A_{1u}	1	1	1	1	−1	−1	−1	−1		
A_{2g}	1	1	1	−1	1	1	1	−1		R_z
A_{2u}	1	1	1	−1	−1	−1	−1	1	z	
E_{1g}	2	2 cos 72°	2 cos 144°	0	2	2 cos 144	2 cos 72°	0	(xz, yz)	(R_x, R_y)
E_{1u}	2	2 cos 72°	2 cos 144	0	−2	−2 cos 144	−2 cos 72°	0	(x, y)	
E_{2g}	2	2 cos 144°	2 cos 288°	0	2	2 cos 288°	2 cos 144°	0	$(x-y, xy)$	
E_{2u}	2	2 cos 144°	2 cos 288°	0	2	−2 cos 288°	−2 cos 144°	0		

T_h	E	$3C_2$	$4C_3$	$4C_3^2$	i	$3\sigma_h(=iC_2)$	$4S_6$	$4S_6^5$		
A_g	1	1	1	1	1	1	1	1	$(x^2+y^2+z^2)$	
A_u	1	1	1	1	−1	−1	−1	−1		
E_g	1	1	ω	ω^2	1	1	ω^2	ω	$(x^2-y^2), (2z^2-x^2+y^2)$	
	1	1	ω^2	ω	1	1	ω	ω^2		
E_u	1	1	ω	ω^2	−1	−1	$-\omega^2$	$-\omega$		
	1	1	ω^2	ω	−1	−1	$-\omega$	$-\omega^2$		
T_g	3	−1	0	0	3	−1	0	0	(xy, xz, yz)	(R_x, R_y, R_z)
T_u	3	−1	0	0	−3	1	0	0	(x, y, z)	

$$\omega = \exp\frac{2\pi i}{3}$$

T_d	E	$8C_3$	$3C_2$	$6\sigma_d$	$6S_4$		
A_1	1	1	1	1	1	$x^2+y^2+z^2$	
A_2	1	1	1	−1	−1		
E	2	−1	2	0	0	$(x^2-y^2, 2z^2-(x^2+y^2))$	
T_1	3	0	−1	−1	1		(R_x, R_y, R_z)
T_2	3	0	−1	1	−1	$\{(x, y, z)$, $(xy, xz, yz)\}$	

Octahedral Groups

O	E	$8C_3$	$3C_2(=C_4^2)$	$6C_2'$	$6C_4$		
A_1	1	1	1	1	1	$x^2+y^2+z^2$	
A_2	1	1	1	−1	−1		
E	2	−1	2	0	0	$(x^2-y^2, 2z^2-(x^2+y^2))$	
T_1	3	0	−1	−1	1	(x, y, z)	(R_x, R_y, R_z)
T_2	3	0	−1	1	−1	(xy, xz, yz)	

O_h	E	$8C_3$	$3C_2$	$6C_2'$	$6C_4$	i	$8S_6$	$3\sigma_h$	$6\sigma_v$	$6S_4$		
A_{1g}	1	1	1	1	1	1	1	1	1	1	$(x^2+y^2+z^2)$	
A_{1u}	1	1	1	1	1	−1	−1	−1	−1	−1		
A_{2g}	1	1	1	−1	−1	1	1	1	−1	−1		
A_{2u}	1	1	1	−1	−1	−1	−1	−1	1	1		
E_g	2	−1	2	0	0	2	−1	2	0	0	$(x^2-y^2, 2z^2-(x^2+y^2))$	
E_u	2	−1	2	0	0	−2	1	−2	0	0		
T_{1g}	3	0	−1	−1	1	3	0	−1	−1	1		(R_x, R_y, R_z)
T_{1u}	3	0	−1	−1	1	−3	0	1	1	−1	(x, y, z)	
T_{2g}	3	0	−1	1	−1	3	0	−1	1	−1	(xy, xz, yz)	
T_{2u}	3	0	−1	1	−1	−3	0	1	−1	1		

Icosahedral Groups

P	E	$15C_2$	$20C_3$	$12C_5$	$12C_5^2$		
A	1	1	1	1	1	$x^2+y^2+z^2$	
T_1	3	−1	0	m	n	(x, y, z)	(R_x, R_y, R_z)
T_2	3	−1	0	n	m		
U	4	0	1	−1	−1		
V	5	1	−1	0	0	$\left\{\begin{matrix}(xy, xz, yz, x^2-y^2,\\ 2z^2-(x^2+y^2))\end{matrix}\right\}$	

$m = 1+2\cos 72° = \frac{1}{2}(\sqrt{5}+1) = +1{\cdot}618$
$n = 1+2\cos 144° = -\frac{1}{2}(\sqrt{5}-1) = -0{\cdot}618$

$P_h \equiv P_i$	E	$15C_2$	$20C_3$	$12C_5$	$12C_5^2$	i	$15\sigma_h(=iC_2)$	$20S_6$	$12S_{10}^3$	$12S_{10}$		
A_g	1	1	1	1	1	1	1	1	1	1	$(x^2+y^2+z^2)$	
A_u	1	1	1	1	1	−1	−1	−1	−1	−1		
T_{1g}	3	−1	0	m	n	3	−1	0	m	n		(R_x, R_y, R_z)
T_{1u}	3	−1	0	m	n	−3	1	0	$-m$	$-n$	(x, y, z)	
T_{2g}	3	−1	0	n	m	3	−1	0	n	m		
T_{2u}	3	−1	0	n	m	−3	1	0	$-n$	$-m$		
U_g	4	0	1	−1	−1	4	0	1	−1	−1		
U_u	4	0	1	−1	−1	−4	0	−1	1	1		
V_g	5	1	−1	0	0	5	1	−1	0	0	$(xy, xz, yz, x^2-y^2, 2z^2-(x^2+y^2))$	
V_u	5	1	−1	0	0	−5	−1	1	0	0		

$m = 1+2\cos 72° = \frac{1}{2}(\sqrt{5}+1) = +1{\cdot}618$
$n = 1+2\cos 144° = -\frac{1}{2}(\sqrt{5}-1) = -0{\cdot}618$

Continuous Groups

C_∞	E	$\overrightarrow{C}_x$	$\overleftarrow{C}_x$		
$A\ (\Sigma)$	1	1	1	z, z^2, x^2+y^2	R_z
$E_1\ (\Pi)$	1	ω	ω^{-1}	$(x, y), (xy, yz)$	(R_x, R_y)
	1	ω^{-1}	ω		
$E_2\ (\Delta)$	1	ω^2	ω^{-2}	(x^2-y^2, xy)	
	1	ω^{-2}	ω^2		
⋮	⋮	⋮	⋮		
E_k	1	ω^k	ω^{-k}		
	1	ω^{-k}	ω^k		
⋮	⋮	⋮	⋮		

$x = \dfrac{2\pi}{\theta} \qquad \omega = \exp \dfrac{2\pi i}{\theta}$

$C_{\infty v}$	E	$2C_x$	σ_v		
$A_1\ (\Sigma^+)$	1	1	1	z, z^2, x^2+y^2	
$A_2\ (\Sigma^-)$	1	1	-1		R_z
$E_1\ (\Pi)$	2	$2\cos\theta$	0	$(x, y), (xz, yz)$	(R_x, R_y)
$E_2\ (\Delta)$	2	$2\cos 2\theta$	0	(x^2-y^2, xy)	
⋮	⋮	⋮	⋮		
E_k	2	$2\cos k\theta$	0		
⋮	⋮	⋮	⋮		

$x = \dfrac{2\pi}{\theta}$

D_∞	E	$2C_x$	C_2'		
$A_1\ (\Sigma^+)$	1	1	1	z^2, x^2+y^2	
$A_2\ (\Sigma^-)$	1	1	-1	z	R_z
$E_1\ (\Pi)$	2	$2\cos\theta$	0	$(x, y), (xz, yz)$	(R_x, R_y)
$E_2\ (\Delta)$	2	$2\cos 2\theta$	0	(x^2-y^2, xy)	
⋮	⋮	⋮	⋮		
E_k	2	$2\cos k\theta$	0		
⋮	⋮	⋮	⋮		

$x = \dfrac{2\pi}{\theta}$

$D_{\infty h}$	E	$2C_x$	$\infty C'_2$	i	$2iC_x$	(σ_h)	$\infty\sigma_v$		
$A_{1g}(\Sigma_g^+)$	1	1	1	1	1	1	1	z^2, x^2+y^2	
$A_{1u}(\Sigma_u^-)$	1	1	1	-1	-1	-1	-1		
$A_{2g}(\Sigma_g^-)$	1	1	-1	1	1	1	-1		R_z
$A_{2u}(\Sigma_u^+)$	1	1	-1	-1	-1	-1	1	z	
$E_{1g}(\Pi_g)$	2	$2\cos\theta$	0	2	$2\cos\theta$	-2	0	(xz, yz)	(R_x, R_y)
$E_{1u}(\Pi_u)$	2	$2\cos\theta$	0	-2	$-2\cos\theta$	2	0	(x, y)	
$E_{2g}(\Delta_g)$	2	$2\cos 2\theta$	0	2	$2\cos\theta$	2	0	(x^2-y^2, xy)	
$E_{2u}(\Delta_u)$	2	$2\cos 2\theta$	0	-2	$2\cos 2\theta$	-2	0		
$\vdots$	$\vdots$	$\vdots$	$\vdots$	$\vdots$	$\vdots$	$\vdots$	$\vdots$		
E_{kg}	2	$2\cos k\theta$	0	2	$2\cos k\theta$	$2(-1)^k$	0		
E_{ku}	2	$2\cos k\theta$	0	-2	$-2\cos k\theta$	$2(-1)^{k+1}$	0		
$\vdots$	$\vdots$	$\vdots$	$\vdots$	$\vdots$	$\vdots$	$\vdots$			

$x = \dfrac{2\pi}{\theta}$ $\quad\sigma_h$ is a special case of $2iC_x$, $x = 2$

INDEX

A 4
B 5
C 6
D 7
E 8
F 9
G 0
H 1
I 2
J 3